Luke Robinson

CCEA AS
APPLIED MATHEMATICS

COLOURPOINT EDUCATIONAL

© Luke Robinson and
Colourpoint Creative Ltd 2022

First Edition
Second Impression, 2024

Print ISBN: 978 1 78073 345 6
eBook ISBN: 978 1 78073 346 3

Layout and design: April Sky Design
Printed by: GPS Colour Graphics Ltd, Belfast

All rights reserved. No part of this publication may be reproduced, stored in a retrieval system or transmitted in any form or by any means, electronic, mechanical, photocopying, scanning, recording or otherwise, without the prior written permission of the copyright owners and publisher of this book.

Copyright has been acknowledged to the best of our ability. If there are any inadvertent errors or omissions, we shall be happy to correct them in any future editions.

The Author
Luke Robinson took a mathematics degree, followed by an MSc and PhD in meteorology. He taught at Northwood College in London before becoming a freelance mathematics tutor and writer. He now lives in County Down with his wife and son.

Colourpoint Educational
An imprint of Colourpoint Creative Ltd
Colourpoint House
Jubilee Business Park
21 Jubilee Road
Newtownards
County Down
Northern Ireland
BT23 4YH

Tel: 028 9182 0505
E-mail: sales@colourpoint.co.uk
Web site: www.colourpointeducational.com

Note: This book has been written to meet the AS Mathematics specification from CCEA. While the authors and Colourpoint Creative Limited have taken all reasonable care in the preparation of this book, it is the responsibility of each candidate to satisfy themselves that they have covered all necessary material before sitting an examination or attempting coursework based on the CCEA specification. The publishers will therefore accept no legal responsibility or liability for any errors or omissions from this book or the consequences thereof.

Contents

Introduction 5

Mechanics

1. Concepts in Mechanics 6
2. Kinematics: Constant Acceleration 14
3. Motion Graphs 28
4. Forces 36
5. Newton's Laws 52
6. Friction 63
7. Connected Bodies 76

Statistics

8. Statistical Sampling 96
9. Data Presentation and Interpretation 107
10. Central Tendency and Variation 122
11. Correlation and Regression 133
12. Data Cleaning 145
13. Probability 156
14. Binomial Distribution 172

Answers 182

Introduction

This book covers the revised specification for Unit AS 2: Applied Mathematics (Mechanics and Statistics) for CCEA, which was available for teaching from September 2018 onwards.

Accuracy

It is important to remember that all answers should be given either exact, or rounded to 3 significant figures. This advice is printed on the front page of all A Level Mathematics papers. Answer marks can be lost for rounding to any other level of accuracy.

Modelling

An important part of Applied Mathematics is **modelling**. Modelling questions may be set in relation to all topics in AS Applied Mathematics.

What does a modelling question look like?
A modelling question typically involves several of the following features, but not necessarily all of them:

- There may be a requirement to make simplifications. The question will ask what simplifications or assumptions have been made.
- The candidate may be required to discuss the limitations of the model used.
- There may also be a requirement to refine or adapt the model or to consider different models.

The Modelling Cycle
The **Modelling Cycle** is outlined in the diagram. From the wording of the problem, the student should devise a way to model the situation. Simplifications and assumptions may be required.

The model should be applied to obtain a solution and this solution is interpreted and evaluated. At this point, it may become clear that certain assumptions were inappropriate, wrong, or not needed. It may be the case that different assumptions are required. In this way the model can be refined, and this modified version of the model is applied to the problem.

The final report should detail results, conclusions, any assumptions made and any limitations of the model being used. In AS Level Mathematics, the report will comprise the solution to the problem.

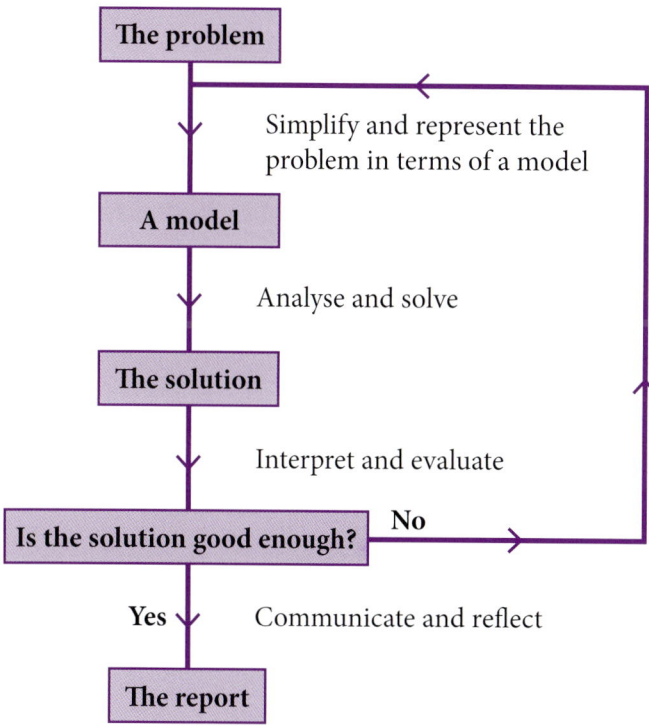

The Modelling Cycle

Chapter 1
Concepts in Mechanics

1.1 Introduction

This chapter is an introduction to some of the concepts used in AS Mechanics.

You will learn to use mathematical **models**. Models make assumptions to simplify a complex situation or process. You will learn when modelling assumptions are valid.

You will learn about the difference between **vectors** (such as velocity and force) and **scalars** (such as mass and time).

You will also learn about the SI units with which we measure many of these physical properties.

Key words
- **Model**: A mathematical representation of a real-life situation.
- **Vector**: A quantity with both size and direction, such as force or velocity.
- **Magnitude**: The length or size of a vector.
- **Resultant**: The sum of more than one vector.
- **Scalar**: A quantity that can be expressed as a single number. Mass, length and time are all scalar quantities.
- **Kilogram**: Unit of mass.
- **Metre**: Unit of length.
- **Second**: Unit of time.
- **Newton**: Unit of force.

Before you start
You should know:
- How to use vectors to represent translations.
- How to use Pythagoras' Theorem.
- How to use trigonometry in right-angled triangles.

What you will learn
In this chapter you will learn about:
- Modelling and modelling assumptions in Mechanics.
- Using vectors in Mechanics.
- The SI system of units.

In the real world...

In about 500 BC, the ancient Greeks were the first to present a mathematical model of the universe. Eudoxus' model placed a spherical Earth at the centre of the universe. The Sun, planets, and stars were then placed on giant transparent spheres surrounding it. A model of the universe that has the Earth at the centre is known as a geocentric model of the universe.

Aristotle extended Eudoxus' model of the universe in the 4th century BC. Aristotle's model of the universe was also geocentric, with the Sun, Moon, planets, and stars all orbiting the Earth inside Eudoxus' spheres. Aristotle believed the universe to be finite in space but to exist eternally in time.

Since then, astronomers have refined our model of the universe many times. Perhaps the most important change was an understanding that the Earth does not lie at the centre.

The concept of the Big Bang followed in the 20th century, the idea that all of matter exploded from a single point in space, and that the universe has been expanding ever since.

But the existing model of the universe is still incomplete. Observations suggest it is expanding more quickly than the model predicts. At the same time, the model suggests that the stars in galaxies should fly apart, but they do not. For these reasons, physicists have introduced the idea of dark matter and dark energy, but the exact nature of these concepts remains a mystery.

Many new ideas are being worked on, in an attempt to explain these anomalies, and others. One idea is that different rules of physics apply in different parts of the universe. Another, called string theory, is that the universe comprises sub-atomic strings oscillating in 11 dimensions. Of one thing we can be sure: our model of the universe will continue to evolve and develop as our understanding improves.

1: CONCEPTS IN MECHANICS

Exercise 1A (Revision)

1. Copy the diagram below. Draw shape X after each of the transformations given below. Label each of the images.

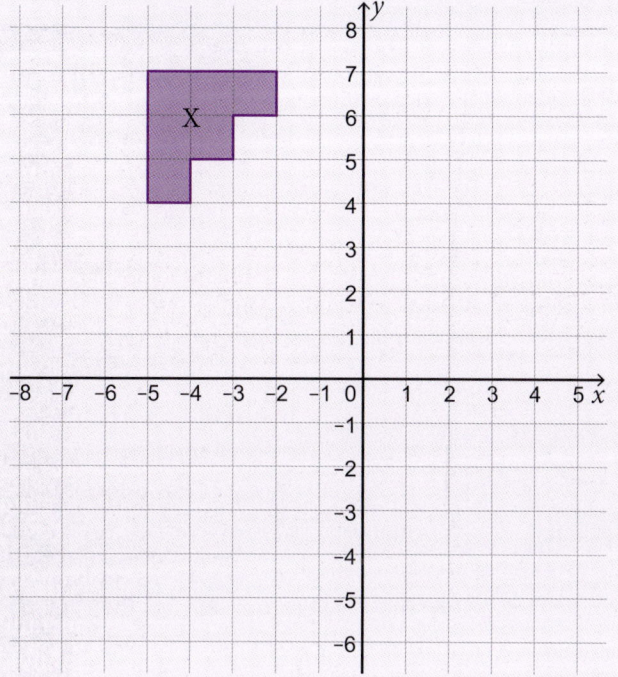

		Translation vector
(a)	X→Y	$\begin{pmatrix} 3 \\ 1 \end{pmatrix}$
(b)	X→Z	$\begin{pmatrix} 2 \\ -6 \end{pmatrix}$
(c)	X→P	$\begin{pmatrix} -2 \\ -5 \end{pmatrix}$
(d)	Z→Q	$\begin{pmatrix} -4 \\ -4 \end{pmatrix}$
(e)	Z→R	$\begin{pmatrix} 5 \\ 1 \end{pmatrix}$
(f)	R→S	$\begin{pmatrix} -2 \\ -4 \end{pmatrix}$

Exercise 1A...

2. For each of the triangles (a) to (d) below:
   (i) Find the length of the hypotenuse.
   (ii) Find the angle between the hypotenuse and the horizontal. In the first two triangles the angle is marked.
   Round your answers to one decimal place where necessary.

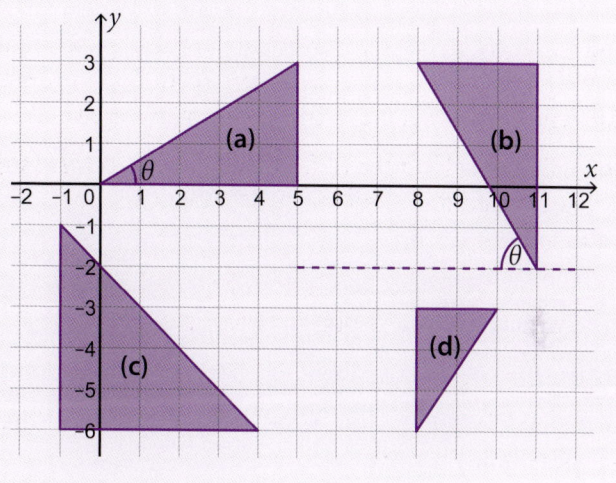

1.2 Modelling

Modelling of a real-life situation is a key part of Mechanics. The modelling cycle was discussed in the introduction to this book.

Modelling assumptions

In all models, **assumptions** are made in order to simplify the problem and to allow you to use known mathematical technique to analyse real-life situations. You need to understand the significance of different modelling assumptions and how they affect the calculations in a particular problem.

Modelling assumptions can affect the validity of a model. For example, when modelling the landing of an aeroplane, it would not be appropriate to ignore the effects of the wind or the air resistance.

The following table shows some common modelling assumptions used in Mechanics.

Modelling assumption	What it means
An object is a **particle**	- The mass of the object is concentrated at a single point. - The object has no size, i.e. it is dimensionless. - Hence, air resistance can be ignored. - Rotational forces can also be ignored.
An object is a **rod**	- The object has a length, but no width or height, i.e. it is one-dimensional. - The mass is concentrated along a line. - The object is **rigid**, i.e. it does not bend or break.
An object is a **wire**	- A wire is a metal **rod**.
An object is **uniform**	- The mass of the object is distributed evenly. - This is equivalent to assuming that the weight force acts at a single point at the centre of the object, the centre of gravity.
An object is **light**	- The mass of the object is small when compared with other masses in the question. - The mass of the object can be neglected and will not appear in any equations. - The tension in a light string is the same throughout the string.
A string is **inextensible**	- The string does not stretch. - If two objects are connected by a taut, inextensible string, they will move with the same velocity and will have the same acceleration.
A surface is **smooth**	- There is no friction between the surface and any object in contact with it.
A surface is **rough**	- There is friction between the surface and any object in contact with it. - A friction force acts upon the object.
An object is a **bead**	- A bead is a particle that has a hole through it. - A bead is usually threaded **smoothly** onto a piece of string or a **wire**. - The tension in the string or wire is the same on either side of the bead.

Modelling assumption	What it means
A support is a **peg**	- A peg is a support. An object can rest on or be suspended from a peg. - A peg is dimensionless, i.e. it has no size. - It can be **rough** or **smooth**.
Air resistance	- The resistance force experienced by an object as it passes through the air. - Air resistance is usually assumed to be small compared with other forces, so is not included in the equations. - If an object is modelled as a **particle**, air resistance is always ignored, since a particle has no size.
Gravity	- Gravity is a force of attraction between all objects, but most importantly between an object and the Earth. - The acceleration due to the force of gravity is always vertically downwards. - The acceleration due to gravity is usually taken to be a constant 9.8 m s^{-2} and is denoted by g.

Worked Example

1. When a space capsule returns to Earth, parachutes are opened to allow the capsule to touch down, or splash down in the ocean, gently. This situation is to be modelled so that the time of touchdown can be estimated. Which of the following assumptions are valid?
   (a) The capsule is a particle.
   (b) The parachute is a particle.
   (c) The strings connecting the capsule to the parachute are light and inextensible.

   (a) The capsule could be modelled as a particle. A particle has mass but no size; in this way, the effects of air resistance on the capsule would be ignored.
   (b) It would not be appropriate to treat the parachute as a particle because the model should not neglect the effects of air resistance on the parachute.
   (c) The strings can be considered light since their mass is negligible compared with the mass of the capsule. The amount they stretch will be small, so they can also be considered inextensible.

Exercise 1B

1. In modelling each of the following situations state at least one valid assumption that could be made.
   (a) When a football is kicked by a goalkeeper, a model is used to work out its maximum height and its range (the distance travelled).
   (b) An art teacher slides a pot of glue across a table towards a student. The model is to be used to calculate whether the pot stops on the table in front of the student, or whether it hits her.

2. Are the modelling assumptions listed valid when modelling each of the problems described?
   (a) A pole vaulter attempts to set a new world record. A mathematical model is used to determine whether he clears the bar.
   Modelling assumptions used:
   (i) The man is a particle.
   (ii) His pole is a rod.
   (iii) Gravity is a constant 9.8 m s^{-2}.

   (b) A toboggan carrying two children slides down a snowy hill and onto a flat field, which is also covered in snow. A mathematical model is used to determine whether the toboggan crashes into a tree, which is standing in the field.
   Modelling assumptions used:
   (i) The toboggan and the two children are modelled as a single particle.
   (ii) The hill and the field are both smooth.
   (iii) Gravity is a constant 9.8 m s^{-2}.

   (c) A medical scientist is studying the effects of space flight on the human body. A mathematical model is used to determine the stresses placed on an astronaut's leg bones while she is orbiting the earth.
   Modelling assumptions used:
   (i) An astronaut's leg bones are rods.
   (ii) Gravity is a constant 9.8 m s^{-2}.

1.3 Quantities and Units in Mechanics

The International System of Units (the SI system) is the modern form of the metric system.

Fundamental quantities and units in the SI system: length, time, mass

The following base SI units are commonly used in Mechanics.

Quantity	Unit	Symbol
Mass	kilogram	kg
Length, distance and displacement	metre	m
Time	second	s

> **Warning: Weight** and **mass** are not the same thing.
>
> **Mass** is a measure of the amount of metal, wood, plastic, water, etc and is measured in **kilograms**.
>
> **Weight** is a **force** acting on the object due to its mass and is measured in newtons (N).
>
> For example, a mass of 2 kg experiences a downwards weight force of 19.6 N.
>
> If this object is moved to the moon, its mass would remain as 2 kg, but the downwards weight force would be smaller because of the lower gravity on the moon.

Derived quantities and units: velocity, acceleration, force, weight

The table below shows some **derived** quantities. Derived units are compound units built from the base units.

Quantity	Unit	Symbol
Speed and velocity	metres per second	m s^{-1}
Acceleration	metres per second per second (or metres per second squared)	m s^{-2}
Forces, including weight	newton	N

AS 2: APPLIED MATHEMATICS

Worked Example

2. Write the following quantities using SI units.
   (a) 5.6 km   (b) 25 g
   (c) 900 km h^{-1} (d) 630 g cm^{-3}

   (a) Multiply by 1000 to convert kilometres to metres:
       5.6 km = 5600 m
   (b) Divide by 1000 to convert grams to kilograms:
       25 g = 0.025 kg
   (c) To convert kilometres per hour to metres per second: multiply by 1000, then divide by 3600:
       $900 \times 1000 \div 3600 = 250$ m s^{-1}
   (d) To convert grams per cubic centimetre to kilograms per cubic metre: divide by 1000, then multiply by 1 000 000:
       $\frac{630}{1000} \times 1\,000\,000 = 630\,000$ kg m^{-3}

Exercise 1C (Revision)

1. Write the following quantities using SI units.
   (a) 30 km   (b) 26 000 g
   (c) 50 km h^{-1} (d) 25 g cm^{-3}
   (e) 45 cm per minute (f) 120 g m^{-2}
   (g) 0.97 km s^{-1} (h) 5.4×10^{-7} days

1.4 An Introduction to Vectors

In previous work you have used the quantities **distance**, **speed**, **time** and **mass**. These are all examples of **scalar** quantities. Scalars have a size, or **magnitude**, only.

In this book you will work with scalars, but also with **vectors**. The chapters on kinematics and forces both use vectors extensively.

A vector is a quantity which has both magnitude and direction. The following quantities are vectors:

- **Displacement**: The distance travelled in a particular direction.
- **Velocity**: The speed in a particular direction.
- **Force**: Force is described both by its size and its direction.

In addition, the word **acceleration** can be used to refer to a scalar quantity (the rate of change of the speed) or to a vector quantity (the rate of change of the velocity).

In Pure Mathematics you may have learnt about **unit vectors**. A unit vector is any vector of length (or **magnitude**) 1.

The special unit vectors **i** and **j** run parallel to the positive x and y axes respectively, as shown in the diagram.

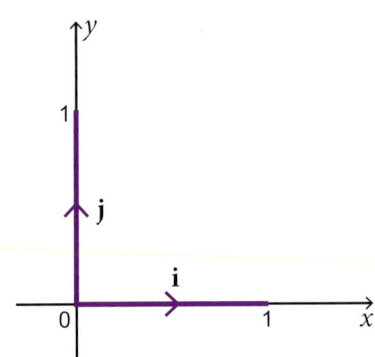

Any two-dimensional vector can be built up using the unit vectors **i** and **j**.

Worked Example

3. The vector **p**, from the point C(1, 1) to the point D(3, 2), is shown below. Express this vector using the unit vectors **i** and **j**.

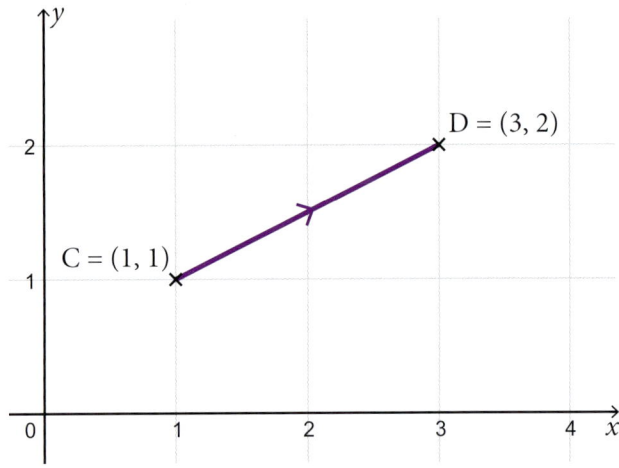

The vector **p** moves 2 units in the horizontal and 1 in the vertical:

p = 2**i** + **j**

Alternatively, this can be written using column vector notation:

$\mathbf{p} = \begin{pmatrix} 2 \\ 1 \end{pmatrix}$

> **Note:** In your work you can use **i-j** notation or column vector notation. They are exactly equivalent. Both notations are used throughout this book.

Note: Vectors can have three components, although this is beyond the scope of the current A Level specification. Vectors with three components are used in the study of motion and forces in three-dimensional space. Vectors can also have more than three components in the abstract study of higher dimensions!

Worked Examples

4. The velocity of a particle is given by $\mathbf{v} = 6\mathbf{i} + 3\mathbf{j}$ m s^{-1}. Find:
   (a) The speed of the particle.
   (b) The angle the direction of travel makes with the horizontal.

It is a good idea to sketch the vector:

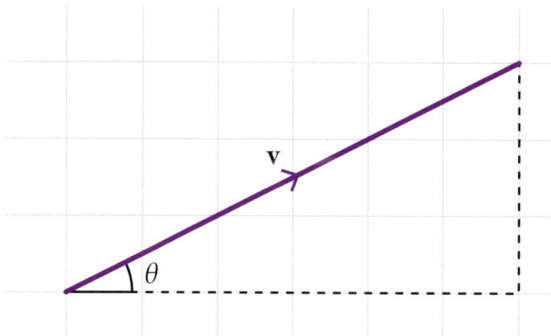

A right-angled triangle has been formed.

(a) The speed is the magnitude (or size) of the velocity vector. This can be found using Pythagoras' Theorem:
$$|\mathbf{v}| = \sqrt{6^2 + 3^2} = 6.71 \text{ m s}^{-1}$$

(b) The angle that the vector makes with the horizontal can be found using trigonometry:
$$\tan\theta = \frac{3}{6}$$
$$\theta = 26.6°$$

Note: There are other ways this question could have been phrased. For example, it could have asked for the angle between the vector and the x-axis, or the angle between the vector $\mathbf{v}$ and the vector $\mathbf{i}$. Alternatively, it could have asked for the direction of the vector $\mathbf{v}$. If you are asked for the direction of a vector, find the angle between the vector and the positive x-axis, as in this example.

5. In the school playground, a girl walks from point A to point B, then to point C.
   Her displacement from A to B is $\binom{5}{3}$ metres.
   Her displacement from B to C is $\binom{2}{-4}$ metres.
   (a) Find her displacement from A to C.
   (b) Find the distance from A to C.
   (c) Find the total distance the girl walks.

Sketch the girl's journey:

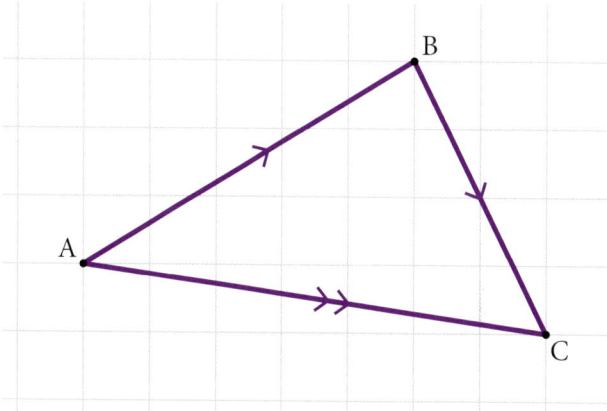

(a) Use the triangle law for vector addition:
$$\overrightarrow{AC} = \overrightarrow{AB} + \overrightarrow{BC}$$

Note: Vector $\overrightarrow{AC}$ is called the **resultant** of vectors $\overrightarrow{AB}$ and $\overrightarrow{BC}$.

$$\overrightarrow{AC} = \binom{5}{3} + \binom{2}{-4} = \binom{7}{-1}$$

Note: You can add vectors on your calculator to check your answer to part (a).

(b) The distance is the magnitude of the displacement vector:
$$\text{Distance} = |\overrightarrow{AC}| = \sqrt{7^2 + (-1)^2} = 7.07 \text{ m (3 s.f.)}$$

(c) The total distance the girl walks is the sum of the two distances AB and BC. Find the magnitudes of both vectors $\overrightarrow{AB}$ and $\overrightarrow{BC}$:
$$|\overrightarrow{AB}| = \sqrt{5^2 + 3^2} = 5.831 \text{ m}$$
$$|\overrightarrow{BC}| = \sqrt{2^2 + (-4)^2} = 4.472 \text{ m}$$

Total distance she walks = 5.831 + 4.472
= 10.3 m (3 s.f.)

You may be asked to find a vector in its component form, given the vector's magnitude and direction.

Worked Example

6. A vector has a magnitude of 20. The angle between the vector and the horizontal is 53.1°, as shown. Find the vector in its component form correct to 2 significant figures.

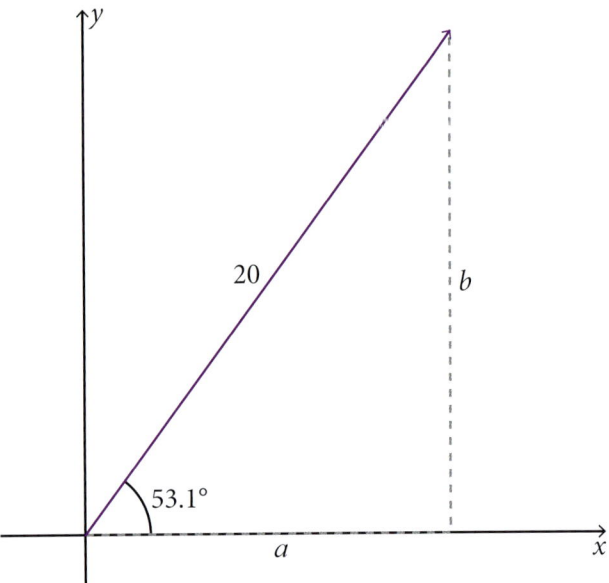

Using trigonometry in the right-angled triangle:

$\cos 53.1° = \dfrac{a}{20}$

$\Rightarrow a = 20 \cos 53.1° = 12$ (2 s.f.)

$\sin 53.1° = \dfrac{b}{20}$

$\Rightarrow b = 20 \sin 53.1° = 16$ (2 s.f.)

Therefore, the vector is **12i + 16j**

Note: −12i − 16j also has the required magnitude and angle, but it is clear from the diagram that both components of the vector are positive.

Exercise 1D

1. Are these quantities scalars or vectors?
(a) Height
(b) Mass
(c) Speed
(d) Time
(e) Velocity
(f) Distance from home
(g) Displacement from a fixed point
(h) Force

Exercise 1D...

2. Solve the following.
(a) A train is travelling with a velocity of −50**i** + 120**j** km h^{-1}. Find the speed of the train.
(b) The acceleration of a particle is −2**i** + 7**j** m s^{-2}. Find the magnitude of this acceleration.
(c) A force of −15**i** − 20**j** N is acting on a particle. Find the magnitude of the force.
(d) A ship has a displacement vector of 30**i** − 27**j** km from its harbour H. Find the ship's distance from H.

3. A ball is thrown with velocity **v** = −8**i** + 10**j** m s^{-1}. Find:
(a) the speed of the ball;
(b) the angle that the direction of the ball's motion makes with the horizontal.

4. A boy cycles from point A to point B and then to point C. The displacement from A to B is (−9**i** + 4**j**) km. The displacement from B to C is (5**i** − 3**j**) km. Find:
(a) the displacement of C from A;
(b) the distance of C from A;
(c) the total distance the boy cycles.

5. A vector has a magnitude of 13. The angle between the vector and the horizontal is 22.6°, as shown.

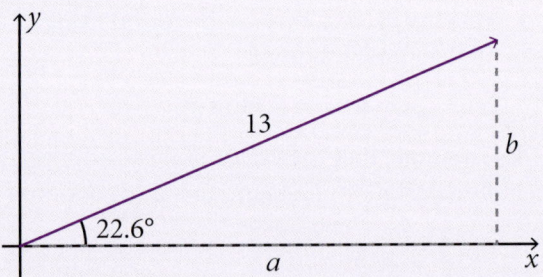

Find the vector in its component form, giving each component to the nearest integer.

1.5 Summary

Mathematical **models** can be used to simulate real-life situations.

Modelling **assumptions** can be made to simplify the calculations.

The **base SI units** used in mechanics are:

- The kilogram (symbol kg), used to measure mass.
- The metre (symbol m), used to measure length, distance or displacement.
- The second (symbol s), used to measure time.

A **vector** is a quantity which has both magnitude and direction.

Vectors can be given using two components (sometimes called the **i** and **j** components), which represent the horizontal and vertical parts of the vector.

Distance is the magnitude of the displacement vector.

Speed is the magnitude of the velocity vector.

Chapter 2
Kinematics: Constant Acceleration

2.1 Introduction

Kinematics is the study of the motion of bodies without considering the forces that cause them to move. Therefore, in this chapter, there will be no references to forces.

This chapter relates to objects moving with **constant acceleration**. In A2 Mathematics, motion with variable acceleration is considered.

Key words
- **Displacement**: The change in position of an object, expressed as a vector.
- **Distance**: The magnitude of the displacement vector.
- **Velocity**: The speed of an object in a given direction. Velocity is a vector quantity.
- **Speed**: The magnitude of the velocity vector.
- **Acceleration**: The rate of change of speed, or the rate of change of velocity.

Before you start
You should know:
- That average speed = $\dfrac{\text{distance travelled}}{\text{time taken}}$
- How to re-arrange a formula and substitute numbers into a formula.

What you will learn
In this chapter you will learn:
- How to model motion in two dimensions with constant velocity.
- About the constant acceleration formulae.
- How to use the constant acceleration formulae with 2D vectors.
- About vertical motion in one dimension.
- How to use the constant acceleration formulae for the motion of two objects.

In the real world...
When a rocket is launched, its precise location at any time can be calculated using the constant acceleration formulae, which you will learn about in this chapter.

The Apollo moon landings of the 1960s and 1970s relied on these formulae, as well as equations involving forces, which you will learn about later in this book.

Exercise 2A (Revision)

1. A car travels 75 km in 2.5 hours. Find the car's average speed for the journey.

2. Re-arrange each formula to make the letter in bold the subject.
   (a) $C = \pi \mathbf{d}$
   (b) $A = \dfrac{1}{2}(\mathbf{a} + b)h$
   (c) $a^2 = \mathbf{b}^2 + c^2$

3. Answer the following.
   (a) If $a = 4$, $b = -2$ and $t = 3$, find the value of C, given that $C = a + bt$
   (b) If $S = \dfrac{1}{2}xy^2$, find the value of S when $x = 10$ and $y = 6$
   (c) Find p given that $p^2 = q^2 + 2rs$, $q = -3$, $r = 2.16$ and $s = 2$

2.2 Constant Velocity With Vectors

If an object travels with **constant velocity**, the following vector formula can be used to describe its motion:

$\mathbf{r} = \mathbf{r_0} + \mathbf{v}t$

where $\mathbf{r}$ is the object's final position vector, $\mathbf{r_0}$ is its initial position vector, $\mathbf{v}$ is its velocity vector and t is the time taken.

If velocity is constant, the acceleration is also constant, being zero. However, it is **not** necessary to use the constant acceleration formulae (the 'suvat' formulae), introduced in the next section.

2: KINEMATICS: CONSTANT ACCELERATION

Worked Examples

1. A particle is initially at a point A with position vector $(7\mathbf{i} - 6\mathbf{j})$ m and moves with constant velocity $(2\mathbf{i} - \mathbf{j})$ m s^{-1}.
   (a) Find the particle's position vector $\mathbf{r}$ after 3 seconds.
   (b) Find the time at which the particle reaches point B with coordinates $(17, -11)$.

(a) $\mathbf{r} = \mathbf{r_0} + \mathbf{v}t$
$= (7\mathbf{i} - 6\mathbf{j}) + 3(2\mathbf{i} - \mathbf{j})$
$= (13\mathbf{i} - 9\mathbf{j})$ m

(b) Point B has position vector $(17\mathbf{i} - 11\mathbf{j})$.
Consider the journey from A to B:
$\mathbf{r} = \mathbf{r_0} + \mathbf{v}t$
$(17\mathbf{i} - 11\mathbf{j}) = (7\mathbf{i} - 6\mathbf{j}) + t(2\mathbf{i} - \mathbf{j})$

Comparing the **i**-components:
$17 = 7 + 2t$
$t = 5$ seconds

Comparing the **j**-components would give the same answer.

2. At 0900 a ship sets sail from port A, which lies at the position with vector $(6\mathbf{i} - 2\mathbf{j})$ km relative to fixed point O. It sails with constant velocity $(-\mathbf{i} + \mathbf{j})$ km h^{-1}.
   (a) Find the position of the ship at midday, relative to O.

   A lighthouse is positioned at the point with position vector $(2\mathbf{i} - 9\mathbf{j})$ km.
   (b) Find the time at which the ship is due north of the lighthouse.
   (c) Find how far the ship is from the lighthouse at this time.
   (d) Port B has position vector $(-\mathbf{i} + 5\mathbf{j})$ km relative to O. At what time does the ship reach port B?

The initial position relative to O is $\mathbf{r_0} = (6\mathbf{i} - 2\mathbf{j})$ km.
The constant velocity $\mathbf{v} = (-\mathbf{i} + \mathbf{j})$ km h^{-1}.

(a) The journey takes 3 hours, so $t = 3$:
$\mathbf{r} = \mathbf{r_0} + \mathbf{v}t$
$\mathbf{r} = (6\mathbf{i} - 2\mathbf{j}) + 3(-\mathbf{i} + \mathbf{j})$
$\mathbf{r} = (3\mathbf{i} + \mathbf{j})$ km

(b) For the ship to be due north of the lighthouse at $(2\mathbf{i} - 9\mathbf{j})$ km, the ship's position vector must have an **i**-component of 2. Let the ship's position vector be $(2\mathbf{i} + y\mathbf{j})$ km. So:
$\mathbf{r} = \mathbf{r_0} + \mathbf{v}t$
$(2\mathbf{i} + y\mathbf{j}) = (6\mathbf{i} - 2\mathbf{j}) + t(-\mathbf{i} + \mathbf{j})$

Comparing the **i**-components:
$2 = 6 - t$
$t = 4$
It takes 4 hours for the ship to reach this point. So the ship arrives at this point at 1300.

(c) $\mathbf{r} = \mathbf{r_0} + \mathbf{v}t$
When $t = 4$:
$\mathbf{r} = (6\mathbf{i} - 2\mathbf{j}) + 4(-\mathbf{i} + \mathbf{j})$
$\mathbf{r} = (2\mathbf{i} + 2\mathbf{j})$ km
So the ship is 11 km north of the lighthouse at $(2\mathbf{i} - 9\mathbf{j})$ km.

(d) The journey is from A to B. The initial position is $\mathbf{r_0} = (6\mathbf{i} - 2\mathbf{j})$ and the final position is $(-\mathbf{i} + 5\mathbf{j})$:
$\mathbf{r} = \mathbf{r_0} + \mathbf{v}t$
$-\mathbf{i} + 5\mathbf{j} = 6\mathbf{i} - 2\mathbf{j} + t(-\mathbf{i} + \mathbf{j})$
$-7\mathbf{i} + 7\mathbf{j} = t(-\mathbf{i} + \mathbf{j})$
$t = 7$
The ship takes 7 hours for this journey, so it arrives at 1600.

Questions may involve two objects travelling with constant velocity.

Worked Example

3. James and Padraig are on a sports field. James starts at the point O. Padraig stands at the point $16\mathbf{i}$ m, relative to point O.

   James runs across the sports field with a constant velocity of $(6\mathbf{i} + \mathbf{j})$ m s^{-1}.

   Two seconds after James leaves O, Padraig kicks a football across the field with a horizontal velocity of $(-2\mathbf{i} + 5\mathbf{j})$ m s^{-1}.

   (a) Show that the football hits James and find the time of the collision in seconds since James started running.
   (b) How far in metres is James from his starting point when the football hits him? Give your answer to 3 significant figures.

(a) Let the time of the collision be T seconds after James leaves O.
For James:
$\mathbf{r} = \mathbf{r_0} + \mathbf{v}t$
$\mathbf{r} = \begin{pmatrix} 0 \\ 0 \end{pmatrix} + \begin{pmatrix} 6 \\ 1 \end{pmatrix}T$
$\mathbf{r} = \begin{pmatrix} 6T \\ T \end{pmatrix}$

For the football:
$\mathbf{r} = \mathbf{r_0} + \mathbf{v}t$

The time taken for the ball's journey is
$T - 2$ seconds. So:

$$\mathbf{r} = \begin{pmatrix} 16 \\ 0 \end{pmatrix} + \begin{pmatrix} -2 \\ 5 \end{pmatrix}(T - 2)$$

$$\mathbf{r} = \begin{pmatrix} 16 - 2(T-2) \\ 5(T-2) \end{pmatrix}$$

$$\mathbf{r} = \begin{pmatrix} 20 - 2T \\ 5T - 10 \end{pmatrix}$$

At the time of collision, James' position vector and the position vector of the ball will be equal:

$$\begin{pmatrix} 6T \\ T \end{pmatrix} = \begin{pmatrix} 20 - 2T \\ 5T - 10 \end{pmatrix}$$

Considering the **i**-components:
$6T = 20 - 2T$
$8T = 20$
$T = 2.5$ s

Considering the **j**-components:
$T = 5T - 10$
$4T = 10$
$T = 2.5$ s

The **i**-components are equal at the same time as the **j**-components are equal. Therefore, there is a collision, when $T = 2.5$.

The ball hits James 2.5 seconds after he starts running.

> **Note:** If these two solutions for T were different, this would indicate that no collision takes place.

(b) At the time of collision, James' position vector is:
$$\mathbf{r} = \begin{pmatrix} 6T \\ T \end{pmatrix} = \begin{pmatrix} 15 \\ 2.5 \end{pmatrix}$$

Distance from starting point
$= |\mathbf{r}| = \sqrt{15^2 + 2.5^2}$
$= 15.2$ m

Exercise 2B

1. A particle P starts at the point with position vector $\mathbf{r_0}$. P moves with velocity $\mathbf{v}$ m s^{-1}. After t seconds, P is at the point with position vector $\mathbf{r}$.
   (a) Find $\mathbf{r}$ if $\mathbf{r_0} = 3\mathbf{i}$ m, $\mathbf{v} = 2\mathbf{i} + 4\mathbf{j}$ m s^{-1} and $t = 5$ s.
   (b) Find $\mathbf{r_0}$ if $\mathbf{r} = -4\mathbf{i} + 7\mathbf{j}$ m, $\mathbf{v} = -\mathbf{i} - 5\mathbf{j}$ m s^{-1} and $t = 10$ s.
   (c) Find $\mathbf{v}$ if $\mathbf{r} = \begin{pmatrix} 6 \\ -1 \end{pmatrix}$ m, $\mathbf{r_0} = \begin{pmatrix} 9 \\ -3 \end{pmatrix}$ m and $t = 2$ s.

Exercise 2B...

   (d) Find t if $\mathbf{r} = \begin{pmatrix} 7 \\ 1 \end{pmatrix}$ m, $\mathbf{v} = \begin{pmatrix} 2 \\ 0 \end{pmatrix}$ m s^{-1} and $\mathbf{r_0} = \begin{pmatrix} 2 \\ 1 \end{pmatrix}$ m.
   (e) Find the **speed** of P if $\mathbf{r_0} = \begin{pmatrix} 17 \\ 16 \end{pmatrix}$ m, $\mathbf{r} = \begin{pmatrix} 2 \\ -4 \end{pmatrix}$ m and $t = 5$ s.
   (f) Find the time taken for the journey if $\mathbf{r} = \mathbf{i} - 8\mathbf{j}$ m, $\mathbf{r_0} = 8\mathbf{i} - 9\mathbf{j}$ m and $\mathbf{v} = -21\mathbf{i} + 3\mathbf{j}$ m s^{-1}.

2. A ball, starting with position vector $\begin{pmatrix} -3 \\ -4 \end{pmatrix}$ m, rolls across the ground with constant velocity. It takes 6 seconds to reach the point with position vector $\begin{pmatrix} 21 \\ -22 \end{pmatrix}$ m. Find the ball's speed.

3. A particle P starts at the point with position vector $-20\mathbf{i} - 35\mathbf{j}$ m. It moves parallel to the vector $5\mathbf{i} + 12\mathbf{j}$ with speed 26 m s^{-1}. Find the time taken for the journey to the point with position vector $\mathbf{r} = 10\mathbf{i} + 37\mathbf{j}$ m.

> **Note:** The following questions involve two particles or objects.

4. At time $t = 0$, a football player kicks a ball on a football field from the point A with position vector $(4\mathbf{i} + 20\mathbf{j})$ m relative to a fixed point O. The ball is modelled as a particle moving horizontally with velocity $(6\mathbf{i} - 4.5\mathbf{j})$ m s^{-1}.
   (a) Find the speed of the ball.
   (b) Find the position vector $\mathbf{r}$ of the ball after t seconds, giving your answer in the form $\mathbf{r} = (a + bt)\mathbf{i} + (c + dt)\mathbf{j}$ m where a, b, c and d are constants to be found.

   The point B on the field has position vector $(22\mathbf{i} + 12.5\mathbf{j})$ m.
   (c) Find the time at which the ball is due south of B.
   (d) At time $t = 0$, another player starts running due south from B and moves with constant speed v. Given that he intercepts the ball, find v.

5. An orange plane flies with a constant velocity of $3\mathbf{i} + 35\mathbf{j}$ m s^{-1} from airport O. A blue plane flies with a constant velocity of $-\mathbf{i} + 45\mathbf{j}$ m s^{-1} from airport A. Both planes leave at midnight. Airport A lies 28.8 km east and 15 km north of O. The vectors $\mathbf{i}$ and $\mathbf{j}$ point east and north respectively. If t is the time in seconds after midnight:

Exercise 2B...

(a) Find the position vector of the blue plane from airport O in terms of t.
(b) Find at which time the blue plane is due north of the orange plane.
(c) Find the **distance** of the blue plane from airport O at this time.

6. Two ships A and B set out from a port O at midnight. Ship A travels due north at 18 km h^{-1}, while ship B travels due east at 12 km h^{-1}.
   (a) Find the position vector of ship A at time t.
   (b) Find the position vector of ship B at time t.
   (c) Find the vector $\overrightarrow{BA}$ in terms of t.
   (d) Find how far apart the two ships are at 1:00 am.
   (e) Find the bearing of ship B from ship A at this time.

7. At noon, a ferry F is 490 m due north of an observation point O. It is moving with a constant velocity $(5\mathbf{i} + 5\mathbf{j})$ m s^{-1}. A speedboat S is 630 m due east of O, moving with velocity $(-4\mathbf{i} + 12\mathbf{j})$ m s^{-1}.
   (a) Write down the position vectors of F and S at time t seconds after noon.
   (b) Show that F and S collide.
   (c) Find the position vector of the point of collision.

8. At 9 a.m. two ships A and B have position vectors $(2\mathbf{i} + 5\mathbf{j})$ km and $(6\mathbf{i} - 3\mathbf{j})$ km from harbour H. Their velocities are $(3\mathbf{i} - 2\mathbf{j})$ km h^{-1} and $(-2\mathbf{i} + 6\mathbf{j})$ km h^{-1} respectively.
   (a) Write down the position vectors of A and B t hours later.
   (b) Show that t hours after 9 a.m. the position vector of B relative to A is given by $(4 - 5t)\mathbf{i} + (-8 + 8t)\mathbf{j}$ km.
   (c) Show that the ships do **not** collide.
   (d) Find the distance between A and B at 11 a.m.

2.3 The Constant Acceleration Formulae

The variables involved in this section are:

s – the displacement
u – the initial velocity
v – the final velocity
a – the acceleration
t – time

There are four formulae that link these variables. These are known as the **constant acceleration formulae** or sometimes the **suvat formulae**.

$$v = u + at$$
$$s = ut + \frac{1}{2}at^2$$
$$s = \frac{1}{2}t(u + v)$$
$$v^2 = u^2 + 2as$$

Note: There is a fifth formula $s = vt - \frac{1}{2}at^2$, but this was not required in the current CCEA AS Mathematics specification at the time of writing.

It is important to note that these formulae can only be used when the acceleration is constant. If the acceleration changes at any point during the object's motion, then the motion should be split up into separate phases with constant acceleration in each one.

In A2 Mathematics you will cover cases where the acceleration is a function of time and is therefore constantly changing. In these cases, the constant acceleration formulae cannot be used; an alternative approach will be demonstrated.

Worked Example

4. A train moves along a straight track with constant acceleration. Three telegraph poles are set at equal intervals beside the track at points A, B and C, where AB = 50 m and BC = 40 m. The train passes A with speed 21 m s^{-1}, and 2 s later it passes B. Find:
   (a) The acceleration of the train.
   (b) The speed of the train when it passes C.
   (c) The time that elapses between the train passing points B and C.

(a) Consider the train's motion from A to B. Draw up a table of the variables s, u, v, a and t. Use a question mark for the variable to find (in this case a). Leave a variable blank if it is unknown.

From A to B:
$s = 50$
$u = 21$
$v = $
$a = ?$
$t = 2$

The formula linking s, u, a and t is required:
$$s = ut + \frac{1}{2}at^2$$
$$50 = 21 \times 2 + \frac{1}{2}a \times 2^2$$
$$50 = 42 + 2a$$
$$a = 4 \text{ m s}^{-2}$$

(b) Draw up a new suvat table to consider the train's motion from A to C:

From A to C:
$s = 90$
$u = 21$
$v = ?$
$a = 4$
$t =$

The formula linking s, u, v and a is needed:
$$v^2 = u^2 + 2as$$
$$v^2 = 21^2 + 2 \times 4 \times 90$$
$$v^2 = 1161$$
$$v = 34.07345...$$
$$v = 34.1 \text{ m s}^{-1} \text{ (3 s.f.)}$$

(c) Firstly, find the time for the journey from A to C. The suvat table is the same as in part (b), but this time t is the unknown:

From A to C:
$s = 90$
$u = 21$
$v = 34.07345...$
$a = 4$
$t = ?$

So we use:
$$s = \frac{1}{2}(u + v)t$$
$$90 = \frac{1}{2}t(21 + 34.07345...)$$
$$t = 3.27 \text{ s}$$

Since the train takes 2 seconds to travel from A to B, the time taken from B to C is 1.27 seconds.

Note: Other formulae could have been used in part (c). Ideally the formula that does not involve v would be used, since the calculation of v in part (b) could be wrong. However, finding t using $s = ut + \frac{1}{2}at^2$ is slightly more complicated, since it involves solving a quadratic equation.

The **deceleration** of an object is the rate at which it slows down. If an object is slowing down, its acceleration is negative. To find the deceleration, change the sign of the acceleration.

Note: The deceleration is sometimes referred to as the **retardation**.

Worked Example

5. A body moves with a velocity of 16 m s⁻¹. Five seconds later it is moving with velocity 2.8 m s⁻¹. Find the body's deceleration.

$s =$
$u = 16$
$v = 2.8$
$a = ?$
$t = 5$

$$v = u + at$$
$$2.8 = 16 + 5a$$
$$a = \frac{2.8 - 16}{5} = -2.64 \text{ m s}^{-2}$$

The acceleration is negative because the body is slowing down.

The acceleration is -2.64 m s⁻²

The deceleration is 2.64 m s⁻².

Exercise 2C

1. A body accelerates uniformly at 1.5 m s⁻² for 2 s, reaching a velocity of 10 m s⁻¹. Find:
   (a) The body's initial velocity.
   (b) The distance travelled.

2. A particle accelerates uniformly from 9.9 m s⁻¹ to 30.7 m s⁻¹ in 6.5 s. Find:
   (a) The particle's acceleration.
   (b) The distance travelled.

3. An object travels 24 m in coming to rest uniformly in 2 seconds. Find:
   (a) The object's initial velocity.
   (b) Its deceleration.

4. During takeoff, an aircraft moves on a straight runway AB of length 1.4 km. Point C lies on the runway between points A and B. The aircraft moves from A with initial speed 4 m s⁻¹. It moves with constant acceleration and 23 s later it is at point C with speed 73 m s⁻¹. Find:

Exercise 2C...

    (a) (i) The acceleration of the aircraft.
        (ii) The distance AC.
        (iii) The distance BC.
    (b) The aircraft must abort its takeoff. It decelerates uniformly to rest from C to B. Find its deceleration.

5. A car travels along a straight road. It passes a post office with speed 5 m s^{-1}. The distance from the post office to the end of the road is 850 m.
    (a) Find the car's acceleration between the post office and the school.

The car then moves with constant acceleration and 25 s later it passes a school with speed 15 m s^{-1}.

    (b) Find the distance from the school to the end of the road.
    (c) As the car passes the school, the driver brakes. The car decelerates uniformly between the school and the end of the road, where it comes to rest. Find the deceleration of the car during this part of its journey.
    (d) Find the total time taken for the car's journey from the post office to the end of the road.

6. Objects A and B both move from point X to point Y along a straight line. Object A starts from rest, has an acceleration of 5 m s^{-2} and reaches point Y with a velocity of 10 m s^{-1}. Object B starts with a velocity of 1 m s^{-1} and accelerates at 4 m s^{-2}.
    (a) How far is it from point X to point Y?
    (b) Which object reaches Y first? Show your working.

2.4 Vertical Motion

You can use the constant acceleration formulae to model an object moving vertically under gravity.

If you ignore the effects of air resistance the object's acceleration is constant. The acceleration does not depend on the object's mass. For example, in a vacuum, an apple and a feather would both accelerate downwards at the same rate. On earth, the acceleration due to gravity is represented by the letter g. In A-Level Mathematics the approximation $g = 9.8 \text{ m s}^{-2}$ is most often used. Some questions may require a different approximation, e.g. $g = 10 \text{ m s}^{-2}$ or $g = 9.81 \text{ m s}^{-2}$.

Note: In reality g is not quite constant. It depends on the object's position on Earth and also its height above sea level.

When solving problems involving vertical motion, you can choose the positive direction to be upwards or downwards. Acceleration due to gravity is always downwards, so if the positive direction is upwards, then the acceleration will be $-g$ or -9.8 m s^{-2}.

Worked Examples

6. A parcel falls from a shelf, which is 1.3 m above the floor. Find:
    (a) The time the parcel takes to hit the floor.
    (b) The speed at which the parcel hits the floor.
Assume that the acceleration due to gravity is 9.8 m s^{-2} and give your answers to an appropriate degree of accuracy.

    (a) Model the parcel as a particle moving in a straight line with constant acceleration of 9.8 m s^{-2}. As the parcel is moving downwards throughout its motion, take downwards as the positive direction. This means $a = 9.8 \text{ m s}^{-2}$. $u = 0$ since the parcel is initially at rest on the shelf.

From the shelf to the floor:
$s = 1.3$
$u = 0$
$v = $
$a = 9.8$
$t = ?$

The formula linking s, u, a and t is required:
$$s = ut + \frac{1}{2}at^2$$
$$1.3 = 0 + \frac{1}{2} \times 9.8 \times t^2$$
$$1.3 = 4.9t^2$$
$$t^2 = \frac{1.3}{4.9} = 0.265306$$
$$t = 0.515079 \ldots \text{ s}$$
$$t = 0.52 \text{ s (2 s.f.)}$$

(b) Use the same suvat table, since the same journey, from shelf to floor, is being considered:

From the shelf to the floor:
$s = 1.3$
$u = 0$
$v = ?$
$a = 9.8$
$t =$

Choose the formula without t:
$v^2 = u^2 + 2as$
$v^2 = 0^2 + 2 \times 9.8 \times 1.3$
$v^2 = 25.48$
$v = 5.047772...$
$v = 5.0 \text{ m s}^{-1}$ (2 s.f.)

> **Note:** Both answers have been given to 2 significant figures. It would be inappropriate to give any greater accuracy, since the value of g has been given to 2 significant figures.

7. A ball is thrown vertically upwards with a speed of 2 m s^{-1} from a bedroom window. It takes 1.1 seconds to reach the garden below. Find the height of the bedroom window to a suitable degree of accuracy.

It is important to remember that s is the displacement from the starting position.

It is easiest to consider the whole journey.

Method 1: Upwards is positive.

This means the initial velocity is positive. The acceleration due to gravity is downwards, therefore negative.

For the ball's journey from the window to the ground:
$s = ?$
$u = 2$
$v =$
$a = -9.8$
$t = 1.1$

Choose the formula without v:
$s = ut + \frac{1}{2}at^2$
$s = 2 \times 1.1 + \frac{1}{2} \times (-9.8) \times (1.1)^2$
$s = -3.729 = -3.7$ m (2 s.f.)

The ball ends its journey with a displacement of −3.7 metres, which means 3.7 metres below the starting position. The window is 3.7 metres above the garden.

> **Note:** The answer has been given to 2 significant figures, since the value of g used is accurate to 2 significant figures.

Method 2: Downwards is positive.

This means the initial velocity is negative. The acceleration due to gravity is downwards, therefore positive.

For the ball's journey from the window to the ground:
$s = ?$
$u = -2$
$v =$
$a = 9.8$
$t = 1.1$

Choose the formula without v:
$s = ut + \frac{1}{2}at^2$
$s = -2 \times 1.1 + \frac{1}{2} \times (9.8) \times (1.1)^2$
$s = 3.729 = 3.7$ m (2 s.f.)

The positive value of the displacement indicates a final position below the starting position. The window is 3.7 metres above the garden.

Method 3

The journey could be separated into two: the upwards part and the downwards part. This is left as an exercise for the student.

Exercise 2D

Note: Throughout this exercise, use a value of $g = 9.8$ m s^{-2} unless otherwise instructed. Give your answers to an appropriate degree of accuracy.

1. A party popper is fired vertically upwards with speed 13 m s^{-1} from a point h metres above the ground. The popper hits the ground 2.8 s later. Assuming no air resistance, find, giving your answers to a suitable level of accuracy:
   (a) The value of h.
   (b) The speed of the popper as it hits the ground.

2. **Use $g = 9.81$ m s^{-2} in this question.**
   A missile is projected vertically upwards with speed 21 m s^{-1} from a point A, which is 1.2 m

Exercise 2D...

above the ground. After launch, the missile moves freely under gravity until it reaches the ground. Modelling the missile as a particle, find:
(a) The greatest height above A reached by the missile.
(b) The speed of the missile as it reaches the ground.
(c) The time between the instant when the missile is launched from A and the instant when it reaches the ground.

3. A stone is thrown vertically upwards with speed 18 m s^{-1} from a point above the ground. The stone hits the ground 6 s later. Find:
(a) How far above the ground the stone was thrown.
(b) The speed of the stone as it hits the ground.

4. **Use $g = 9.81$ m s^{-2} in this question.**
A firework is fired vertically upwards with an initial speed of u m s^{-1} from a point 7 m above horizontal ground. The firework reaches a maximum height of 46.2 m above the ground.
(a) Find u.
(b) Find the total time it takes for the firework to reach the ground.

5. A bird carries a twig in its beak. As the bird flies across a field, it drops the twig onto the field. It takes 4 seconds for the twig to land in the field.
(a) Find how far the twig falls.
(b) Find the speed with which the twig hits the field.
(c) What modelling assumptions have been made?

6. At time $t = 0$ seconds a cricket ball is thrown vertically upwards with speed u m s^{-1} from ground level. When $t = 2$ the ball returns to ground level.
(a) Find u.
(b) Find the greatest height above ground level reached by the ball.

7. A stone is thrown vertically upwards from ground level with a speed of 12 m s^{-1}.
(a) Find the maximum height above the ground reached by the stone.
(b) Find the time taken for the stone to reach this maximum height.

Exercise 2D...

(c) Find the distance travelled by the stone during the first two seconds of its motion.
(d) For how long is the stone more than a metre above ground level?

8. **Use $g = 10$ m s^{-2} in this question.**
A rocket is launched from rest at ground level and moves vertically upwards. It rises 100 m in the first 5 seconds of its motion. Model the rocket as a particle moving with constant acceleration.
(a) Show that the acceleration of the rocket is 8 m s^{-2}.
(b) The rocket runs out of fuel when it has reached a height of 100 m. It can now be modelled as a particle moving vertically under gravity. Find the maximum height above the ground reached by the rocket.
(c) Find the total time taken for the rocket to reach its maximum height from the moment it is launched.

9. A stone is thrown vertically downwards from the top of a cliff towards the sea below with a speed of 2 m s^{-1}. The top of the cliff is 100 m above the sea.
(a) Find the speed of the stone as it hits the sea.
(b) Find the time taken for the stone to reach the sea.
(c) State one assumption that has been made in modelling the stone's trajectory.

10. A ball is thrown vertically upwards with initial velocity of 8 m s^{-1} from a point 2 m above the ground.
(a) Find the ball's maximum height.
(b) Find how long the ball is in the air before it hits the ground for the first time.
(c) The ball hits the ground and rebounds with half the speed it had prior to impact. Find the greatest height it reaches after this first bounce.

AS 2: APPLIED MATHEMATICS

Exercise 2D...

11. A diver launches himself from a 10-metre diving board, giving himself an initial speed of 2 m s⁻¹ upwards.

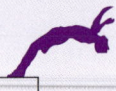

10 m

(a) How long does it take for the diver to reach the water below?
(b) What is his speed when he hits the water?
(c) When the diver hits the water, his body decelerates at 40 m s⁻². Find how far beneath the surface of the water he moves.

2.5 Two Objects

The constant acceleration formulae can be used to model the motion of two particles in motion at the same time. These two objects may be moving vertically under the influence of gravity. The task is often to work out where and when they collide or pass each other. This often involves solving simultaneous equations involving s and t.

Worked Example

8. Objects A and B move towards each other from opposite ends of a straight line XY, as shown in the diagram. The distance XY is 10 m.

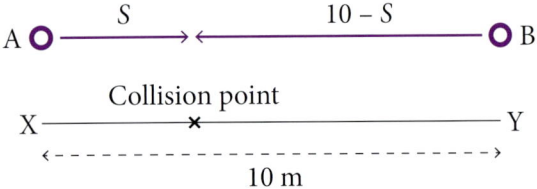

At time $t = 0$, object A moves from point X with initial speed 1.5 m s⁻¹ and acceleration 4 m s⁻².

Also at $t = 0$, object B moves from point Y with initial speed 0.5 m s⁻¹ and acceleration 12 m s⁻².

(a) After how long do the two objects collide?
(b) How far from point X do they collide?

For object A, consider the journey from X to the collision point. Let the distance travelled be S m and the time taken T s.

For object B, consider the journey from Y to the collision point. The distance travelled is $(10 - S)$ m and the time taken is T s.

Object A

$s = S$
$u = 1.5$
$v =$
$a = 4$
$t = T$

Object B

$s = (10 - S)$
$u = 0.5$
$v =$
$a = 12$
$t = T$

(a) For object A:
$$s = ut + \frac{1}{2}at^2$$
$$S = 1.5T + \frac{1}{2}(4)T^2$$
$$S = 1.5T + 2T^2 \quad (1)$$

For object B:
$$s = ut + \frac{1}{2}at^2$$
$$10 - S = 0.5T + \frac{1}{2}(12)T^2$$
$$10 - S = 0.5T + 6T^2 \quad (2)$$

Add equations (1) and (2) to eliminate S:
$$10 = 2T + 8T^2$$
$$8T^2 + 2T - 10 = 0$$
$$4T^2 + T - 5 = 0$$
$$(T - 1)(4T + 5) = 0$$
$$T = 1 \text{ or } T = -\frac{5}{4} \text{ s}$$

Reject the negative time.
The two objects collide after 1 second.

(b) To find the distance travelled by object A, substitute T into equation (1):
$$S = 1.5(1) + 2(1)^2 = 3.5 \text{ m}$$

The two objects collide 3.5 metres from X.

Exercise 2E

Note: Many of the questions in this exercise involve vertical motion. In these questions, use a value of $g = 9.8$ m s^{-2} unless otherwise instructed. Give your answers to an appropriate degree of accuracy.

1. A red car travelling at a constant velocity of 20 m s^{-1} overtakes a stationary police car. The red car's speed is above the speed limit. The police car sets off in pursuit, accelerating at a uniform 5 m s^{-2}.
   (a) How long does the police car take to catch up with the red car?
   (b) How far does the police car travel before catching up with the red car?

2. At time $t = 0$ seconds, a stone A is thrown vertically upwards from ground level, with speed u m s^{-1}. The greatest height above the ground reached by A is 3 m.
   (a) Find u.
   (b) When $t = 1$, a stone B is thrown vertically upwards from ground level, with speed 4 m s^{-1}. Find t when A and B collide.

3. Two boys, Jake and Leo, run towards each other from opposite ends of a straight road, which is 60 metres long. At time $t = 0$ seconds, Jake enters the road from the west, running with a speed of 1 m s^{-1} and runs eastwards with a uniform acceleration of 0.1 m s^{-2}. Also at $t = 0$, Leo enters the road from the east, and runs westwards with a constant speed of 1.5 m s^{-1}. Giving answers to 3 significant figures, find:
   (a) the value of t at which Jake and Leo meet;
   (b) how far Jake has run, after entering the road, at the time they meet.

4. **Use g = 9.81 m s^{-2} in this question.**
   A ball A is dropped from the top of a tower 10 m high. At the same time as A is dropped, a second ball B is launched vertically upwards from the base of the tower with speed 11 m s^{-1}. The balls collide. Find the distance of the point where the balls collide from the base of the tower.

5. A particle P is projected vertically upwards from a point O with speed 10 m s^{-1}. One second later, another particle Q is projected vertically upwards from O with speed 15 m s^{-1}. Find:

Exercise 2E...

   (a) How long after P is launched the two objects collide.
   (b) The distance above O of the point at which the two objects collide.

6. Three towns Avonford, Beckbridge and Cloverfield lie along a straight road, in that order. The distance between Avonford and Beckbridge is 2 km and the distance between Beckbridge and Cloverfield is 3 km. A motorbike, initially at rest, leaves Avonford, heading towards Cloverfield with a constant acceleration of 0.1 m s^{-2}. At the same time, a car starts from rest at Beckbridge and heads towards Cloverfield with a constant acceleration of 0.05 m s^{-2}.
   (a) How far from Avonford are the two vehicles when the motorbike catches up with the car?
   (b) The speed limit on the road from Beckbridge to Cloverfield is 60 miles per hour, or roughly 26.8 m s^{-1}. Are either of the vehicles breaking the speed limit when the motorbike catches up with the car? Show all your working.

7. At time $t = 0$ two balls A and B are projected vertically upwards. The ball A is given an initial speed of 5 m s^{-1} from a balcony 50 m above the horizontal ground. The ball B is launched from the ground with an initial speed of 20 m s^{-1}. At time $t = T$ seconds, the two balls are at the same vertical height, h metres above the ground. The balls are both modelled as particles moving freely under gravity.
   (a) Show that $T = \dfrac{10}{3}$ s.
   (b) Find the exact value of h.

8. Two particles A and B begin at rest 1 metre apart and move towards each other. Object A has constant acceleration a m s^{-2}, while object B has constant acceleration $2a$ m s^{-2}.

   Show that they collide when $T = \sqrt{\dfrac{2}{3a}}$ and find the distance of the collision from A's starting position.

AS 2: APPLIED MATHEMATICS

Exercise 2E...

9. At $t = 0$ seconds, particle A is dropped from a height of H metres. At the same time, particle B is thrown vertically upwards with a speed of u m s^{-1}. The particles collide.
   (a) Show that the time of collision is given by $\dfrac{H}{u}$ seconds.
   (b) In terms of H, g and u, find the height above the ground of the two particles when they collide, where g is the acceleration due to gravity.

10. Towards the end of a boat race the red boat is x metres from the finish line and is being rowed at a constant speed of 9 m s^{-1}. From here the red boat finishes the race in T seconds. The blue boat is 15 m behind the red boat and is being rowed at 7.5 m s^{-1}. It accelerates uniformly at 1.5 m s^{-2} until it crosses the finish line 0.75 seconds after the red boat.
   Find T and x.

2.6 The Constant Acceleration Formulae in Vector Form

In section 2.3 the constant acceleration formulae were introduced. Three of these formulae can be generalised to a vector form. In the formulae below, **s**, **u**, **v** and **a** are all vectors, whereas t remains a scalar quantity (there is no such thing as vector time).

$$\mathbf{v} = \mathbf{u} + \mathbf{a}t$$

$$\mathbf{s} = \mathbf{u}t + \frac{1}{2}\mathbf{a}t^2$$

$$\mathbf{s} = \frac{1}{2}t(\mathbf{u} + \mathbf{v})$$

Note: There is a vector form for the fourth formula, but it is beyond the scope of the current CCEA AS Mathematics specification.

The following example demonstrates that, for two objects to be travelling in parallel, their velocity vectors must be multiples of each other.

Worked Example

9. A fighter plane is on a training mission. As it passes over RAF Abercorn it is flying with velocity $(30\mathbf{i} + 25\mathbf{j})$ m s^{-1}. It begins to accelerate for a short time at $(2\mathbf{i} + \mathbf{j})$ m s^{-2}. Both vectors are relative to a fixed point O.
   (a) Find the plane's velocity during its acceleration in terms of t.

   The plane's acceleration continues for 2 seconds.
   (b) Find the plane's **speed** after its acceleration.

   RAF Abercorn has position vector $(-18\mathbf{i} + 32\mathbf{j})$ m relative to O.
   (c) Find the plane's position after its acceleration, giving your answer as a vector relative to the point O.
   (d) A second plane is flying with a constant velocity of $16\mathbf{i} + 13\mathbf{j}$ m s^{-1}. Find the time at which the two planes are flying parallel to each other.

(a) $\mathbf{s} =$
$\mathbf{u} = (30\mathbf{i} + 25\mathbf{j})$
$\mathbf{v} = ?$
$\mathbf{a} = (2\mathbf{i} + \mathbf{j})$
$t = t$

Using: $\mathbf{v} = \mathbf{u} + \mathbf{a}t$
$= (30\mathbf{i} + 25\mathbf{j}) + (2\mathbf{i} + \mathbf{j}) \times t$
$= (30 + 2t)\mathbf{i} + (25 + t)\mathbf{j}$

(b) When $t = 2$, $\mathbf{v} = (34\mathbf{i} + 27\mathbf{j})$
So speed $= \sqrt{34^2 + 27^2} = 43.4$ m s^{-1}

(c) $\mathbf{s} = ?$
$\mathbf{u} = (30\mathbf{i} + 25\mathbf{j})$
$\mathbf{v} =$
$\mathbf{a} = (2\mathbf{i} + \mathbf{j})$
$t = 2$

Using:
$\mathbf{s} = \mathbf{u}t + \frac{1}{2}\mathbf{a}t^2$
$= 2(30\mathbf{i} + 25\mathbf{j}) + \frac{1}{2}(2\mathbf{i} + \mathbf{j}) \times 2^2$
$= (64\mathbf{i} + 52\mathbf{j})$ m

This is the displacement vector of the plane relative to RAF Abercorn. To find its position vector relative to O:
$(-18\mathbf{i} + 32\mathbf{j}) + (64\mathbf{i} + 52\mathbf{j}) = (48\mathbf{i} + 84\mathbf{j})$ m

(d) The planes are flying parallel to each other when their velocity vectors are multiples of each other.

$$\binom{16}{13} = k\binom{30 + 2t}{25 + t}$$

i-components: $16 = 30k + 2kt$ (1)
j-components: $13 = 25k + kt$ (2)
$(2) \times 2 \Rightarrow 26 = 50k + 2kt$ (3)
$(3) - (1) \Rightarrow 20k = 10$
$$k = 0.5$$
Sub in (2): $13 = 25(0.5) + 0.5t$
$$0.5t = 0.5$$
$$t = 1 \text{ second}$$

The following example features two objects in motion. To work out whether the two particles collide, consider the position vectors of the two particles, in terms of t.

If the particles collide, the **i**-components must be equal at the same time as the **j**-components are equal.

Worked Example

10. A particle A starts at the point with position vector $\binom{-5}{-1}$ m. It has initial velocity $\binom{2}{-1}$ m s^{-1} and constant acceleration $\binom{1}{2}$ m s^{-2}.

    Another particle B starts at a point with position vector $\binom{-1}{7}$ m. The initial speed of B is $2\sqrt{5}$ m s^{-1} in the direction of the vector $\binom{1}{-2}$. B has constant acceleration $\binom{-1}{1}$ m s^{-2}.

    (a) Show that particles A and B collide and find the time at which this happens.
    (b) Find the position vector of the point at which they collide.
    (c) The speeds of the two particles when they collide.

(a) **For particle A**
$$\mathbf{u} = \binom{2}{-1} \text{ and } \mathbf{a} = \binom{1}{2}$$

The displacement of A from its starting position is given by:
$$\mathbf{s} = \mathbf{u}t + \frac{1}{2}\mathbf{a}t^2$$

$$\mathbf{s} = \binom{2}{-1}t + \frac{1}{2}\binom{1}{2}t^2$$

$$\mathbf{s} = \binom{2t + 0.5t^2}{-t + t^2}$$

Since particle A starts at the point with position vector $\binom{-5}{-1}$, its position vector from the origin is:

$$\mathbf{r_A} = \binom{-5}{-1} + \binom{2t + 0.5t^2}{-t + t^2} = \binom{-5 + 2t + 0.5t^2}{-1 - t + t^2}$$

For particle B

The unit vector in the direction of $\binom{-1}{2}$ is $\frac{1}{\sqrt{5}}\binom{1}{-2}$.

The initial velocity is $2\sqrt{5} \times \frac{1}{\sqrt{5}}\binom{1}{-2}$. So:

$$\mathbf{u} = 2\binom{1}{-2} = \binom{2}{-4}$$

The displacement of B from its starting position is given by $\mathbf{s} = \mathbf{u}t + \frac{1}{2}\mathbf{a}t^2$. So:

$$\mathbf{s} = \binom{2}{-4}t + \frac{1}{2}\binom{-1}{1}t^2$$

$$\mathbf{s} = \binom{2t - 0.5t^2}{-4t + 0.5t^2}$$

Since particle B starts at the point with position vector $\binom{-1}{7}$, its position vector from the origin is:

$$\mathbf{r_B} = \binom{-1}{7} + \binom{2t - 0.5t^2}{-4t + 0.5t^2} = \binom{-1 + 2t - 0.5t^2}{7 - 4t + 0.5t^2}$$

Find the time at which the **i**-components are equal:
$$-5 + 2t + 0.5t^2 = -1 + 2t - 0.5t^2$$
$$t^2 = 4$$
$$t = 2 \text{ (since } t = -2 \text{ is rejected)}$$

Find the time at which the **j**-components are equal:
$$-1 - t + t^2 = 7 - 4t + 0.5t^2$$
$$\frac{1}{2}t^2 + 3t - 8 = 0$$
$$t^2 + 6t - 16 = 0$$
$$(t - 2)(t + 8) = 0$$
$$t = 2 \text{ (since } t = -8 \text{ is rejected)}$$

The **i**-components are equal at the same time as the **j**-components are equal. Therefore, the two particles collide at time $t = 2$ seconds.

(b) When $t = 2$:
$$\mathbf{r_B} = \binom{-1 + 2(2) - 0.5(2)^2}{7 - 4(2) + 0.5(2)^2} = \binom{1}{1} \text{ m}$$

This can be checked using the position vector of particle A.

AS 2: APPLIED MATHEMATICS

(c) For particle A

$\mathbf{v} = \mathbf{u} + \mathbf{a}t$

$\mathbf{v} = \begin{pmatrix} 2 \\ -1 \end{pmatrix} + \begin{pmatrix} 1 \\ 2 \end{pmatrix} t$

When $t = 2$:

$\mathbf{v} = \begin{pmatrix} 2 \\ -1 \end{pmatrix} + \begin{pmatrix} 1 \\ 2 \end{pmatrix}(2) = \begin{pmatrix} 4 \\ 3 \end{pmatrix}$

Speed $= \sqrt{4^2 + 3^2} = 5$ m s^{-1}

For particle B

$\mathbf{v} = \mathbf{u} + \mathbf{a}t$

$\mathbf{v} = \begin{pmatrix} 2 \\ -4 \end{pmatrix} + \begin{pmatrix} -1 \\ 1 \end{pmatrix} t$

When $t = 2$:

$\mathbf{v} = \begin{pmatrix} 2 \\ -4 \end{pmatrix} + \begin{pmatrix} -1 \\ 1 \end{pmatrix}(2) = \begin{pmatrix} 0 \\ -2 \end{pmatrix}$

Speed $= \sqrt{0^2 + (-2)^2} = 2$ m s^{-1}

Exercise 2F

1. At time $t = 0$ seconds a particle is at a fixed point O and has an initial velocity of $(4\mathbf{i} - 2\mathbf{j})$ m s^{-1}. The particle has a constant acceleration of $(\mathbf{i} + 3\mathbf{j})$ m s^{-2}.
   (a) Find the velocity of the particle when $t = 3$.
   (b) Find the **distance** of the particle from O when $t = 10$.

2. At time $t = 0$ seconds, the particle P is at the point A with position vector $(3\mathbf{i} + 7\mathbf{j})$ m relative to a fixed origin O. It is moving with velocity $\mathbf{u}$ m s^{-1} and constant acceleration $(2\mathbf{i} - \mathbf{j})$ m s^{-2}. At time $t = 4$ seconds, P is at the point B with position vector $(-15\mathbf{i} + 10\mathbf{j})$ m. Find $\mathbf{u}$.

3. At time $t = 0$ seconds a particle P is passing through a fixed point O with a velocity of $(6\mathbf{i} - 4\mathbf{j})$ m s^{-1}. P has a constant acceleration of $(\mathbf{i} + 3\mathbf{j})$ m s^{-2} for $0 \leq t \leq 4$.
   (a) Find the velocity of P when $t = 2$.
   (b) Find the displacement of P when $t = 4$.

The following questions involve two particles or objects.

4. Two particles A and B start from a fixed point O at time $t = 0$ seconds. Particle A moves with an initial velocity of $(\mathbf{i} - \mathbf{j})$ m s^{-1} and a constant acceleration of $(\mathbf{i} + 2\mathbf{j})$ m s^{-2}. Particle B has a constant velocity of $(2\mathbf{i} + 2.5\mathbf{j})$ m s^{-1}.
   (a) Find the velocity of A at any time t.

Exercise 2F...

   (b) Find the time at which particles A and B are moving parallel to each other.
   (c) Find the speed of A when $t = 7$.

5. A particle A starts at a point with position vector $\begin{pmatrix} 9 \\ 9 \end{pmatrix}$ m. The initial velocity of A is $\begin{pmatrix} 3 \\ -2 \end{pmatrix}$ m s^{-1} and it has constant acceleration $\begin{pmatrix} -1 \\ 1 \end{pmatrix}$ m s^{-2}. Another particle B has initial velocity $\begin{pmatrix} 0 \\ 1 \end{pmatrix}$ m s^{-1} and constant acceleration $\begin{pmatrix} 1 \\ 0 \end{pmatrix}$ m s^{-2}. After 4 seconds the two particles collide. Find:
   (a) The position vector of the point at which they collide.
   (b) The position vector of B's starting position.
   (c) The **speeds** of the two particles when they collide. Give both answers in simplified surd form.

6. Caitlyn and Sonya are in boats on a boating lake. At time $t = 0$ seconds, Caitlyn is at a point with position vector $(7\mathbf{i} + 5\mathbf{j})$ m relative to a fixed point O on the lake. Sonya is at the point with position vector $(5\mathbf{i} + 18\mathbf{j})$ m relative to O. Caitlyn is moving with a constant speed of 0.25 m s^{-1} towards the point with position vector $(15\mathbf{i} + 11\mathbf{j})$ m.
   (a) Show that Caitlyn's velocity is $(0.2\mathbf{i} + 0.15\mathbf{j})$ m s^{-1}.
   (b) Find Caitlyn's position vector at any time t.

   Sonya's boat has initial velocity $(0.1\mathbf{i} - 0.3\mathbf{j})$ m s^{-1} and moves with acceleration $(0.02\mathbf{i} - 0.02\mathbf{j})$ m s^{-2}.
   (c) Find Sonya's velocity at any time t.
   (d) Show that Sonya's position vector relative to O at any time t is given by:
   $(5 + 0.1t + 0.01t^2)\mathbf{i} + (18 - 0.3t - 0.01t^2)\mathbf{j}$ m
   At $t = 20$ seconds, Sonya is at the point P.
   (e) Find the position vector of point P.
   (f) Show that the two boats collide at point P.

2.7 Summary

If an object moves with constant velocity in 2 dimensions, this vector formula can be used to describe its motion:

r = **r**$_0$ + **v**t

If an object moves with constant acceleration in one dimension, the following **constant acceleration formulae** can be used:

$v = u + at$

$s = ut + \frac{1}{2}at^2$

$s = \frac{1}{2}t(u + v)$

$v^2 = u^2 + 2as$

Three of these formulae can be used in a vector form:

v = **u** + **a**t

s = **u**t + $\frac{1}{2}$**a**t^2

s = $\frac{1}{2}t$(**u** + **v**)

Questions involving two particles or objects often require the solution of two simultaneous equations.

Chapter 3
Motion Graphs

3.1 Introduction

Key words
- **Velocity-time graph**: A graph showing an object's velocity on the y-axis plotted against time on the x-axis.
- **Displacement-time graph**: A graph showing an object's displacement on the y-axis plotted against time on the x-axis.

Before you start
You should know:
- How to use the formulae for constant acceleration (the 'suvat' formulae).
- That average speed $= \dfrac{\text{distance travelled}}{\text{time taken}}$

In this chapter you will learn about
- Displacement-time graphs.
- Velocity-time graphs.

In the real world...
A speed camera on a busy road can record a car's speed as it passes and also take a photograph if the car is travelling above the speed limit.

Sometimes a more sophisticated system is in place: average speed cameras are in place on some roads. These measure a car's speed at two points on a stretch of road and calculate its average speed. Number plate recognition technology is used to correctly pair up the two speed readings.

Drivers breaking the speed limit may be fined. If the case ends up in court, the police can use evidence from technology such as average speed cameras to give a detailed picture of the car driver's behaviour.

Exercise 3A (Revision)

1. A family travels by car to see relatives.
   (a) For the first 100 metres the car accelerates from rest at 2 m s^{-2}. Find the velocity it reaches.
   (b) The car then maintains this velocity for 20 minutes. Find the distance covered during this time.
   (c) When the car parks it comes to rest uniformly in 5 seconds. Find the deceleration.
   (d) Find the car's average speed for the entire journey.

3.2 Displacement-Time Graphs

A **displacement-time graph** shows the displacement of an object plotted against time.

Key points
- The gradient of a straight-line section of the graph gives the velocity of the object during this phase of its motion.
- If a section of the graph has a negative gradient, the object has a negative velocity. The direction of travel has changed.
- A horizontal line on a displacement-time graph represents a velocity of zero, i.e. the object is stationary.

Worked Example

1. A man has just started a new job and he is lost inside the office building. He leaves Room 30, looking for the stairs on a long straight corridor. He walks east along the corridor for 2 minutes at a constant speed. He then realises he has come the wrong way. He walks back along the corridor at the same speed, passes room 30 and continues walking for another 2 minutes until he reaches the stairs. The man walks at 1 m s^{-1}.

(a) How far has the man walked by the time he returns to Room 30?
(b) Draw a displacement-time graph of the man's journey to the stairs, showing his displacement from Room 30.

Take east as the positive direction and west as negative.

(a) The man walks at 1 m s^{-1}, so in 2 minutes (120 seconds) the man walks 120 m. He then walks another 120 m to return to Room 30. In total he has walked 240 m by the time he returns to Room 30.

(b) The man starts with a displacement of 0 m. He walks 120 m in the positive direction in 120 s. He then returns to his start point and then 120 m in the negative direction. So we can draw the graph:

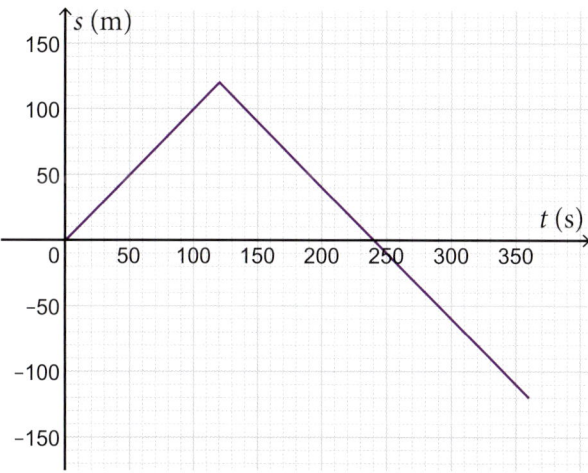

Exercise 3B

1. A car travels 1600 metres along a straight road. It travels at a constant velocity for the whole journey, which takes 80 s.
   (a) Sketch the car's displacement-time graph.
   (b) What is the car's velocity?

2. The displacement-time graph that follows shows the journey of a delivery man one morning. The y-axis shows his displacement from his home in kilometres.
   (a) Between which times was he travelling fastest?
   (b) How far from home was he at 10:30 am?
   (c) Describe what the delivery man was doing at 10:00 a.m.
   (d) Describe what the delivery man was doing between 11:00 and 11:30 a.m.

Exercise 3B...

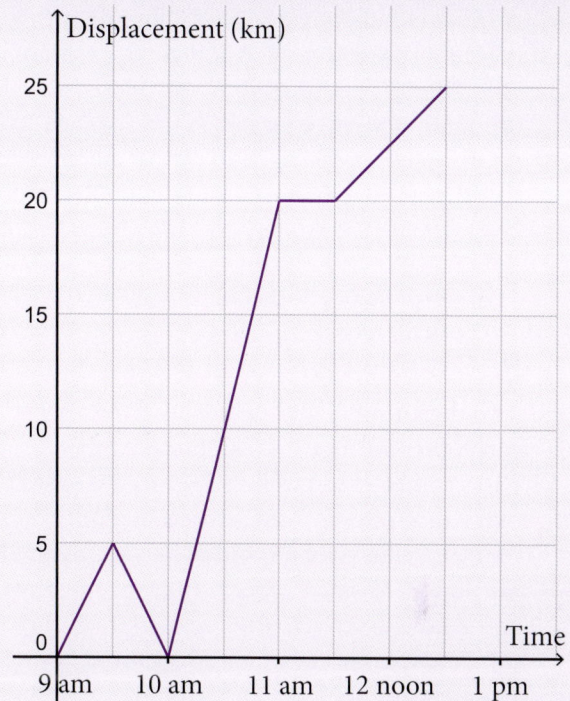

3. A ball is dropped from a height of 2 metres. It bounces twice and is then caught as it reaches the highest point on its trajectory after the second bounce. Sketch a displacement-time graph for the ball's motion.

4. The diagram shows a displacement-time graph for a cyclist on a short journey. The labels O, A, B, C and D represent points on his journey.

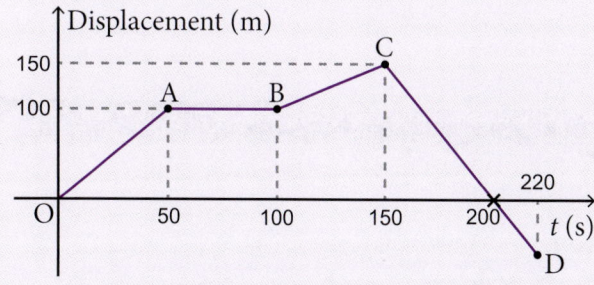

(a) Find the velocity of the cyclist during the first 50 seconds of his journey, from point O to point A.
(b) The cyclist travels at a constant velocity as he travels from C to D. Find the displacement of D from O.
(c) Find the cyclist's average speed for the entire journey.

AS 2: APPLIED MATHEMATICS

Exercise 3B...

5. A snooker ball moves in a straight line with a constant speed of u m s^{-1}. It hits the cushion after a time T seconds.
   (a) Find, in terms of u and T, the distance travelled to the cushion by the snooker ball.
   (b) The snooker ball rebounds along the same path with a constant speed of $\left(\dfrac{u}{2}\right)$ m s^{-1}. Find, in terms of T, the time taken for the return journey.
   (c) Sketch the displacement-time graph.

6. Two dogs Amber and Benny head towards each other. At time $t = 0$ seconds, Amber passes point A and walks at a constant speed of 0.2 m s^{-1}. Benny passes through point B at $t = 0$ and runs with a speed of 1 m s^{-1}. Benny accelerates towards Amber at 0.1 m s^{-2}. Points A and B are 9 metres apart.
   (a) Find the time at which Amber and Benny meet and their distance from point A at this time.
   (b) On the same diagram, sketch the displacement–time graphs for Amber and Benny from $t = 0$ to the time at which they meet.

7. A police car chases a suspect's car along a straight road. At time $t = 0$, the suspect is 200 m ahead. The police car is travelling with a speed of u m s^{-1}. The suspect is travelling at $\left(\dfrac{u}{2}\right)$ m s^{-1}.
   (a) Find, in terms of u, how long it takes for the police car to catch up with the suspect.
   (b) Find the distance travelled by the police car in catching up with the suspect.
   (c) Draw a displacement-time graph for the motion of both vehicles on the same diagram.

8. The displacement-time graph for a particle is shown below.

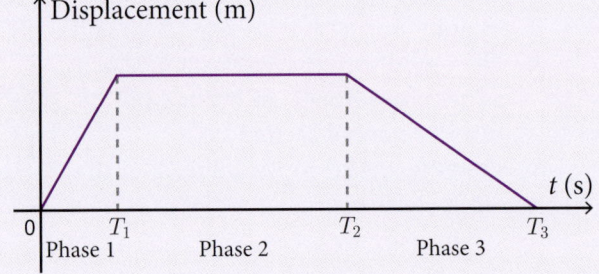

Exercise 3B...

The speed of the particle during Phase 1 of its motion is three times larger than its speed during Phase 3. The particle is in motion for 18 seconds longer than it is stationary. The total time for the particle's journey is 1 minute and 18 seconds.
   (a) Find the times T_1, T_2 and T_3.
   (b) The speed during Phase 1 is 2.5 m s^{-1}. Find the greatest distance the particle moves from its starting position.

3.3 Velocity-Time Graphs

A **velocity-time graph** shows the velocity of an object plotted against time.

Key points
If velocity is plotted against time for an object's journey:

- The distance travelled can be found by calculating the area between the line and the time axis.
- The gradient of a line represents the acceleration of the object.
- A horizontal line represents travel at a constant velocity. The object is still moving.
- A part of the graph with a negative gradient represents the object slowing down, or decelerating. It is still moving.

Worked Example

2. A ball is thrown vertically upwards from ground level with an initial speed of u m s^{-1}. The ball rises and then returns to ground level.
   (a) Find the time taken, in terms of u and g, for the ball to reach the top of its trajectory.
   (b) Sketch a velocity-time graph for the ball's motion.
   (c) The maximum height reached by the ball is 10 m. Find u. Use $g = 9.8$ m s^{-2}.

(a) Use the constant acceleration formulae for the upwards journey to find t.
When the ball reaches the top of its trajectory it has a velocity of 0 m s^{-1}.
$u = u$
$v = 0$
$a = -g$
$t = ?$

$v = u + at$
$0 = u - gt$
$t = \dfrac{u}{g}$

(b) The ball begins with a positive (upwards) velocity of u m s^{-1}.

The velocity decreases until, at the top of the trajectory, the velocity becomes zero.

The ball then moves downwards, so its velocity is negative. The ball's velocity becomes $-u$ by the time it reaches the ground. The time taken for the upwards and downwards journeys are equal, so it reaches the ground after $\dfrac{2u}{g}$ seconds. So we sketch:

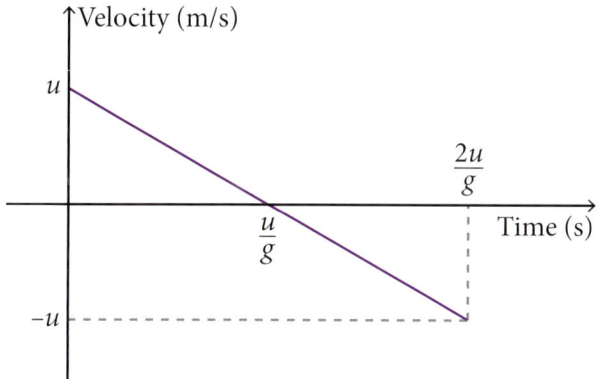

(c) Use the constant acceleration formulae for the upwards journey to find u:
$s = 10$
$u = ?$
$v = 0$
$a = -9.8$

$v^2 = u^2 + 2as$
$\Rightarrow u^2 = v^2 - 2as$
$u^2 = 0^2 - 2 \times (-9.8) \times 10$
$u^2 = 196$
$u = 14$ m s^{-1}

Velocity-time graphs often involve more than one object or person.

Worked Example

3. Aoife walks to school at a constant speed of 2 m s^{-1}. Her brother Max leaves home 2 minutes after Aoife. He cycles to school. Max accelerates from rest for 30 seconds with an acceleration of ⅙ m s^{-2} until he reaches his top speed. He then maintains this speed until he catches up with Aoife. As soon as he catches up with Aoife, Max decelerates uniformly to rest at 0.5 m s^{-2}.

(a) Find Max's top speed.
(b) Find how long it takes Max to decelerate.
(c) Draw a velocity-time graph, showing both Aoife's journey and Max's journey on the same set of axes.
(d) Find out how long it takes Max to catch up with Aoife after he started cycling.
(e) Find how far Max has cycled when he catches up with Aoife.
(f) When Max catches up with Aoife, Aoife continues walking at 2 m s^{-1}. Find out who has travelled further by the time Max comes to rest.

(a) For the first part of Max's journey he has constant acceleration, so the suvat formulae can be used.
$u = 0$
$v = ?$
$a = \frac{1}{6}$
$t = 30$

$v = u + at$
$= 0 + \dfrac{1}{6} \times 30$
$= 5$ m s^{-1}

(b) During the final phase of Max's journey, he has constant acceleration again (since the question says that he decelerates uniformly), so the suvat formulae can be used again.
$u = 5$
$v = 0$
$a = -0.5$
$t = ?$

$v = u + at$
$0 = 5 + (-0.5 \times t)$
$0.5t = 5$
$t = 10$ s

> **Note:** It is important to remember that the constant acceleration formulae can only be used when the acceleration is constant.
>
> So, for example, it is possible to use the formulae for the first phase of Max's journey, while he is accelerating with a constant acceleration. It is also possible during the final phase while he is decelerating, since the acceleration is constant during this phase of the motion.

(c) Let T seconds be the time at which Max catches up with Aoife. So we can draw the graph:

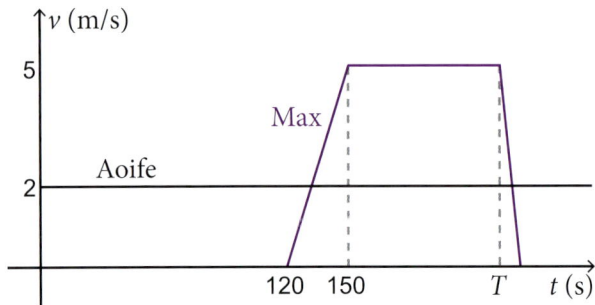

(d) When Max catches up with Aoife at time T, their distances travelled are equal. Therefore, the area below Max's graph is equal to the area below Aoife's graph up to the time T.

Therefore, to find Aoife's distance travelled, find the area of a rectangle with width T and height 2, which is $2T$.

To find Max's distance travelled, find the area of a trapezium. The parallel sides have length $T - 120$ and $T - 150$. The distance between the parallel sides is 5.

Therefore, Max's distance travelled:
$$= \frac{1}{2}(5)\big((T-120) + (T-150)\big)$$
$$= \frac{1}{2}(5)(2T - 270)$$
$$= 5T - 675$$

Since Max and Aoife have travelled the same distance in time T:
$2T = 5T - 675$
$3T = 675$
$T = 225$

This is 105 seconds (or 1 minute 45 seconds) after Max started cycling.

(e) Max's distance travelled, in terms of T, is $5T - 675$. Since $T = 225$:
Max's distance = $5(225) - 675$
$= 450$ m

Alternatively, calculate Aoife's distance travelled, since she has travelled the same distance at this time:
Distance travelled = $2T = 2(225) = 450$ m

(f) After catching up with Aoife, Max decelerates to rest. During these 10 seconds, his additional distance travelled is the area of a triangle, with a base of 10 and a height of 5.

Max's additional distance = $\frac{1}{2} \times 10 \times 5 = 25$ m

Aoife continues walking with constant speed. Her additional distance travelled is the area of a rectangle with base of 10 and height of 2.

Aoife's additional distance = $10 \times 2 = 20$ m

By the time Max comes to rest, he is $25 - 20 = 5$ metres ahead of Aoife.

Exercise 3C

1. A mouse runs in a straight line across a kitchen floor. It accelerates from rest to 2 m s^{-1} in 1.5 seconds. It then runs for 4.5 seconds at a constant speed of 2 m s^{-1} before slowing to rest uniformly in 0.5 seconds.
   (a) Draw a velocity-time graph of the mouse's motion.
   (b) Find the mouse's acceleration.
   (c) Find the mouse's deceleration.
   (d) How far does the mouse travel?

2. Two trains, A and B, run on parallel straight tracks. Initially both are at rest in a station and side by side. At time $t = 0$ seconds, train A starts to move. It moves with constant acceleration for 10 s until it reaches a speed of 24 m s^{-1}. It then moves with a constant speed of 24 m s^{-1}.

   When $t = 50$ s, train B starts to move in the same direction as train A. It accelerates with the same initial acceleration as train A, up to a speed of 36 m s^{-1}. It then moves at a constant speed of 36 m s^{-1}. Train B overtakes train A at $t = T$, after both trains have reached their maximum speed.
   (a) Find the acceleration of each train.
   (b) Sketch, on the same diagram, the velocity-time graphs of both trains for $0 \leq t \leq T$.
   (c) Find the value of T.

3. Two cars are on a straight racetrack. At time $t = 0$, car A starts from rest. It accelerates at 3 m s^{-2} for 10 seconds.
   (a) Find car A's speed after 10 seconds.
   Car A then maintains this speed.
   (b) At time $t = 15$, car B starts from rest. It accelerates at 0.72 m s^{-2} for 50 seconds. Find the speed of car B after this time.
   Car B then maintains this speed and catches up with car A at time T seconds.
   (c) Sketch, on the same diagram, the speed-time graphs for both cars up to T seconds.

Exercise 3C...

(d) Find the value of T.
(e) Find the distance travelled by car B at this time.

4. Alfie is running at a steady speed of 3 m s^{-1} on a straight road. When he passes Caitlyn, who is at rest, he keeps running at the same speed. Twelve seconds after Alfie passes her, Caitlyn gets on her bike and starts to chase after Alfie. She accelerates to a top speed of 5 m s^{-1} in 20 seconds.
   (a) Find Caitlyn's acceleration in m s^{-2}.

   Caitlyn maintains her speed of 5 m s^{-1} until she catches up with Alfie. As soon as she catches up with Alfie, she decelerates uniformly to rest at ⅓ m s^{-2}.
   (b) Find how long it takes Caitlyn to decelerate to rest.
   (c) Draw a velocity-time graph, showing both Alfie's journey and Caitlyn's journey on the same set of axes.
   (d) Find out how long it takes Caitlyn to catch up with Alfie after he started running.
   (e) Find how far Caitlyn has cycled when she catches up with Alfie.
   (f) While Caitlyn is decelerating, Alfie continues to run at 3 m s^{-1} and catches up with Caitlyn. Find how many seconds elapse between Caitlyn catching up with Alfie and Alfie catching up with Caitlyn.
   (g) When Caitlyn finally comes to rest, Alfie is ahead. Find how far ahead he is at this time.

5. Towns A, B and C all lie on the same straight road. Town B is 20 km east of Town A. Town C is 30 km east of Town B.
   A blue car leaves Town A travelling east. The blue car accelerates from rest to a speed of 40 m s^{-1} in 6 seconds. It then maintains this speed.
   A yellow car leaves Town B at the same time as the blue car leaves Town A. It travels east with a constant speed of 20 m s^{-1}.
   (a) Plot a velocity-time graph for each car on the same diagram.
   (b) Find the time each car takes to reach Town C. Which car reaches Town C first?

6. The graph below shows the velocities of particles A and B, plotted against time.

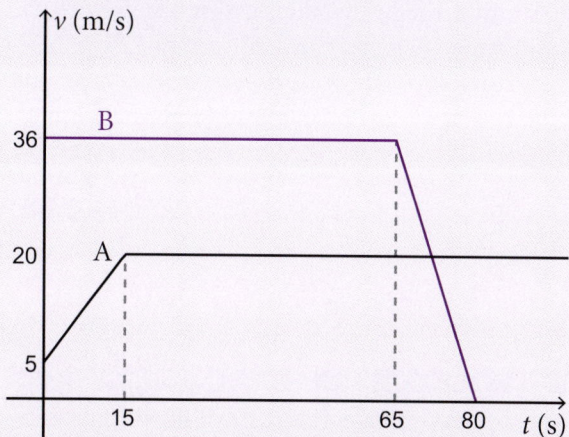

   (a) What speed does particle A have initially?
   (b) What speed does particle B have initially?
   (c) For how long does particle B maintain this speed?
   (d) Calculate the acceleration of particle A during the first 15 seconds of its motion.
   (e) Calculate the deceleration of particle B during the final 15 seconds of its motion.
   (f) Which particle comes to rest? Describe the behaviour of the other particle.
   (g) Which particle travels furthest during the first 80 seconds?
   (h) At what time have the two particles travelled the same distance? Give your answer to the nearest whole number of seconds.

7. A tortoise and a hare are running a race, which is 60 metres long. The tortoise is slow and steady. The hare is fast, but a little bit lazy. At exactly midday the tortoise sets off, walking from the start line towards the finish line at a constant speed of 0.5 m s^{-1}.
   The hare sets off 100 seconds later. He accelerates uniformly for 10 seconds, reaching his top speed of 3 m s^{-1}. The hare maintains this speed for 10 seconds, then decelerates to rest in a further 10 seconds.
   (a) Plot a velocity-time graph for both the tortoise and the hare on the same diagram.
   (b) Who wins the race? Show all of your working.
   (c) How many seconds elapse between the times the two animals cross the finish line?

Exercise 3C...

8. Two trucks A and B move in the same direction along a straight, horizontal road. At time $t = 0$, they are side-by-side, passing a sign S. Truck A is travelling at a constant speed of 27 m s^{-1}. Truck B passes S with a speed of 15 m s^{-1} and moves with a constant acceleration of 3 m s^{-2}. Truck B passes a second sign R, 58.5 metres from S. Find:
   (a) The speed of truck B as it passes sign R.
   (b) The distance of truck A from S when truck B passes sign R.
   (c) The time when truck B overtakes truck A.

9. A car is driven on a straight road at 30 m s^{-1}. The car passes a sign S warning of roadworks ahead. The speed limit through the roadworks is 20 m s^{-1}. After passing the sign, the driver takes 1 second to react and then begins to decelerate. He reduces his speed to 20 m s^{-1} with a constant deceleration of 2 m s^{-2}. He then continues at this constant lower speed.
   (a) Sketch a velocity-time graph to show the motion of the car after it passes the sign S.
   (b) Assuming drivers take 1 second to react to the sign at S, and then decelerate at 2 m s^{-2}, calculate the minimum distance the sign should be from the roadworks to ensure deceleration from 30 m s^{-1} to 20 m s^{-1}.

10. Starting at time $t = 0$ seconds, Particle F moves from rest in a straight line from a fixed point O with a constant acceleration of 0.4 m s^{-2}. At $t = 5$ seconds, Particle G starts to move along the same straight line. It accelerates at 1 m s^{-2} for 10 seconds. Immediately after reaching its top speed, Particle G decelerates to rest, with a deceleration of 1 m s^{-2}.
    (a) Find Particle G's top speed.
    (b) Find the time at which Particle G comes to rest.
    (c) Draw, on the same set of axes, a velocity-time graph for the motion of Particles F and G, up to the time that Particle G comes to rest.
    (d) Find the two times at which Particles F and G have both travelled the same distance from O. Give your two answers in seconds to one decimal place. You may assume that, when the two particles are at an equal distance from O, they do not collide, but pass through each other.

11. A car accelerates uniformly along a straight road. It moves from rest for 8 seconds, reaching a maximum speed of V m s^{-1}. The car then decelerates uniformly for 12 seconds until it comes to rest. The total distance travelled by the car is 200 m.
    As the car starts, a cyclist also starts from rest, at a point 6.25 metres ahead of the car. The cyclist accelerates at 2 m s^{-2} for 6 seconds along the same road in the same direction as the car. He then continues at a constant speed.
    (a) On the same diagram, sketch a velocity-time graph for both the car and the cyclist.
    (b) Find the top speed reached by the car.
    (c) Find the top speed reached by the cyclist.
    (d) The car catches up with the cyclist while both the car and cyclist are accelerating. Find the time at which this happens.
    (e) The cyclist catches up with the car again as the car is decelerating. Find the time at which this happens. Give your answer in seconds to 3 significant figures.

3.4 Mixed Section

The following exercise comprises questions that involve both types of motion graph.

Exercise 3D

1. (a) A displacement-time graph for the motion of an object meets the time axis. What does this indicate about the object's motion?
   (b) If a line on a velocity-time graph meets the time axis, what conclusion could you draw?
   (c) A section of a displacement-time graph lies below the time axis. What does this indicate about the object's motion?
   (d) If a line on a velocity-time graph lies below the time axis, what conclusion could you draw?
   (e) If the line on an object's displacement-time graph is horizontal, describe the object's behaviour during this time.
   (f) If the line on an object's velocity-time graph is horizontal, describe the object's behaviour during this time.

Exercise 3D...

2. A cat jumps from ground level. The cat's displacement-time graph is shown in the following diagram.

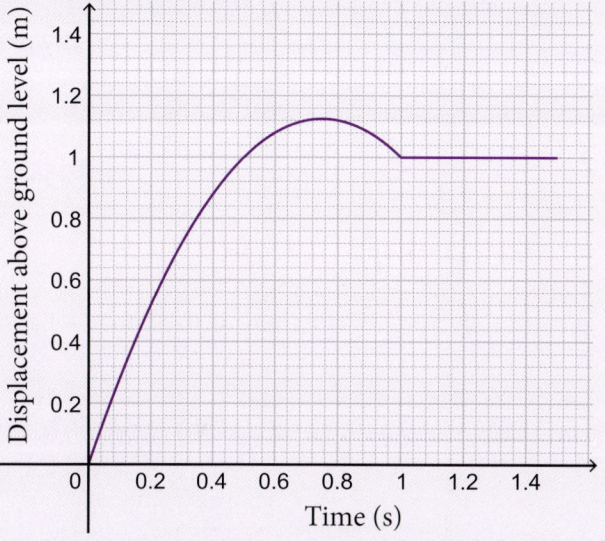

   (a) Describe the cat's motion between 0 and 1.5 seconds.
   (b) Sketch a velocity-time graph for the cat's journey.

3.5 Summary

A **displacement-time graph** shows the displacement of an object plotted against time. The gradient of a straight-line section of the graph gives the velocity of the object during this phase of its motion.

A **velocity-time graph** shows the velocity of an object plotted against time. The gradient of a straight-line section of the graph gives the acceleration of the object during this phase of its motion. The distance travelled can be found by calculating the area between the line and the time axis.

Chapter 4
Forces

4.1 Introduction

Key words
- **Force**: An influence that can change the motion of an object, e.g. weight, friction or tension in a string. Force is a vector quantity.
- **Equilibrium**: An object is in equilibrium if it is stationary, or motionless. In this case there are either no forces acting on the object, or all the forces cancel each other out.
- **Resultant**: The resultant of two or more forces is the sum of the force vectors.
- **Magnitude**: The magnitude of a vector is its size or length.

Before you start
You should know about:
- Pythagoras' Theorem and trigonometry.
- Adding and subtracting vectors.
- The acceleration due to gravity.

What you will learn
In this chapter you will learn about:
- The concept of a force and different types of force.
- Resolving forces.
- Equilibrium.
- Finding resultant forces.

In the real world...

An aeroplane needs a force called lift, or upward thrust, to get it into the air.

An aeroplane wing is designed in a very special way, so that the air flows faster around the top of the wing than around the bottom of it.

The slower-moving air beneath the wing provides a greater force upon the wing than the faster moving air above it. If the plane is moving fast enough, this difference is big enough to overcome the plane's own weight and the plane takes off.

We say there is a **resultant** force upwards.

Exercise 4A (Revision)

1. Calculate:
   (a) $(\mathbf{i} + 3\mathbf{j}) + (6\mathbf{i} - 2\mathbf{j})$
   (b) $\begin{pmatrix} 2 \\ -5 \end{pmatrix} - \begin{pmatrix} -3 \\ 4 \end{pmatrix}$

2. The diagram shows a right-angled triangle.

   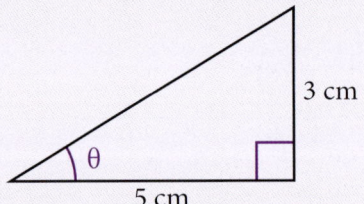

   Find:
   (a) the length of the hypotenuse,
   (b) the size of the angle θ.

3. (a) What is the value of the constant g, the acceleration due to gravity on Earth?
   (b) Two objects, an apple and a feather, are dropped from 10 metres in a vacuum tube, with both objects starting from rest. Which one reaches the ground first?
   (c) Why would the experiment described in part (b) give different results if the vacuum tube were not used?

4.2 The Concept of a Force

In your study of mechanics, you will encounter force of various types, some of which are described below.

Forces are vectors, so in a force diagram they are shown using arrows. Draw all force arrows coming out of the object, even if the object is being pushed.

The weight force
Weight should not be confused with mass. The **weight is a force** equal to the object's mass multiplied by the acceleration due to gravity, g.

$$W = mg$$

Remember that g is a constant, with a value of roughly 9.8 m s^{-2}.

The weight force always acts vertically downwards.

The normal reaction

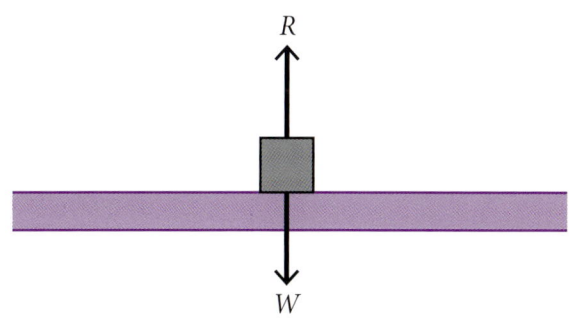

In the diagram, W is the block's weight force.

R is called the **normal reaction**. There is always a reaction force when an object rests upon or against a surface.

The word 'normal' is used because the reaction force acts at right angles to the surface.

This block is not moving; we say it is in **equilibrium**. The forces acting upwards must be equal to the forces acting downwards, so $R = W$.

Tension in a string and the friction force

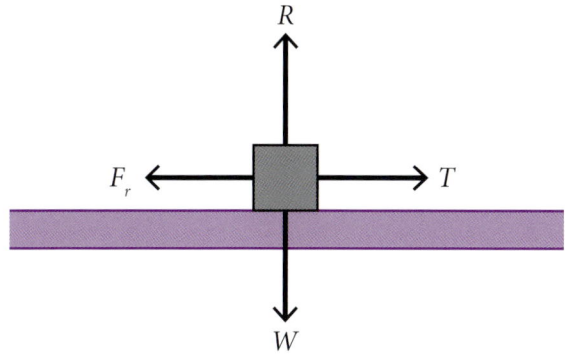

In this diagram, in addition to the weight and normal reaction forces, there are two more forces acting on the block.

The block is at rest (in equilibrium) on a **rough** surface, but it is being **pulled** to the right by a string.

T is called the **tension** in the string.

F_r is the **friction** force. There is friction because the surface is rough.

Since this block is in equilibrium, $T = F_r$.

The friction force always opposes the motion. If the tension force was large enough to make the block move, it would move to the right. Therefore, the friction force acts to the left.

Friction is discussed further in Chapter 6.

Thrust or compression in a rod

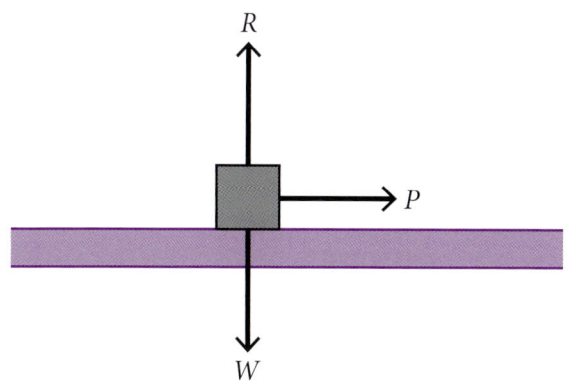

If the block is **pushed** from left to right (for example by a rod), there is a **thrust** or **compression** force in the rod. In this example, the block is being pushed from the left by a force of magnitude P newtons. Since forces are drawn as arrows coming **out** of the block, the thrust force is an arrow from the block to the right.

In this example, the surface is **smooth**, so there is no friction force opposing the motion. Since the forces acting to the right are greater than the forces acting to the left, this block is not in equilibrium. We say there is a **resultant** force in the horizontal.

Buoyancy

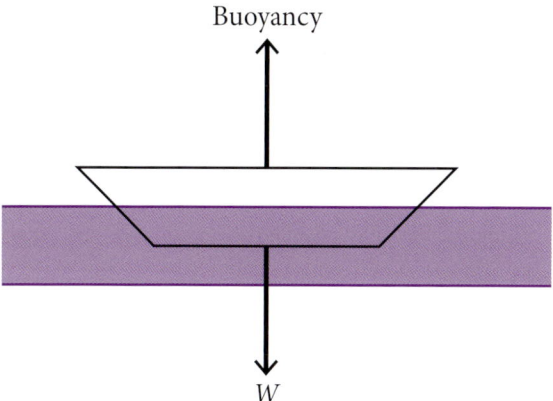

This diagram shows a boat on water. The boat is in equilibrium.

Its weight force acting vertically downwards is counteracted by a **buoyancy** force from the water.

Forward thrust and water resistance

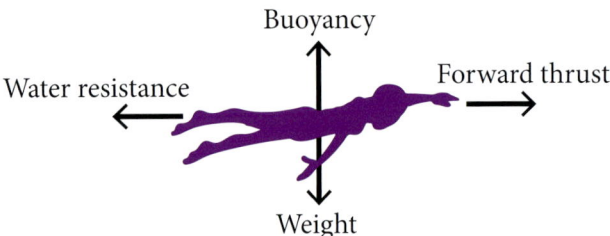

There are four forces acting on this swimmer. In addition to the swimmer's weight force (vertically downwards) and the buoyancy (vertically upwards), the swimmer is providing **forward thrust,** which propels her through the water.

There is also a resistance force from the water, opposing the motion.

Object on a rough inclined plane

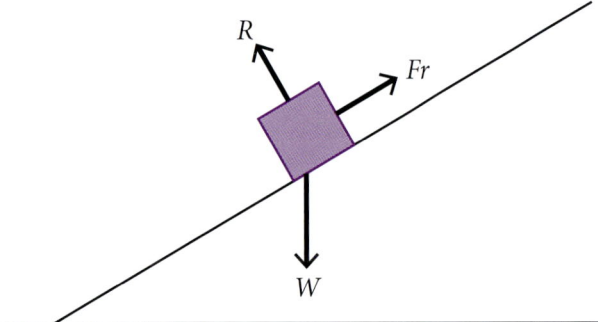

This block rests on a **rough, inclined plane**. There are three forces acting on the block. Its weight force acts vertically downwards. The normal reaction force acts at right angles to the plane. There is a friction force, since the slope is rough. The friction force always acts in the opposite direction to the motion.

Friction will be discussed further in Chapter 6.

Other examples

The following are other examples of forces acting on objects in motion:

- An aircraft is given **forward thrust** by its engines. The aircraft is also given an upwards force called **lift** (or **upward thrust**).
- A car is given a forward force by its engine, sometimes called a **tractive force**.
- A lift is pulled upwards by **tension** in the lift cable. It also experiences its own weight force downwards.
- A downhill skier experiences three forces: her weight force, a normal reaction from the ski slope and a friction force opposing the motion.

Exercise 4B

1. Write down the names of the forces acting on these objects.
   (a) A rescue helicopter heading out to sea.
   (b) The same rescue helicopter hovering above the sea.
   (c) A car travelling at a constant speed on a motorway.
   (d) A box being dragged across a rough floor by a horizontal rope.
   (e) A toboggan sliding down an icy slope.
   (f) A ball kicked vertically upwards into the air, rising and then falling.

4.3 Resolving Forces in Two Dimensions

Splitting a force into horizontal and vertical components

Any force in a two-dimensional plane can be split into two components, one horizontal, the other vertical. This is called **resolving** a force.

Worked Example

1. A force of 4 N acts on a particle P at an angle of 65° to the horizontal, as shown in the diagram. Resolve this force into its horizontal and vertical components.

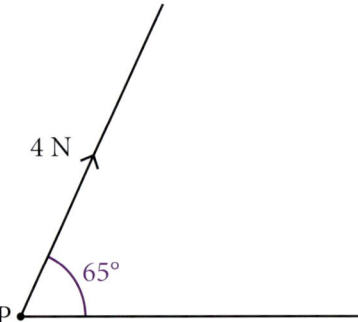

Resolve the 4 N force into its two components as shown in the next diagram.

Starting from the position of the particle, draw the horizontal component R_H and then the vertical component R_V to form a right-angled triangle.

4: FORCES

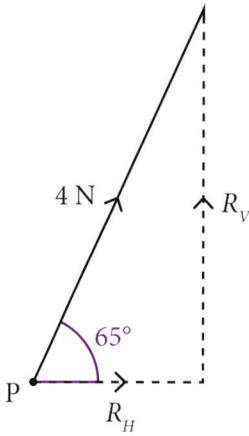

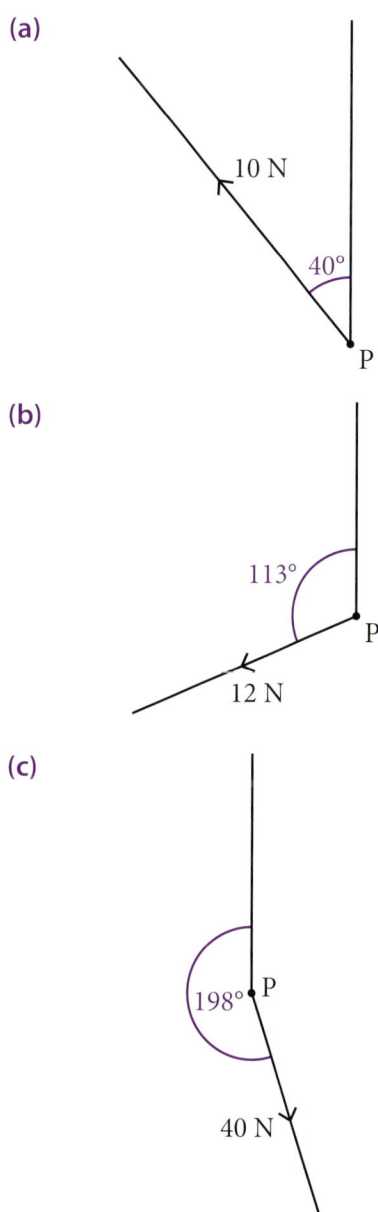

Using trigonometry in this triangle:

$\cos 65° = \dfrac{R_H}{4}$

$\Rightarrow R_H = 4 \cos 65° = 1.69 \text{ N}$

$\sin 65° = \dfrac{R_V}{4}$

$\Rightarrow R_V = 4 \sin 65° = 3.63 \text{ N}$

Shortcut

To move from the direction of the force to the horizontal involves crossing the angle 65°. In this case, to find the horizontal component, multiply the size of the force by cos 65°:

$\therefore R_H = 4 \cos 65° = 1.69 \text{ N}$

To find the vertical component, multiply the force by sin 65°:

$R_V = 4 \sin 65° = 3.63 \text{ N}$

Note: It is conventional to use a solid line for the original force, but to use a dashed or dotted line for each component.

When resolving a force, if you begin by constructing a triangle, it may then be necessary to find the size of one of the interior acute angles.

Worked Example

2. The three diagrams that follow each show a single force acting on a particle P. In each case, resolve the force into its horizontal and vertical components.

In each case, draw a diagram showing the force vector split up into its horizontal and vertical components. For each one, start from the position of P, draw the horizontal component first, then draw the vertical component.

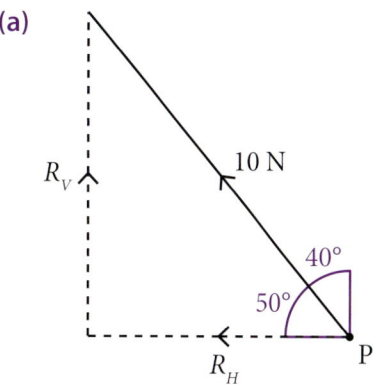

39

To move from the direction of the 10 N force to the horizontal, the angle 50° is crossed.

∴ $R_H = 10 \cos 50° = 6.43$ N
$R_V = 10 \sin 50° = 7.66$ N

(b)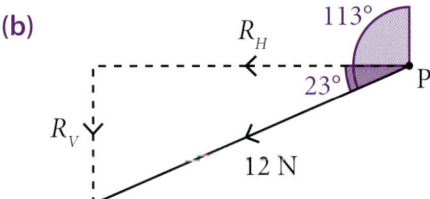

From the direction of the force to the horizontal, an angle of 23° is crossed.

∴ $R_H = 12 \cos 23° = 11.0$ N
$R_V = 12 \sin 23° = 4.69$ N

(c)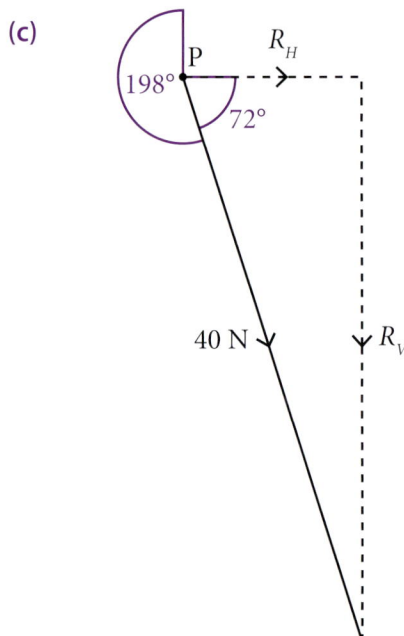

From the direction of the force to the horizontal, an angle of 72° is crossed.

∴ $R_H = 40 \cos 72° = 12.4$ N
$R_V = 40 \sin 72° = 38.0$ N

Note: All answers have been given to 3 significant figures.

Splitting a force into two components parallel and perpendicular to a slope

In the previous set of examples, forces were resolved into their horizontal and vertical components. In the following example, a weight force is resolved into two components parallel and perpendicular to a slope.

Worked Example

3. The diagram shows a 5 kg block on a slope inclined at 30° to the horizontal.
   (a) Find the size of the block's weight force.
   (b) Resolve the weight force into two components, one parallel to the slope, the other perpendicular to the slope.

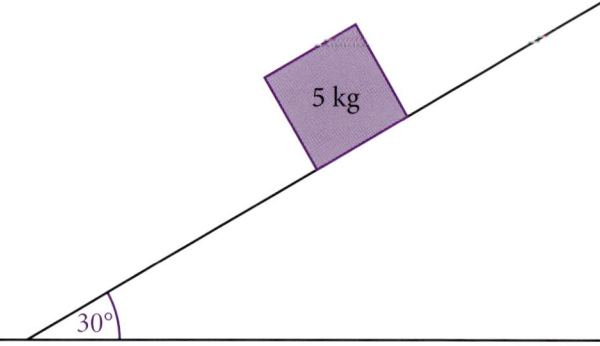

(a) The weight force is $5g$ N or 49 N acting vertically downwards.

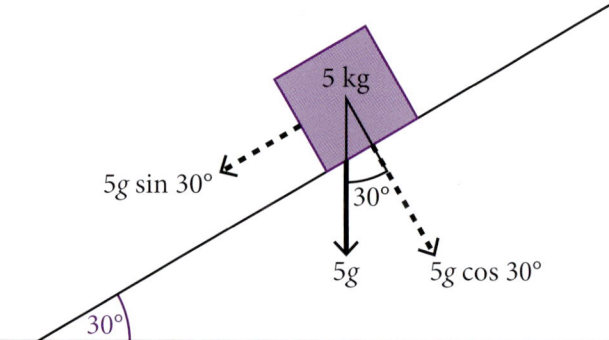

Note: In this diagram all other forces acting on the block, for example the normal reaction, have been omitted for clarity.

(b) The angle between the vertical and the perpendicular is the same as the angle of the slope, 30° in this case. (It is left as an optional exercise to show that these two angles are always equal.)

The $5g$ force has a component $5g \cos 30°$ in the perpendicular direction, since to move from the direction of the force to the perpendicular direction an angle of 30° is crossed.

The component in the direction parallel to the slope is $5g \sin 30°$.

Perpendicular component:
$5g \cos 30° = 42.4$ N (3 s.f.)

Parallel component:
$5g \sin 30° = 24.5$ N

Exercise 4C

1. In each diagram below, a single force is acting on the particle P. Resolve the force shown into its horizontal and vertical components.

   (a)

   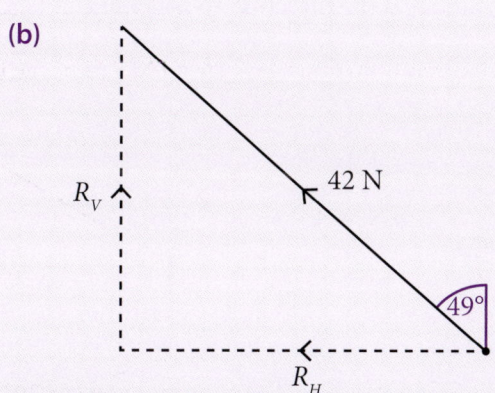

   (b)

   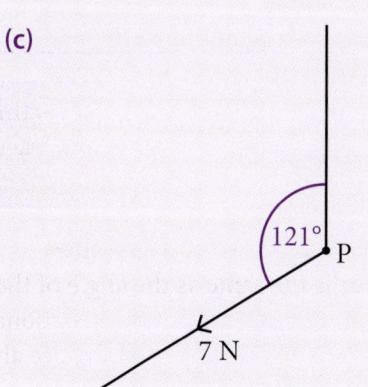

   (c)

   (d)

   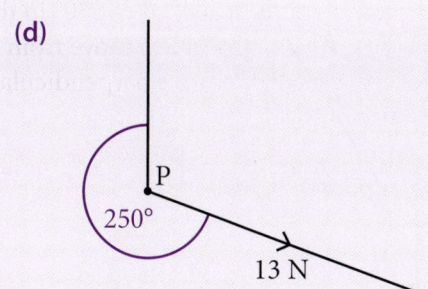

Exercise 4C...

   (e)

   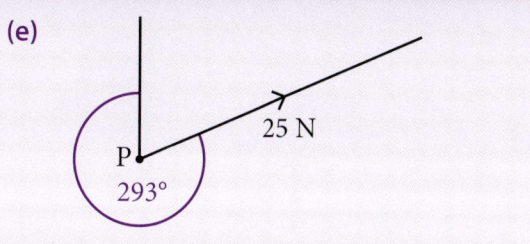

   (f)

   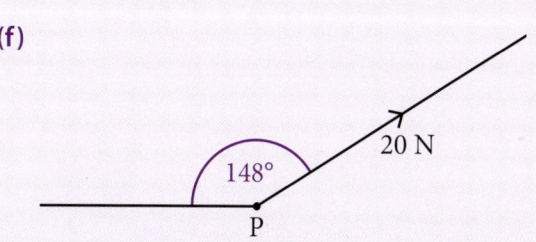

2. Each of the diagrams below shows an object resting on an inclined plane. For each one:
   (i) Copy the diagram, showing the object's weight force. Also show the two components of the weight force, one parallel and one perpendicular to the plane.
   (ii) Calculate the size of the two components, giving your answers to 3 significant figures where appropriate.

   (a)

   (b)

   (c)

   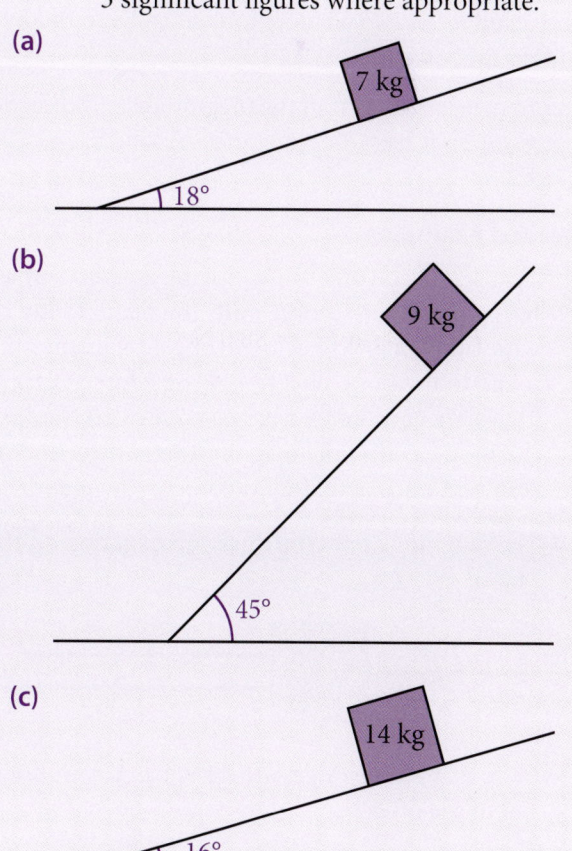

Exercise 4C...

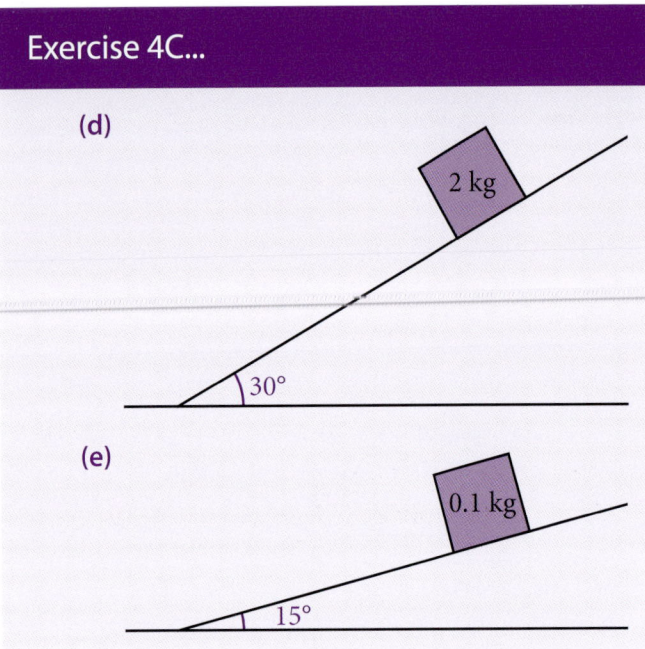

(d) 2 kg, 30°

(e) 0.1 kg, 15°

3. A particle of mass 6 kg rests on a slope inclined at 27° to the horizontal. Find the component of the particle's weight force acting down the slope, parallel to it.

4.4 The Resultant of a System of Forces

The resultant of two or more forces can be found by adding the force vectors.

You may be asked to find the magnitude and direction of the resultant force.

Worked Examples

4. Two removal workers are moving a piano across a horizontal floor. George **pulls** the piano to the right with a force of 80 N using a rope. Mel **pushes** the piano in the same direction using a force of 100 N. There is a constant resistance force of 50 N opposing the motion. Find the magnitude and direction of the resultant horizontal force acting on the piano.

The diagram shows the three forces acting on the piano.

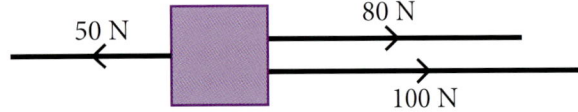

Note that all the forces are shown as arrows coming out of the object.

The three forces all act in the horizontal. To find the resultant, add the two forces acting to the right and subtract the force acting to the left:

80 + 100 − 50 = 130

The resultant force acting on the object is 130 N to the right.

> **Note:** The question only asks for the resultant horizontal force. It is not necessary to consider the weight force or the normal reaction, which act in the vertical.

5. A rectangular frame comprises four rigid struts. Forces act along each strut, as shown in the diagram. Find the magnitude of the resultant force acting on the frame and the direction in which it acts.

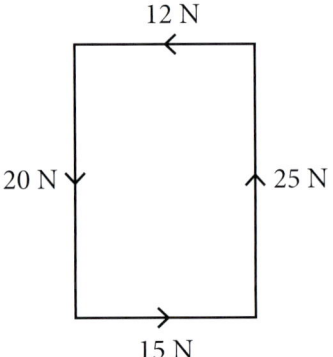

Consider the forces acting in the horizontal:
15 − 12 = 3 N to the right.

Consider the forces acting in the vertical:
25 − 20 = 5 N upwards

Sketch the resultant force:

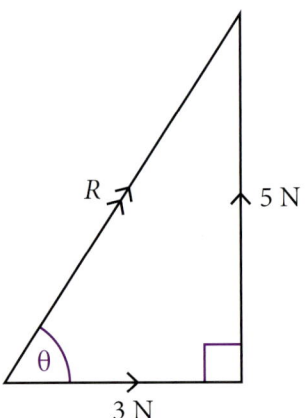

$|R| = \sqrt{5^2 + 3^2} = \sqrt{34}$
$|R| = 5.83 \text{ N}$ (3 s.f.)

To find the angle θ use trigonometry in this right-angled triangle:

$\tan \theta = \dfrac{5}{3}$

$\theta = \tan^{-1}\left(\dfrac{5}{3}\right) = 59.0°$ (3 s.f.)

The resultant force has a magnitude of 5.83 N and acts at an angle 59.0° above the horizontal.

Force vectors may be given in component form. The resultant is the sum of all the forces acting. Addition of vectors in component form involves adding the **i** components and adding the **j** components.

Worked Example

6. Two forces A and B act on a particle P, where $A = \begin{pmatrix} 1 \\ -3 \end{pmatrix}$ N and $B = \begin{pmatrix} -7 \\ 1 \end{pmatrix}$ N.

Find the magnitude and direction of the resultant force.

The two forces acting on the particle P are shown.

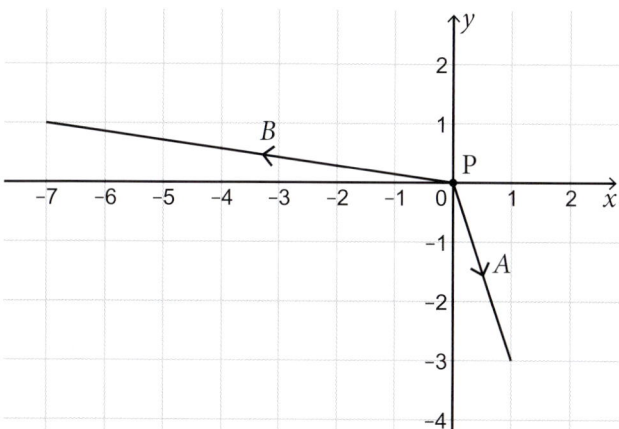

The resultant R is given by:
$R = A + B$
$= \begin{pmatrix} 1 \\ -3 \end{pmatrix} + \begin{pmatrix} -7 \\ 1 \end{pmatrix}$
$= \begin{pmatrix} -6 \\ -2 \end{pmatrix}$ N

This vector addition is shown in the next diagram. Force vector A is drawn first from (0, 0) to the point (1, −3). From this point the force vector B is drawn, 7 units to the left and one upwards. The resultant R is also shown.

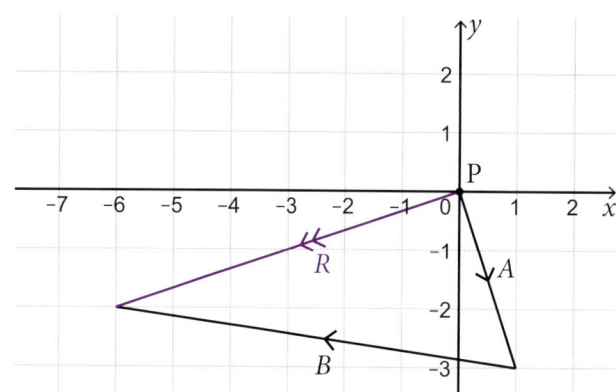

This graphical representation of vector addition is called a **vector triangle**.

To find the magnitude of R:

$R = \begin{pmatrix} -6 \\ -2 \end{pmatrix}$

$|R| = \sqrt{(-6)^2 + (-2)^2} = \sqrt{40}$
$= 6.32$ N (3 s.f.)

The direction is the angle between the resultant and the horizontal. It is marked θ in the diagram below.

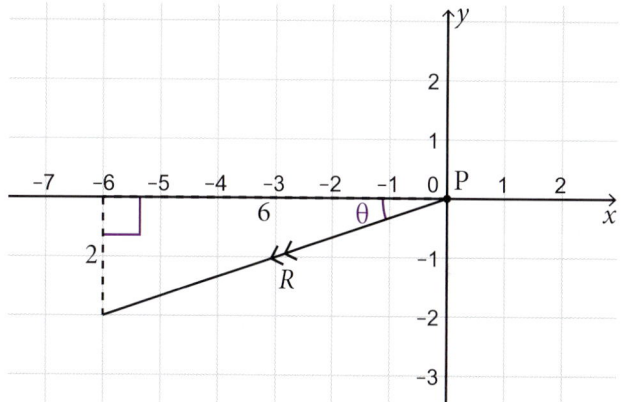

$\tan \theta = \dfrac{2}{6}$

$\theta = \tan^{-1}\left(\dfrac{2}{6}\right) = 18.4°$ (3 s.f.)

The resultant has a magnitude of 6.32 N and acts at 18.4° below the negative x-axis.

Note: When answering a question involving the addition of vectors to find the resultant, it is not necessary to draw all three diagrams as above. In this example the diagrams are included to aid with the explanation. In the calculation of the direction, the third diagram may be required to visualise the opposite and adjacent sides in the right-angled triangle.

AS 2: APPLIED MATHEMATICS

Force vectors may also be given in terms of their magnitude and direction. If there are two forces acting, it may be possible to find the resultant using trigonometry.

Worked Example

7. Two forces of 8 N and 5 N act on the particle P, as shown in the diagram. The 8 N force acts horizontally and the angle between the two forces is 30°. Find the magnitude and direction of the resultant force.

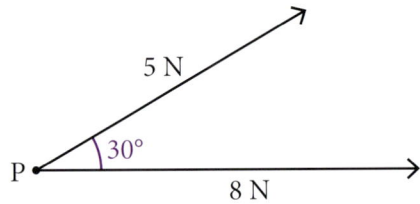

Draw a diagram showing addition of the two force vectors. The 8 N force has been drawn from P. The 5 N force is then drawn from the end point of the 8 N vector. The resultant R is also shown.

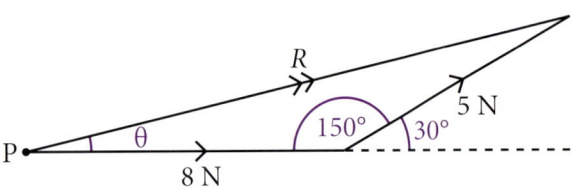

The 5 N force acts at an angle of 30° to the horizontal. The obtuse angle inside the vector triangle is therefore 150°.

In this triangle two of the sides are known, as well as the angle between them. The magnitude of R can be found using the cosine rule:

$|R|^2 = 8^2 + 5^2 - 2(8)(5) \cos 150°$
$|R|^2 = 158.28 ...$
$|R| = 12.581 ...$
$|R| = 12.6$ N (3 s.f.)

The sine rule can be used to find the direction:

$\dfrac{\sin \theta}{5} = \dfrac{\sin 150°}{12.581 ...}$

$\sin \theta = 0.1987 ...$
$\theta = 11.5°$ (3 s.f.)

The resultant force has a magnitude of 12.6 N and acts at 11.5° above the horizontal, with both numbers given to 3 significant figures.

To find the resultant if more than two forces are acting, resolve each force into its horizontal and vertical components.

Worked Example

8. Three coplanar forces act on a particle P, as shown in the diagram below.

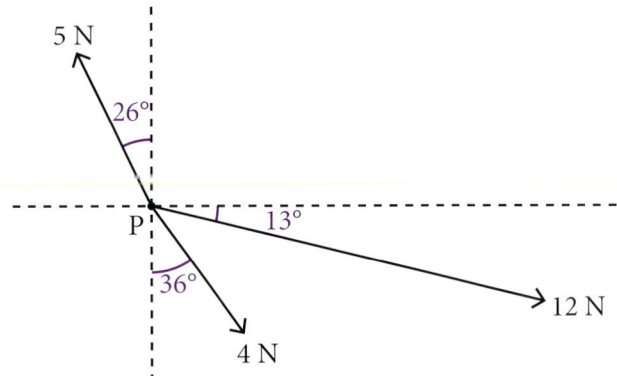

Find the magnitude and direction of the resultant force R acting on the particle. Give both answers to 1 decimal place.

Split the 5 N force into horizontal and vertical components:

Horizontal component: 5 sin 26° = 2.192 N to the left

Vertical component: 5 cos 26° = 4.494 N upwards

In vector form this is $-2.192\mathbf{i} + 4.494\mathbf{j}$ or $\begin{pmatrix} -2.192 \\ 4.494 \end{pmatrix}$ N

Split the 12 N force into horizontal and vertical components:

Horizontal component: 12 cos 13° = 11.692 N to the right

Vertical component: 12 sin 13° = 2.699 N downwards

In vector form this is $11.692\mathbf{i} - 2.699\mathbf{j}$ or $\begin{pmatrix} 11.692 \\ -2.699 \end{pmatrix}$ N

Split the 4 N force into horizontal and vertical components:

Horizontal component: 4 sin 36° = 2.351 N to the right

Vertical component: 4 cos 36° = 3.236 N downwards

In vector form this is $2.351\mathbf{i} - 3.236\mathbf{j}$ or $\begin{pmatrix} 2.351 \\ -3.236 \end{pmatrix}$ N

The resultant is the sum of all three vectors:

$R = \begin{pmatrix} -2.192 \\ 4.494 \end{pmatrix} + \begin{pmatrix} 11.692 \\ -2.699 \end{pmatrix} + \begin{pmatrix} 2.351 \\ -3.236 \end{pmatrix}$

$= \begin{pmatrix} 11.851 \\ -1.441 \end{pmatrix}$ N

The resultant R is sketched below.

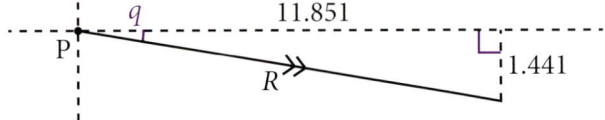

The magnitude of R can be found using Pythagoras' Theorem:

$|R| = \sqrt{(11.851)^2 + (1.441)^2} = 11.9 \text{ N}$ (1 d.p.)

Its direction can be found using trigonometry:

$\tan \theta = \dfrac{1.441}{11.851}$

$\theta = \tan^{-1}\left(\dfrac{1.441}{11.851}\right) = 6.9°$ (1 d.p.)

The resultant has a magnitude of 11.9 N and acts at 6.9° below the positive x-axis.

Find the resultant force acting on an object on an inclined plane by resolving each force into its components parallel and perpendicular to the plane.

Worked Example

9. A block of mass 1.1 kg is on a smooth plane, which is inclined at an angle of 20° to the horizontal. A rope is attached to the block. The tension in the rope is 1.5 N and this acts to pull the block along the line of greatest slope up the plane, parallel to it. The block also experiences a normal reaction force, which acts perpendicular to the plane.
   (a) Sketch a diagram showing all the forces acting on the block.
   (b) Find the resultant force acting on the block.

(a) The weight force of 1.1g is shown resolved into its two components parallel and perpendicular to the plane.

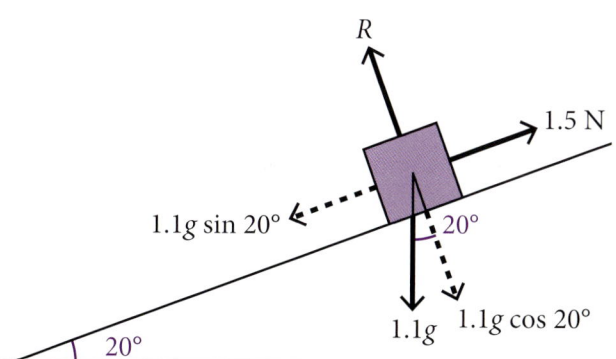

(b) The reaction force R is equal in magnitude to the perpendicular component of the weight force, so there is no resultant force in the perpendicular direction.

Parallel to the slope, the component of the weight force down the plane is greater than the tension of 1.5 N acting up the plane.

$1.1g \sin 20° - 1.5 = 2.19 \text{ N}$ (3 s.f.)

Therefore, the resultant force acting on the block is 1.85 N acting down the plane.

Exercise 4D

1. Four forces act on an object, as shown in the diagram. Find the magnitude and direction of the resultant force.

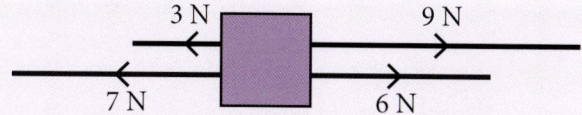

2. Four rods are connected in the shape of a rectangle. Forces are applied to each of the rods, as shown in the diagram.

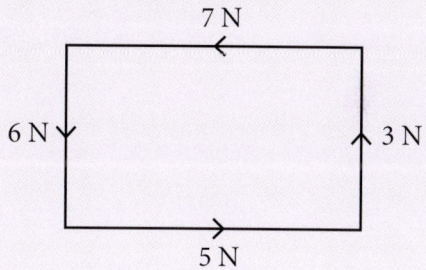

Find the magnitude of the resultant force acting on the rectangular frame and the direction in which it acts.

3. Three coplanar constant forces act on a particle P, as shown in the diagram.

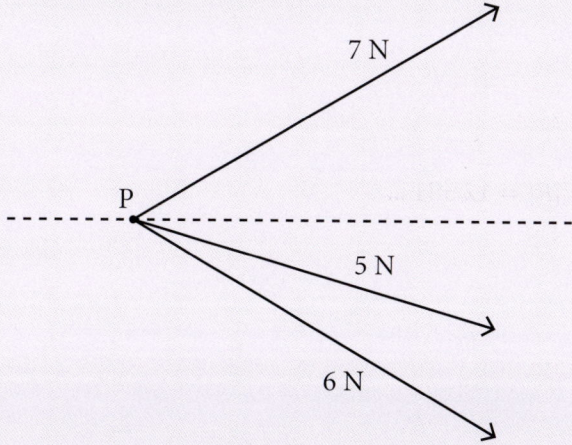

The 7 N force acts at an angle of 40° to the horizontal. The 5 N force acts at an angle of 15° to the horizontal. The 6 N force acts at an angle of 30° to the horizontal.

Find the resultant force acting on the particle and the direction in which it acts.

Exercise 4D...

4. Three tugboats are trying to release a cargo ship C that has become stuck in a canal, as shown in the diagram.

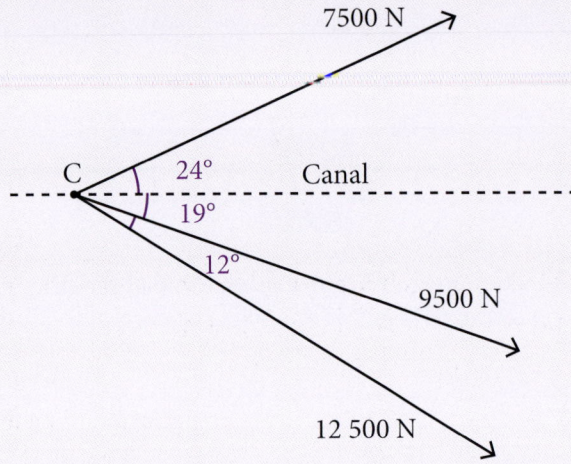

To free the ship, the resultant force must be at least 27 000 N in the direction of the canal (the horizontal in this diagram).

Do the tugboats manage to free the cargo ship? Show all your working.

5. Two forces of 70 N and 90 N act on a particle P. The 90 N force is horizontal and the angle between the two forces is 120° as shown in the diagram.

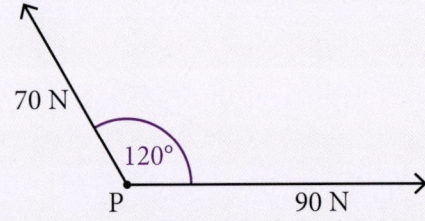

Find the size of the resultant force and the angle at which it acts to the horizontal.

6. A block of mass 17 kg is on a smooth plane, which is inclined at an angle of 15° to the horizontal, as shown in the diagram.

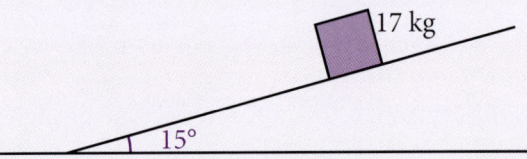

The block is being pushed up the plane by a force of 50 N, which acts along the line of greatest slope up the plane, parallel to it.

Exercise 4D...

The block also experiences a reaction force, which acts perpendicular to the plane.
(a) Copy the diagram, showing all the forces acting on the block.
(b) Find the magnitude of the resultant force acting on the block. State the direction in which the resultant force acts.

7. Two forces of size 25 N and X N act on a particle P, with the 25 N force horizontal and the angle between the two forces 145° as shown in the diagram.

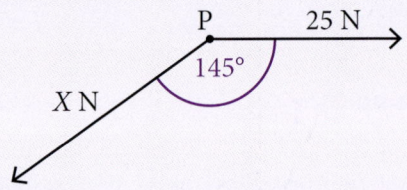

The angle between the 25 N force and the resultant is 100°.
(a) Find the size of the resultant force.
(b) Find X.

4.5 Equilibrium of Forces on a Particle

If a particle is in equilibrium, the resultant force acting on it is zero.

If the forces acting upon the particle have been given in component form, the sum of the **i** components must be zero and the sum of the **j** components must be zero.

Worked Example

10. Three forces $(15\mathbf{i} - 4\mathbf{j})$ N, $(-\mathbf{i} + 7\mathbf{j})$ N and P N act on a particle. Given that the particle is in equilibrium, find P.

Using column vector notation may help to form the equations.

Let $P = \begin{pmatrix} p \\ q \end{pmatrix}$. Then: $\begin{pmatrix} 15 \\ -4 \end{pmatrix} + \begin{pmatrix} -1 \\ 7 \end{pmatrix} + \begin{pmatrix} p \\ q \end{pmatrix} = \begin{pmatrix} 0 \\ 0 \end{pmatrix}$

Using the **i** components: $15 - 1 + p = 0 \Rightarrow p = -14$
Using the **j** components: $-4 + 7 + q = 0 \Rightarrow q = -3$

$\therefore P = \begin{pmatrix} -14 \\ -3 \end{pmatrix}$ N or $(-14\mathbf{i} - 3\mathbf{j})$ N

If each force acting is given by its magnitude and direction, the forces can be resolved into their horizontal and vertical components. The resultant force in the horizontal and the resultant force in the vertical are both zero.

Worked Example

11. Three forces act on a particle P as shown in the diagram. The 7 N force acts at an angle of 63° to the horizontal. The 9 N force acts at an angle of 58° to the vertical. Given that the particle is in equilibrium, find the magnitude and direction of the third force F N.

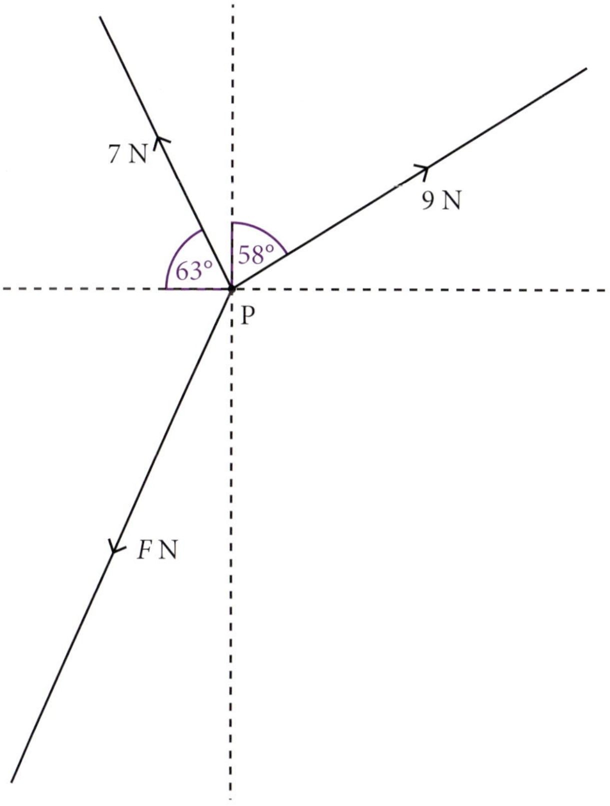

Write the force F in vector form: $F = \begin{pmatrix} F_H \\ F_V \end{pmatrix}$.

Since P is in equilibrium, the sum of the horizontal components is zero:

$$9 \sin 58° - 7 \cos 63° + F_H = 0$$
$$F_H = 7 \cos 63° - 9 \sin 58°$$
$$F_H = -4.454$$

The sum of the vertical components is zero:

$$7 \sin 63° + 9 \cos 58° + F_V = 0$$
$$F_V = -7 \sin 63° - 9 \cos 58°$$
$$F_V = -11.006$$

So we can sketch:

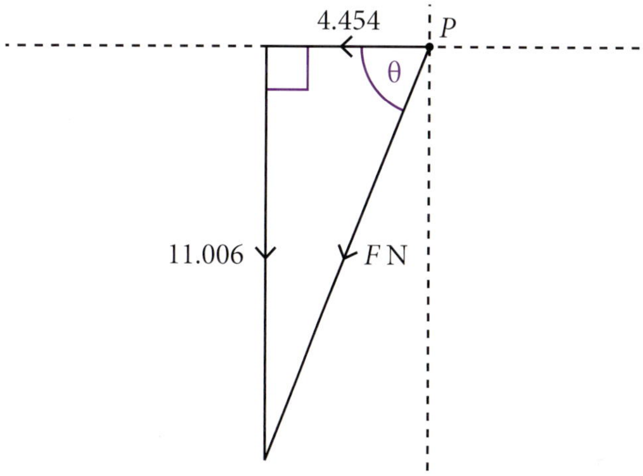

Find the magnitude of F using Pythagoras' Theorem:

$$|F| = \sqrt{(4.454)^2 + (11.006)^2} = 11.873$$
$$= 11.9 \text{ N (3 s.f.)}$$

The direction required is the angle between the force vector and the horizontal, marked θ above.

$$\tan \theta = \frac{11.006}{4.454}$$
$$\theta = \tan^{-1}\left(\frac{11.006}{4.454}\right)$$
$$\theta = 67.967° = 68.0° \text{ (3 s.f.)}$$

below the negative x-axis.

> **Note:** The working uses more than 3 significant figures accuracy, ensuring no loss of accuracy when the final answer is rounded to 3 significant figures.

There may be two or more unknown forces.

Worked Example

12. Consider the three forces shown in the diagram below, which act upon the particle P.

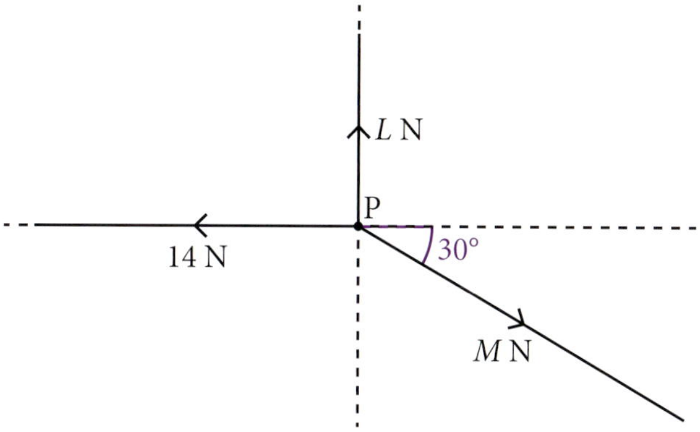

Given that P is in equilibrium, find the magnitudes of the forces L and M.

Since the particle is in equilibrium:

Forces acting to the right are equal to the forces acting to the left. So:
$$M \cos 30° = 14$$
$$M \times 0.8660 = 14$$
$$M = \frac{14}{0.8660} = 16.166\ldots$$
$$M = 16.2 \text{ N (3 s.f.)}$$

In the vertical, the forces upwards are equal to the forces downwards.
$$L = M \sin 30°$$
$$L = 16.166 \times 0.5$$
$$L = 8.08 \text{ N (3 s.f.)}$$

> **Note:** In this example it is important to consider the horizontal first to find M. Considering the vertical components first would result in an equation involving both unknowns, L and M.

In the example above, considering forces in the horizontal results in an equation in one unknown M.

In harder problems, there may be a component of both unknown forces in both the horizontal and the vertical. This type of problem requires simultaneous equations.

Worked Example

13. A particle P is acted upon by three forces, as shown in the diagram.

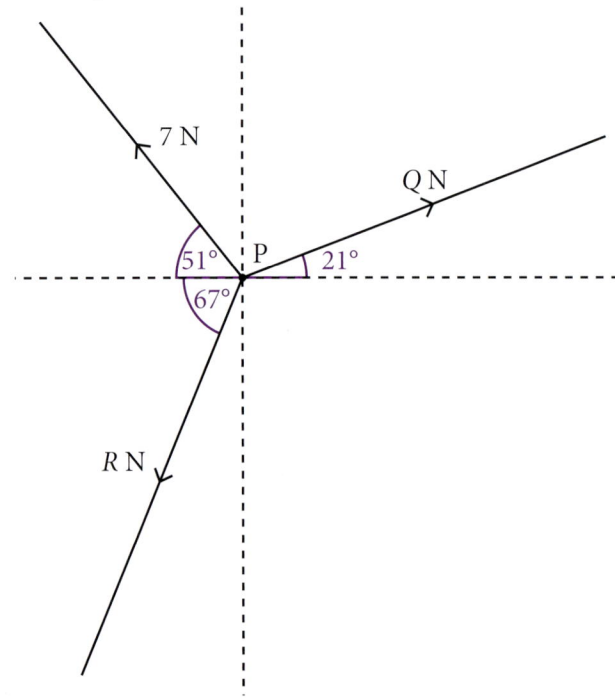

Given that P is in equilibrium under the action of these forces, find the size of the forces marked Q N and R N.

In the horizontal:
$$7 \cos 51° + R \cos 67° = Q \cos 21°$$
$$R \cos 67° = Q \cos 21° - 7 \cos 51°$$
$$R = \frac{Q \cos 21° - 7 \cos 51°}{\cos 67°} \quad (1)$$

In the vertical:
$$7 \sin 51° + Q \sin 21° = R \sin 67° \quad (2)$$

Substitute (1) into (2):
$$7 \sin 51° + Q \sin 21° = \left(\frac{Q \cos 21° - 7 \cos 51°}{\cos 67°}\right) \sin 67°$$
$$7 \sin 51° + Q \sin 21° = \frac{Q \sin 67° \cos 21° - 7 \sin 67° \cos 51°}{\cos 67°}$$
$$7 \sin 51° + Q \sin 21° = Q \tan 67° \cos 21° - 7 \tan 67° \cos 51°$$
$$7 \sin 51° + 7 \tan 67° \cos 51° = Q \tan 67° \cos 21° - Q \sin 21°$$
$$7 \sin 51° + 7 \tan 67° \cos 51° = Q(\tan 67° \cos 21° - \sin 21°)$$
$$Q = \frac{7 \sin 51° + 7 \tan 67° \cos 51°}{(\tan 67° \cos 21° - \sin 21°)}$$
$$Q = 8.5921\ldots$$
$$Q = 8.59 \text{ N (3 s.f.)}$$

Substitute into (1):
$$R = \frac{8.5921 \cos 21° - 7 \cos 51°}{\cos 67°}$$
$$R = 9.2549\ldots$$
$$R = 9.26 \text{ N (3 s.f.)}$$

> **Note:** Decimal values can be used throughout the working as an alternative to the trigonometric ratios cos 21°, etc. If approaching the problem this way, keep enough accuracy in the working to ensure no loss of accuracy when the final answers are rounded to 3 significant figures.

When considering an object in equilibrium on an inclined plane, consider equilibrium in the directions parallel and perpendicular to the slope.

Worked Example

14. A crate of mass 5 kg is at rest on a smooth slope inclined at an angle of 25° to the horizontal as shown in the diagram. A rope is attached to the crate. The tension in the rope is of magnitude P N and acts parallel to the slope, along the line of greatest slope, keeping the crate in equilibrium.

4: FORCES

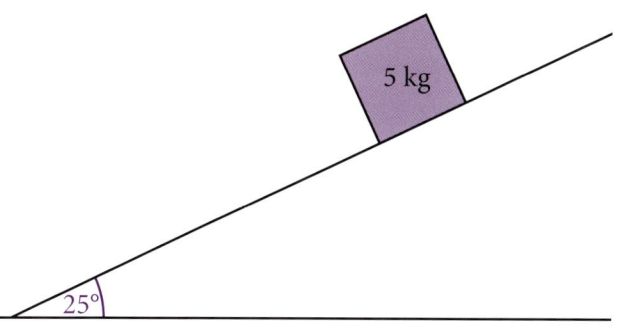

(a) Copy the diagram, showing all of the external forces acting upon the crate. Resolve the weight force into two components acting parallel and perpendicular to the slope.
(b) Find the size of the normal reaction force.
(c) Find the value of P.

(a)

[Diagram showing crate on slope with forces R (normal), P (up slope), 5g (weight) resolved into 5g sin 25° parallel to slope and 5g cos 25° perpendicular to slope, angle 25°]

(b) The crate is in equilibrium in the direction perpendicular to the slope.
∴ $R = 5g \cos 25° = 44.4$ N (3 s.f.)
(c) The crate is also in equilibrium in the direction parallel to the slope.
∴ $P = 5g \sin 25° = 20.7$ N (3 s.f.)

Exercise 4E

1. A rowing boat carrying two people is stationary on a calm boating lake. The boat has a mass of 90 kg and each person has a mass of 60 kg.
   (a) Draw a diagram showing all the forces acting on the boat.
   (b) What is the name of the force that acts vertically upwards on the boat?
   (c) Find the size of this force.

Exercise 4E...

2. The forces given below act on a particle P. Given that the particle is in equilibrium, find the force F in each case. Give your answer in component form.
   (a) $2\mathbf{i} + 3\mathbf{j}$ N and F N
   (b) $7\mathbf{i} - 10\mathbf{j}$ N, $-5\mathbf{i} + 6\mathbf{j}$ N, and F N
   (c) $\begin{pmatrix} -7 \\ 4 \end{pmatrix}$ N, $\begin{pmatrix} 9 \\ -14 \end{pmatrix}$ N, $\begin{pmatrix} 1 \\ -5 \end{pmatrix}$ N and F N

3. Three forces act on a particle P, as shown in the diagram. Two of the forces have magnitudes 8 N and 10 N respectively. The third force is marked F N. If the particle P is in equilibrium, find the magnitude and direction of this third force.

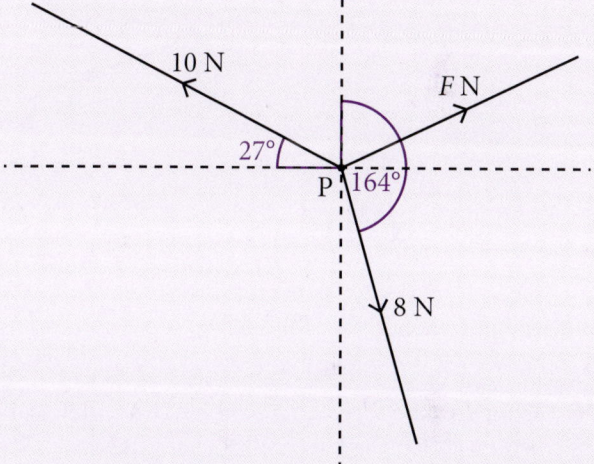

4. Consider a box of mass 7 kg at rest on a smooth slope inclined at 20° to the horizontal, as shown in the diagram.

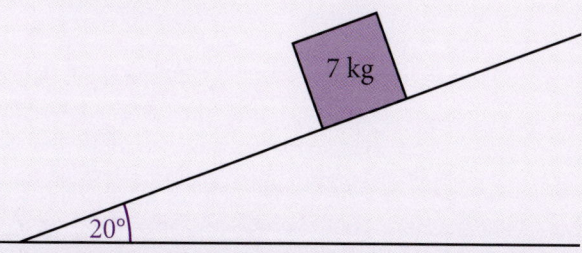

A rope is attached to the box, providing a force of magnitude P N. This force acts up the slope and parallel to it, keeping the box in equilibrium.
(a) Copy the diagram, showing all the external forces acting upon the box.
(b) Find the size of the normal reaction force acting on the box.
(c) Find the size of P.

Exercise 4E...

5. Three children are fighting over a toy. They exert forces on the toy as shown in the diagram.

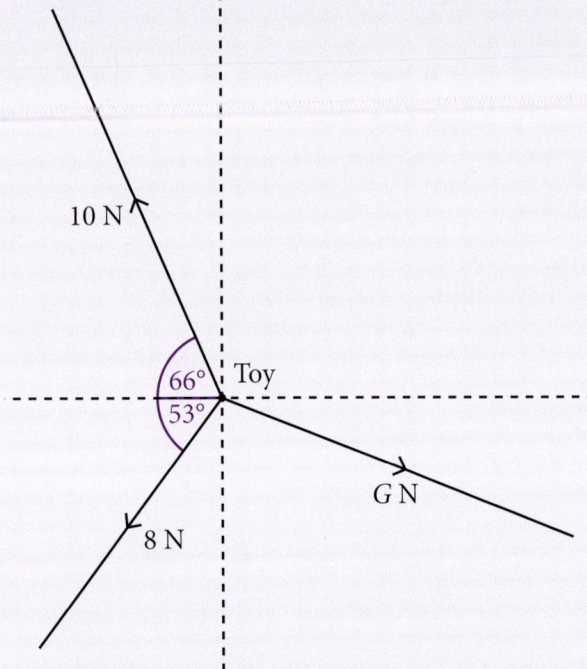

Given that the toy is in equilibrium, find the magnitude and direction of the force marked G N.

6. Consider the three forces shown in the diagram, which act on the particle P. Given that the particle is in equilibrium under the action of these three forces, find the unknown forces marked G N and J N.

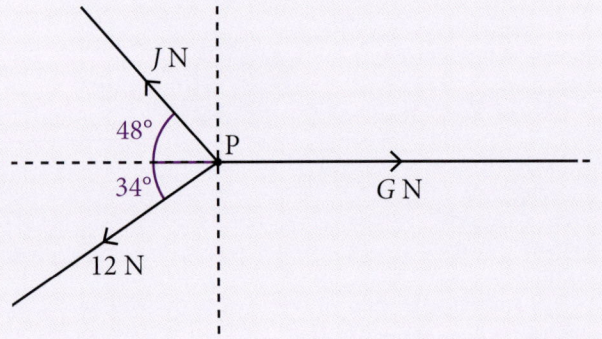

Exercise 4E...

7. The diagram shows three forces S N, T N and 10 N acting on a particle P. Given that P is in equilibrium, find the values of S and T.

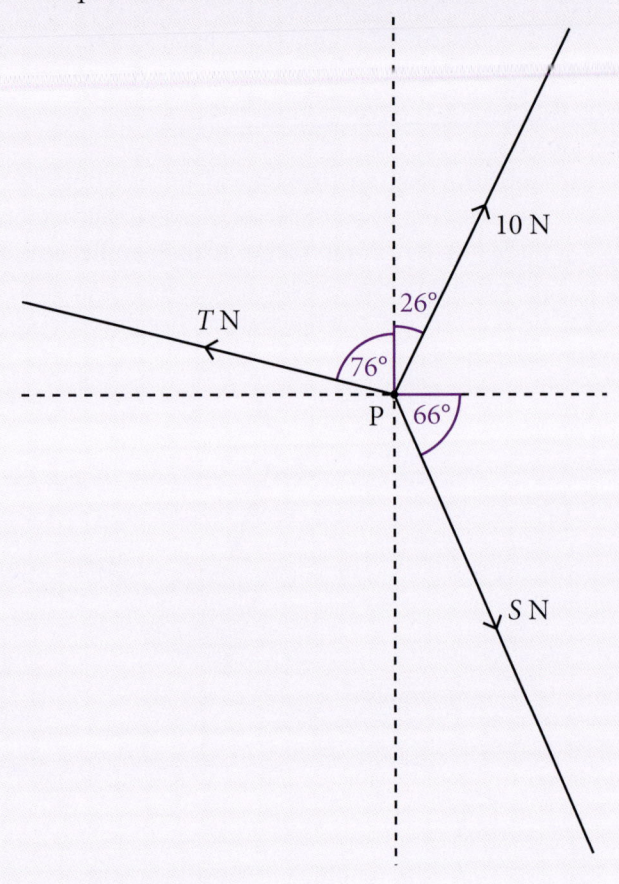

4.6 Summary

Forces upon objects arise in many ways, for example: an object's own **weight**; the **tension** in a rope or string; a **normal reaction** force caused by the surface with which an object is in contact; a **friction** force if the surface is **rough**.

Forces in two dimensions can be resolved into two components, often a vertical and horizontal component, known as the **i** and **j** components. Otherwise, if an object is on an inclined plane it is common to resolve the force into components parallel and perpendicular to the plane.

The **resultant** of two or more forces is the sum of the force vectors. The resultant can be found by adding the **i** components and adding the **j** components. If the magnitude and direction of the forces are given, they can be resolved into their **i** and **j** components before addition.

If an object is in **equilibrium**, then the resultant force is zero. This means that the sum of the **i** components is zero and the sum of the **j** components is zero. Using this, an unknown force can be found in a system of forces acting on a particle.

Chapter 5
Newton's Laws

5.1 Introduction

Sir Isaac Newton first presented his three laws of motion in the *Principia Mathematica Philosophiae Naturalis* in 1686. For over 300 years, these three laws have been used to describe the way in which forces act upon objects.

Key words
- **Force**: An influence that can change the motion of an object, e.g. weight, friction or tension in a string. Force is a vector quantity.
- **Mass**: The quantity of matter in an object. The SI unit of mass is the kilogram.
- **Acceleration**: The rate of change of speed, or the rate of change of velocity.
- **Reaction**: According to Newton's third law of motion, all forces occur in pairs, so that if one object exerts a force on another object, then the second object exerts an equal and opposite **reaction** force on the first.

Before you start
You should know about:
- Kinematics.
- Finding the magnitude and direction of a force.
- Forces in equilibrium.

What you will learn
In this chapter you will learn about:
- Newton's three laws of motion.

In the real world...

Newton's first law states that every object will remain at rest, or in motion with a constant velocity, unless acted upon by some external force. If there is no net force acting on an object (i.e. if no forces are acting, or if all the external forces cancel each other out), then the object will maintain a constant velocity. If that velocity is zero, then the object remains at rest.

There are many examples of Newton's first law that involve aerodynamics.

Two spacecraft named Voyager 1 and Voyager 2 were launched in the 1970s. After giving us spectacular images of the planets in our solar system, they are now heading out into interstellar space. As the distances between the stars are so vast, there are no significant gravitational forces acting on these two spacecraft. Since they left the solar system travelling at a very high speed, they will maintain this speed for many thousands of years, until they encounter the gravitational forces from some other star system.

Exercise 5A (Revision)

1. A speedboat starts from rest and accelerates uniformly at 3 m s^{-2}.
   (a) Work out the speed of the speedboat after 8 seconds.

   After 8 seconds the speedboat stops accelerating and begins to decelerate at a constant rate of 2.5 m s^{-2}.
   (b) Find how long it takes for the speedboat to come to rest.
   (c) Find how far the speedboat travels in total.

2. Two forces P and Q act on a particle. They are defined as follows:
   $P = 15\mathbf{i} - 20\mathbf{j} \text{ N}$
   $Q = -20\mathbf{i} + 4\mathbf{j} \text{ N}$
   Find the magnitude and direction of the following.
   (a) P
   (b) the resultant of P and Q.

3. Three forces F_1, F_2 and F_3 act on a particle P, as shown in the diagram.

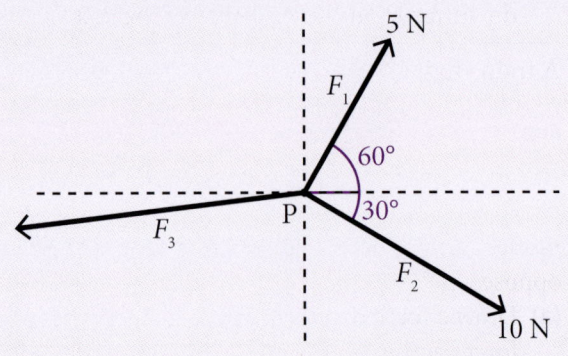

Exercise 5A...

F_1 has magnitude 5 N and acts at an angle of 60° to the horizontal. F_2 has magnitude 10 N and acts at an angle of 30° below the horizontal. The particle is in equilibrium. Find F_3, giving your answer in component form.

5.2 Newton's First Law

Newton's first law states that an object will remain at rest or at a constant velocity unless some external force is applied to the object. If there is no net force (no **resultant** force) acting on an object (i.e. if all the external forces cancel each other out), then the object will maintain its constant velocity. If that velocity is zero, then the object remains at rest.

If, however, an additional external force is applied, the velocity will change because of the force. The object will accelerate in the direction of the resultant force. In the next section, Newton's second law is discussed. This allows calculation of the acceleration in the direction of the resultant force.

Worked Examples

1. A picture hangs on a picture hook attached to a wall. The picture is in equilibrium.
   (a) Describe the forces acting on the picture.
   (b) Suddenly, the picture hook snaps. Describe what happens next, referring to Newton's first law.

   (a) Two forces act on the picture: its weight force acting vertically downwards and a reaction force from the picture hook, acting vertically upwards.
   (b) When the picture hook snaps, the reaction force disappears. The weight force is now the only force acting. Since there is now a resultant force, the velocity of the picture will change according to Newton's first law. The picture will accelerate downwards, in the direction of the resultant force.

2. A train engine travels along a track that is at a slight inclination of $\theta°$ to the horizontal, where $\tan\theta = 0.01$. The engine has a mass of 10 000 kg and travels downhill at a constant speed. The engine produces a tractive force of P N to power its own motion. A constant resistive force of 30 000 N opposes the motion.
   (a) Draw a force diagram showing all the external forces acting on the engine.
   (b) Find P while the engine moves downhill.

The engine returns along the same stretch of track, at the same constant speed. The resistive force remains unchanged.
   (c) Draw a force diagram showing all the external forces acting on the engine while it moves uphill.
   (d) Find the tractive force P while the engine moves uphill.
   (e) Referring to Newton's first law, explain what would happen to the engine if the power was lost and the tractive force disappeared.

(a)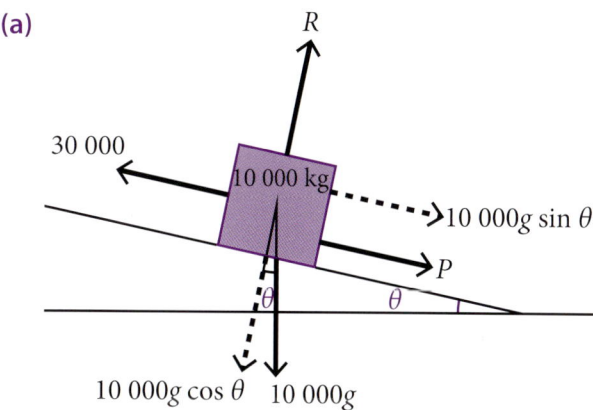

(b) $\tan\theta = 0.01 = \dfrac{1}{100}$

Using a right-angled triangle, $\sin\theta = \dfrac{1}{\sqrt{10001}}$ (this is left as an exercise for the reader).
Since the engine is travelling at a constant velocity, the forces up the plane are equal to the forces down the plane:
$30000 = 10000g \sin\theta + P$
$P = 30000 - 10000g \sin\theta$
$P = 29020.049$
$P = 29000$ N (3 s.f.)

(c)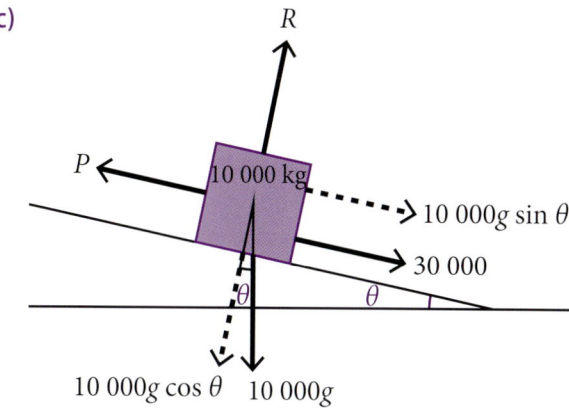

(d) Again, since the engine is travelling at a constant velocity, the forces up the plane are equal to the forces down the plane:
$P = 10000g \sin\theta + 30000$
$P = 30979.951$
$P = 31000$ N (3 s.f.)

(e) If the tractive force disappeared, there would be a resultant force acting down the plane. According to Newton's first law, the engine would begin to move in that direction.

Exercise 5B

1. A model rocket is launched into the atmosphere. Describe how Newton's first law applies to its motion **during the launch**.

2. A kite is flying above a beach.
   (a) The kite is stationary at a point 20 metres above the beach. Describe the forces acting upon the kite.
   (b) Suddenly the kite moves rapidly. Describe why this happens, referring to Newton's first law.

3. A skydiver jumps out of an aeroplane.
   (a) What forces are acting on the skydiver before the parachute is opened?
   (b) Why does the skydiver spread their arms and legs out?
   (c) An object falling in a vacuum continues to accelerate at roughly 9.8 m s^{-2} as it falls. Do you think this happens to the skydiver? Explain your answer.
   (d) Thinking about the forces acting on the skydiver, describe what happens when the parachute is opened.

4. Three forces act on a particle on a horizontal plane. Two forces of 10 N and 15 N are acting on the particle to the right. A single force of Q N acts on the particle to the left.
   (a) The particle travels at a constant speed of 5 m s^{-1} to the right. Find Q.
   (b) The force Q is then removed. Explain the behaviour of the particle, referring to Newton's first law.

5. Three forces $\begin{pmatrix}3\\1\end{pmatrix}$ N, $\begin{pmatrix}1\\-3\end{pmatrix}$ N and **P** N act on a particle. The particle is moving with a constant velocity. Find the vector **P**.

6. A plane's engines give it a constant thrust of 5000**i** − 2400**j** N. A strong wind is acting on the plane, providing an additional force of −400**i** − 600**j** N. In addition, the plane is affected by an air resistance force.

Exercise 5B...

(a) If the plane travels at a constant velocity of 250**i** + 300**j** miles per hour, find the size of the air resistance force.
(b) The wind suddenly stops, but the plane's engines continue to provide the same thrust. The air resistance force also remains the same. Does the plane's velocity remain constant? Explain your answer by referring to Newton's first law.

7. A box of mass 20 kg is on a smooth slope inclined at $\theta°$ to the horizontal. Two sisters Hannah and Stella are helping each other to move the box up the slope. Stella pushes with a force of 12 N and Hannah pulls the box using a rope. The tension in the rope is 15 N. The box moves up the slope with a constant velocity.
   (a) Draw a force diagram, showing all the forces acting on the box.
   (b) Find the angle of the slope, θ, giving your answer in degrees to one decimal place.
   (c) Stella then stops pushing the box. Explain, referring to Newton's first law, what happens next.

8. Sebastian is cutting his grass with a lawnmower. His garden is on a slope at an angle of $\tan^{-1}\left(\frac{1}{20}\right)$ to the horizontal.
   The lawnmower has a mass of 25 kg. When the lawnmower is in motion across the grass, there is a constant resistance force of size 35 N opposing its motion.
   (a) When Sebastian pushes the lawnmower downhill, he keeps the lawnmower moving at a constant speed.

   Draw a force diagram showing all the forces acting on the lawnmower when Sebastian is pushing it downhill. You may assume that the force Sebastian uses to push the lawnmower is parallel to the slope.
   (b) Find the size of the force Sebastian must use to push the lawnmower downhill.

Exercise 5B...

(c) When Sebastian pushes the lawnmower uphill, he maintains the same constant speed.

Draw a new force diagram and calculate the size of the force Sebastian must use now.

(d) State any modelling assumptions you have made.

5.3 Newton's Second Law

Newton's first law, discussed in section 5.2, states that an object maintains its velocity unless acted upon by some external force.

If an external force is applied to an object, the object will accelerate because of the force. The acceleration is determined by Newton's second law of motion, $F = ma$.

Worked Example

3. An object with a mass of 5 kg is at rest on a smooth table. A 10 N force is applied to the object, as shown in the diagram.

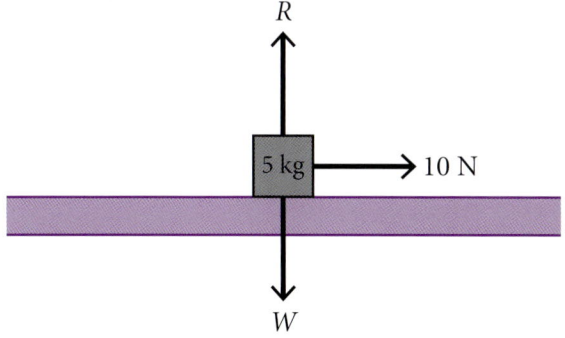

Calculate the object's horizontal acceleration.

In the horizontal there is only one force acting. So:
$F = ma$
$10 = 5a$
$a = \dfrac{10}{5} = 2 \text{ m s}^{-2}$

Newton's second law can be applied to an object on an inclined plane.

The following example requires both Newton's second law, $F = ma$, and the constant acceleration formulae.

Worked Example

4. A basket of mass 3 kg is being pulled up a smooth ramp by a rope. The ramp is 1.5 metres long and is angled at 15° to the horizontal. The tension in the rope is 10 N.
   (a) Draw a diagram showing all the forces acting on the basket.
   (b) Find the basket's acceleration up the ramp.
   (c) The basket starts from rest at the bottom of the ramp. How fast is it moving when it reaches the top of the ramp?
   (d) Find by how much the tension would increase to give an acceleration of 2 m s⁻².

(a)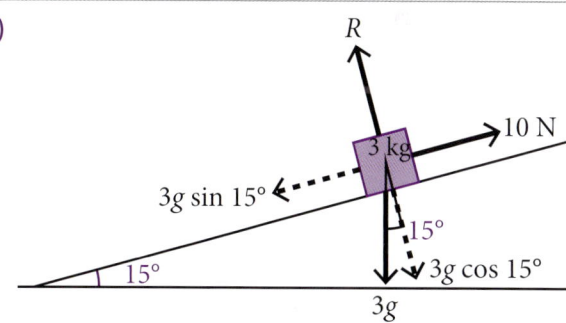

(b) Use $F = ma$ in the direction parallel to the ramp:
$10 - 3g \sin 15° = 3a$
$3a = 2.39072$
$a = 0.796907 = 0.797 \text{ m s}^{-2}$ (3 s.f.)

(c) Calculating the final velocity requires one of the constant acceleration formulae:
$v^2 = u^2 + 2as$
$v^2 = 0^2 + 2 \times 0.796907 \times 1.5$
$v^2 = 2.39072$
$v = 1.55 \text{ m s}^{-1}$ (3 s.f.)

(d) Draw a second diagram. The tension in the rope is unknown.

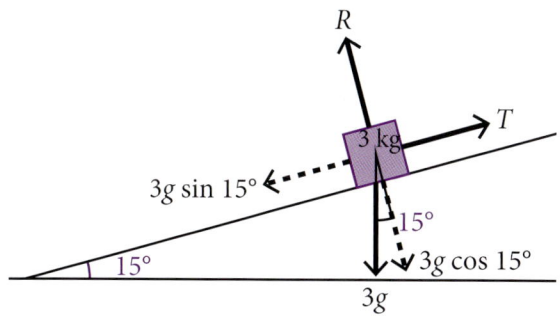

Use $F = ma$ in the direction parallel to the ramp, with $a = 2$:
$T - 3g \sin 15° = 3 \times 2$
$T = 6 + 3g \sin 15°$
$T = 13.609 ...$
The tension increases by 3.61 N (3 s.f.).

Exercise 5C

1. (a) Find the resultant force that would give a mass of 5 kg an acceleration of 2 m s^{-2}.
   (b) Find the acceleration of a body of mass 8 kg when it is acted upon by an external force of 20 N.
   (c) Find the resultant force required to bring a body of mass 16 kg from rest to a velocity of 12 m s^{-1} in 5 seconds.
   (d) Find the magnitude of the resultant force required to bring a mass of 18 kg to rest from 25 m s^{-1} in a distance of 250 m.

2. A bus of mass 9000 kg is travelling at 20 m s^{-1}.
   (a) Find the magnitude of the force required to bring the bus to rest in 2 seconds.
   (b) Find the distance covered during this time.

3. A block of mass 4 kg is acted upon by the forces shown in the diagram below.

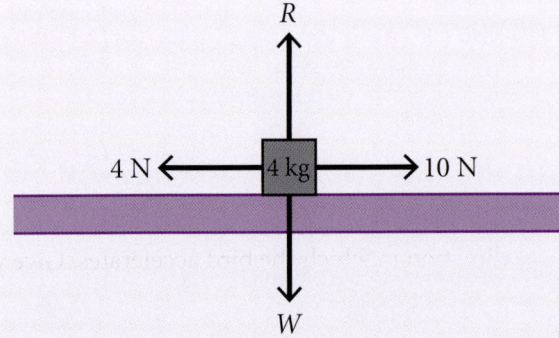

   Find the acceleration of the block.

4. The *Thrust SSC* is a British car powered by jet engines that holds the world land speed record. On 15 October 1997 it achieved a speed of 1228 km h^{-1} (763 mph) and became the first land vehicle to break the sound barrier. The car's mass, when it is full of fuel, is 10 700 kg and the jet engines produced 223 kN of thrust.
   (a) Calculate the initial acceleration of the car.
   (b) Why may the acceleration change as the car moves?
   (c) What assumptions or simplifications have been made in your calculations?

Exercise 5C...

5. At night, when nobody is around, a hedgehog climbs to the top of a children's slide in a playground. It then slides down. The slide is 2 metres long, angled at 30° to the horizontal and is smooth.
   (a) Find the hedgehog's acceleration.
   (b) Find the hedgehog's speed as it reaches the bottom of the slide.

6. A 9 kg object is at rest on a smooth plane, which is inclined at 20° to the horizontal, as shown in the diagram.

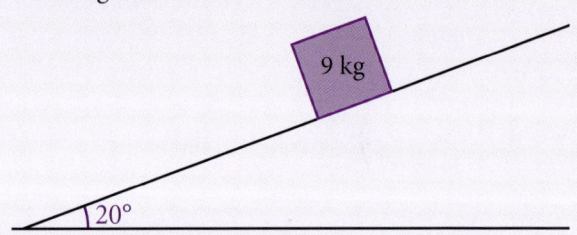

   A rope is attached to the object. The rope is parallel to the plane and the tension force T acts up the plane, parallel to it.
   (a) Copy the diagram, showing all the external forces acting upon the object.
   (b) When $T = 40$ N, the object begins to accelerate up the plane. Find the acceleration.

   An additional force of size P N is added. This force also acts parallel to the plane.
   (c) Find P and the direction of this force if the acceleration:
      (i) is increased by 0.1 m s^{-2}
      (ii) is decreased by 0.1 m s^{-2}.

7. Three dogs are each pulling at a bone B with their teeth. The magnitudes of the forces exerted by the three dogs are shown in the diagram. The mass of the bone is 100 grams. All three dogs exert forces that are constant and act in a horizontal plane.

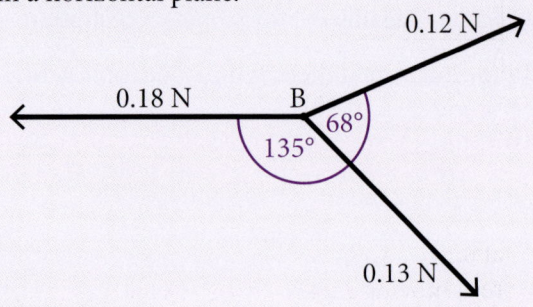

Exercise 5C...

(a) Find the magnitude and direction of the resultant force acting on the bone.
(b) Find the magnitude of the acceleration of the bone.
(c) Find how far the bone moves from rest in 3 seconds.

8. A particle P of mass 5 kg is at rest on a smooth horizontal table. Two horizontal forces of magnitude 5 N and 3 N act on P, as shown in the diagram. The angle between the two forces is 67°.

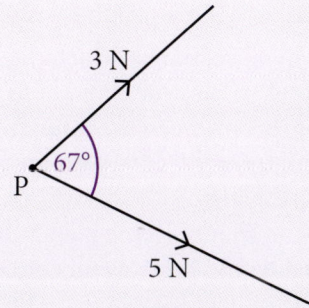

(a) Draw a vector triangle, showing addition of these two forces and the resultant.
(b) Find the magnitude of the resultant force acting on P.
(c) Find the magnitude of the particle's acceleration.
(d) Find the distance travelled by P in the first 2 seconds of its motion.

9. Mr Madill loves his holidays in the sun. It's getting harder to travel to Spain these days, so Mr Madill decides to fly to a different part of the solar system. He looks up his options on the internet.

Destination	Acceleration due to gravity (m s^{-2})
The Moon	1.62
Mercury	3.59
Venus	8.87
Mars	3.71

When he reaches his destination, Mr Madill relaxes by the pool with a refreshing drink. He notices that his glass is slowly sliding off the smooth table, because the table top is angled at 0.5° to the horizontal. The glass slides from rest, moving 15.48 cm in 2 seconds. Which destination did Mr Madill choose for his holiday? Show all your working.

Exercise 5C...

10. A particle of mass M kg is on a smooth inclined plane. A rope, which is parallel to the plane, is attached to the particle so that the particle accelerates up the plane. With the help of a force diagram, show that, for the acceleration to increase by k m s^{-2}, the tension in the rope must increase by Mk newtons.

5.4 Newton's Second Law in Vector Form

A vector form of Newton's second law can be used:
F = m**a**

When used in this form, the force and acceleration are both vectors. Mass is always a scalar quantity.

Remember to use Pythagoras' Theorem if a question requires the **magnitude** of the force or acceleration.

Worked Example

5. A bird of mass 150 grams, taking off from rest, gives itself a thrust force of magnitude $\frac{\sqrt{74}}{5}$ N in the direction of the vector $(5\mathbf{i} + 7\mathbf{j})$.
(a) Show that this thrust force is $(\mathbf{i} + 1.4\mathbf{j})$ N.

A strong wind is blowing, which exerts an additional force of $(0.8\mathbf{i} - 0.2\mathbf{j})$ N upon the bird.
(b) Find the resultant force acting upon the bird.
(c) Find the **magnitude** of the bird's acceleration.
(d) If **i** is the unit vector from west to east, and **j** is the unit vector from south to north, find the direction in which the bird accelerates. Give your answer as a bearing.
(e) Find the distance the bird travels in 2 seconds, assuming the bird maintains the same acceleration throughout this time.

(a) The magnitude of vector $\begin{pmatrix}5\\7\end{pmatrix}$ is $\sqrt{5^2 + 7^2} = \sqrt{74}$

The unit vector in the direction of $\begin{pmatrix}5\\7\end{pmatrix}$ is $\frac{1}{\sqrt{74}}\begin{pmatrix}5\\7\end{pmatrix}$, or $\begin{pmatrix}5/\sqrt{74}\\7/\sqrt{74}\end{pmatrix}$

The vector with magnitude $\frac{\sqrt{74}}{5}$ in this direction is $\frac{\sqrt{74}}{5}\begin{pmatrix}5/\sqrt{74}\\7/\sqrt{74}\end{pmatrix}$, or $\begin{pmatrix}1\\1.4\end{pmatrix}$, or $(\mathbf{i} + 1.4\mathbf{j})$ N.

(b) To find the resultant force, add the two forces acting upon the bird.
$$\mathbf{F} = \begin{pmatrix} 1 \\ 1.4 \end{pmatrix} + \begin{pmatrix} 0.8 \\ -0.2 \end{pmatrix} = \begin{pmatrix} 1.8 \\ 1.2 \end{pmatrix} \text{N}$$

(c) Use $\mathbf{F} = m\mathbf{a}$ in vector form.
$$\begin{pmatrix} 1.8 \\ 1.2 \end{pmatrix} = 0.15\mathbf{a}$$
$$\mathbf{a} = \frac{1}{0.15}\begin{pmatrix} 1.8 \\ 1.2 \end{pmatrix} = \begin{pmatrix} 12 \\ 8 \end{pmatrix}$$

The magnitude of this vector can be found using Pythagoras' Theorem:
$$|\mathbf{a}| = \sqrt{12^2 + 8^2} = 14.4 \text{ m s}^{-2}$$

Note: If the question asked for the acceleration rather than its magnitude, give the vector $\begin{pmatrix} 12 \\ 8 \end{pmatrix}$ m s^{-2}.

(d) The diagram shows the acceleration vector.

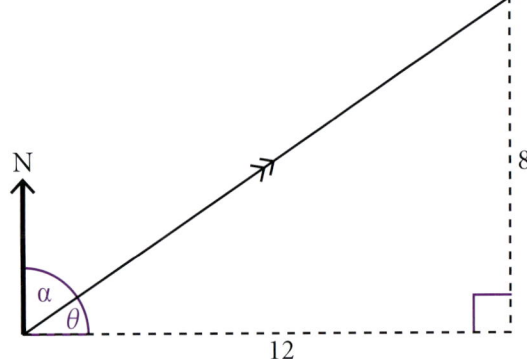

The angle α between the north line and the vector is the bearing.
$$\tan\theta = \frac{8}{12}$$
$$\theta = \tan^{-1}\left(\frac{8}{12}\right) = 33.7°$$
So: $\alpha = 90 - \theta = 56.3°$
The bearing is 056°.

Note: Remember bearings always have three digits and no decimal places.

(e) Use the constant acceleration formula
$\mathbf{s} = \mathbf{u}t + \frac{1}{2}\mathbf{a}t^2$ in vector form:
$$\mathbf{s} = \begin{pmatrix} 0 \\ 0 \end{pmatrix} \times 2 + \frac{1}{2} \times \begin{pmatrix} 12 \\ 8 \end{pmatrix} \times 2^2$$
$$\mathbf{s} = \begin{pmatrix} 24 \\ 16 \end{pmatrix} \text{m}$$
Distance = $\sqrt{24^2 + 16^2} = 28.8$ m (3 s.f.)

Exercise 5D

1. Two forces $F_1 = (6\mathbf{i} - 2\mathbf{j})$ N and $F_2 = (2\mathbf{i} + 8\mathbf{j})$ N act on a particle P of mass 2 kg.
   (a) Find the **magnitude** of the acceleration of P.
   (b) Find the angle between the resultant force acting on P and the vector **i**.

2. During take-off, a small aeroplane is given a thrust force of $(2900\mathbf{i} - 4180\mathbf{j})$ N by its engines. In addition, a constant wind at the airport exerts a force of $(100\mathbf{i} + 180\mathbf{j})$ N on the plane. The plane's mass is 5000 kg.
   (a) Find the acceleration of the plane.
   (b) Find the direction at which the plane takes off, giving your answer as an angle in degrees between the resultant force and the unit vector **i**. Give your answer to one decimal place.
   (c) Find the change in the plane's direction caused by the wind, again giving your answer in degrees to one decimal place.

3. A box of mass 8 kg is in equilibrium under the action of three forces P, Q and R where $P = \begin{pmatrix} 3 \\ 2 \end{pmatrix}$ N and $Q = \begin{pmatrix} -1 \\ -4 \end{pmatrix}$ N.
   (a) Find R.
   (b) If the direction of R is now reversed, find the acceleration given to the box.

4. At time $t = 0$ seconds, a particle of mass 4 kg is at rest at the origin. A constant force F of magnitude 20 N acts on the particle in the direction of the vector $\mathbf{p} = 4\mathbf{i} + 3\mathbf{j}$
   (a) Find the magnitude of **p**.
   (b) Find F.
   (c) Find the speed of the particle when $t = 2$.

5. Two forces F_1 and F_2 act on a particle of mass 6 kg, where $F_1 = (2\mathbf{i} - 5\mathbf{j})$ N. The resultant of these two forces gives the particle an acceleration of $(\mathbf{i} + 7\mathbf{j})$ m s^{-2}.
   (a) Find F_2.
   (b) A third force F_3 now acts on the particle together with F_1 and F_2. The resultant of these three forces causes the particle to move with a constant velocity. Find F_3.

6. A particle P of mass 0.5 kg is acted upon by two constant forces F_1 and F_2. F_1 has magnitude 15 N and acts in the direction of the vector $4\mathbf{i} - 3\mathbf{j}$. F_2

Exercise 5D...

has magnitude 6.5 N and acts in the direction of the vector $12\mathbf{i} - 5\mathbf{j}$.
(a) Find, in vector form, the resultant force acting on P.
(b) Hence, show that the magnitude of the acceleration of P is $5\sqrt{73}$ m s^{-2}.

At time $t = 0$ seconds, P is moving with velocity $12\mathbf{i} + 2\mathbf{j}$ m s^{-1}.
(c) Find the velocity of P at $t = 1$.
(d) Show that at $t = 2$, the particle is moving parallel to the vector $21\mathbf{i} - 11\mathbf{j}$.

7. A particle, P, of mass 4 kg is acted on by a force $(-3\mathbf{i} + \mathbf{j})$ N. Initially, P is at rest at a point A. After 4 seconds, P reaches the point B.
   (a) Find the velocity of P at B.

   A is $(3\mathbf{i} + 12\mathbf{j})$ m from a fixed point O.
   (b) Find the displacement vector $\overrightarrow{OB}$.
   (c) Find the exact unit vector in the direction of P's motion.

8. In this question you may assume that $\mathbf{i}$ is the unit vector from west to east, and $\mathbf{j}$ is the unit vector from south to north.

   A particle of mass 2 kg moves from a point P to a point Q under the action of the constant force $(4\mathbf{i} - 3\mathbf{j})$ N. P has position vector $(2\mathbf{i} + 3\mathbf{j})$ m relative to a fixed point O. The particle's velocity at P is zero. It takes 2 seconds to move from P to Q. Show that Q lies due east of O.

9. Two forces $\begin{pmatrix} 6 \\ -5 \end{pmatrix}$ N and $\begin{pmatrix} 3 \\ c \end{pmatrix}$ N act on a particle P of mass 2.5 kg. The resultant of the two forces is parallel to the vector $\begin{pmatrix} 1 \\ -2 \end{pmatrix}$.
   (a) Find the value of c.
   (b) At time $t = 0$, P is moving with velocity $\begin{pmatrix} 4 \\ -5 \end{pmatrix}$ m s^{-1}. Find the **speed** of P at time $t = 2$ seconds.

10. Two forces $(3\mathbf{i} + 9\mathbf{j})$ N and $(12\mathbf{i} - 6\mathbf{j})$ N act on a body of mass 3 kg.
    (a) Find the acceleration of the body.

    The body starts from rest.
    (b) Find the **speed** of the body after 5 s.
    (c) Find the bearing of the direction in which P accelerates.

5.5 Vertical Motion Under Gravity

In section 4.2 the **weight** force was briefly introduced.

Weight should not be confused with mass. The **weight of an object is a force** equal to the object's mass multiplied by the acceleration due to gravity, g:

$$W = mg$$

On Earth the value of g is roughly 9.8 m s^{-2}. This is almost constant, but does vary slightly depending on geographical location. For example, the value of g is slightly lower near to the equator. It is also lower for objects that are far away from the Earth's surface, such as satellites.

The value most commonly used as an approximation for g is 9.8 m s^{-2}, although some work involving gravity may require a different approximation for g, for example 10 or 9.81.

When giving an answer to a problem, it is inappropriate to give more accuracy than is used for the numbers in the question. For example, if using $g = 9.8$ m s^{-2}, answers should not be given to more than 2 significant figures.

Some problems may require an answer given in terms of g, in which case a value for g is not required.

On the moon, or on a different planet, the value of g is different. The moon is smaller than Earth and the acceleration due to its gravity is only about 1.625 m s^{-2}. An object of mass 2 kg weighs 19.6 N on earth, but only 3.25 N on the moon.

Questions involving vertical motion often involve use of the constant acceleration formulae, discussed in Chapter 2.

Worked Example

6. The weight force experienced by an apple is 0.49 N.
   (a) What is the mass of the apple?
   (b) The apple falls from its tree to the ground in 1.5 seconds. How far does it fall?
   (c) State one modelling assumption that has been made.

(a) $W = mg$
   $0.49 = m \times 9.8$
   $m = \dfrac{0.49}{9.8} = 0.05$ kg $= 50$ grams

(b) $s = ut + \dfrac{1}{2}at^2$
   $s = 0 \times 1.5 + \dfrac{1}{2}(9.8)(1.5)^2$
   $= 11.025$
   $= 11.0$ m (3 s.f.)

(c) One assumption is that there is no air resistance. You could also state that the apple is modelled as a particle. These two statements are roughly equivalent (a particle has no size and therefore experiences no air resistance).

In some questions involving weight, other forces may be involved. When using $F = ma$, F represents the resultant force, i.e. all the forwards forces minus all the backwards forces. Such questions may involve lifts, where there is an upwards force from the lift cable, as well as a weight force. Alternatively, a question may involve a rocket, which has an upwards force provided by combustion of the fuel, as well as its weight force.

Worked Example

7. **Use, as an approximation, $g = 10$ m s^{-2} throughout this question.**
   A lift has a mass of 2000 kg and is supported by a cable.
   (a) Draw a diagram showing the forces acting on the lift when it is empty and stationary.

   Calculate the tension in the cable when the empty lift is:
   (b) at rest,
   (c) ascending with a constant velocity,
   (d) descending with a constant velocity,
   (e) accelerating upwards at 1 m s^{-2},
   (f) accelerating downwards at 1 m s^{-2}.

(a)

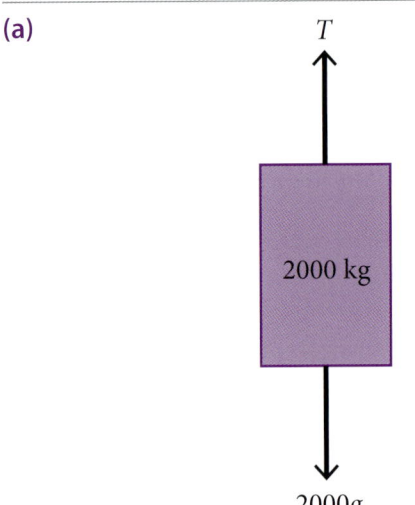

(b) At rest the lift is in equilibrium, so the upwards force is equal to the downwards force:
$T = 2000g = 20\,000$ N

(c) and (d) If the lift is moving with a constant velocity the upwards forces are again equal to the downwards forces, so the tension in the cable is 20 000 N.

(e) If there is acceleration, then use $F = ma$. The acceleration is upwards, so this is the positive direction:
$$T - 2000g = 2000 \times 1$$
$$T = 2000 + 2000g = 22\,000 \text{ N}$$

(f) Acceleration is downwards, so this is the positive direction. Using $F = ma$:
$$2000g - T = 2000 \times 1$$
$$T = 2000g - 2000 = 18\,000 \text{ N}$$

Exercise 5E

1. **You may assume $g = 10$ m s^{-2} throughout this question.**
   A model rocket has a mass of 50 grams. It is propelled into the air by a thrust of 5 N.
   (a) Calculate the initial acceleration of the rocket.
   (b) Assume that the acceleration remains constant for one second after launch. How high does the rocket rise during this time?
   (c) What is the velocity of the rocket after one second?

   After one second the fuel runs out.
   (d) Find the greatest height achieved by the rocket.
   (e) Find the total time of flight.

2. The space shuttle launched on 24 February 2011 had a lift-off mass of 2 million kilograms. The combined thrust from its solid rocket boosters and its engines was 30.1 million newtons.
   (a) Calculate the initial acceleration of the shuttle as it left the launchpad.
   (b) Why was the acceleration of the space shuttle far less than the acceleration of the model rocket in Question 2?

3. A lift has a mass of 600 kg.
   (a) Find the tension in the lift cable when the lift is accelerating downwards at 1.2 m s^{-2} without any passengers.

   The greatest tension the lift cable can support is 12 000 N. The average mass of a person travelling in the lift is 60 kg.

Exercise 5E...

(b) Find the maximum number of people that the lift can safely carry when it is accelerating upwards at 0.8 m s^{-2}.

(c) Give two modelling assumptions you have made in answering part (b).

4. The rover *Perseverance* reached the surface of Mars in February 2021. In the final stages of its descent, it was lowered onto the surface using a 'sky crane'. The sky crane lowered *Perseverance* on nylon cables, which were broken as the rover touched down. The mass of the rover is 1025 kg. The acceleration due to gravity on Mars is 3.71 m s^{-2}. Find the tension in the nylon cables if the rover was decelerating at 0.1 m s^{-2} as it landed.

5. A stone of mass 0.15 kg is thrown vertically downwards from the top of a cliff at a speed of 3 m s^{-1}. The top of the cliff is 100 m above horizontal ground. If air resistance is ignored, find:
   (a) The speed of the stone as it hits the ground.
   (b) The time taken for the stone to fall to the ground.

   If instead air resistance is included as a constant force of magnitude 0.5 N, find:
   (c) The acceleration of the stone as it falls.
   (d) The speed of the stone as it hits the ground.

6. An object with a mass of 200 grams is thrown upwards from the top of a 100 metre building with an initial speed of 21 m s^{-1}.
   (a) What speed does the object have as it hits the ground?
   (b) What modelling assumptions were made in the calculation for part (a)?
   (c) The ground is soft and the object sinks into it. If the ground exerts a constant resistive force of 2500 N, how far does the object sink into the ground?

5.6 Newton's Third Law

According to **Newton's third law** of motion, whenever two objects come into contact, they exert equal and opposite forces on each other. This is often worded as 'every action has an equal and opposite reaction'.

Worked Examples

8. A swimmer pushes with her feet against the side of the swimming pool with a force of 120 N. What is the size of the force that the swimmer experiences to propel her through the water?

By Newton's third law, if the swimmer pushes against the side of the pool with a force of 120 N, the side of the pool exerts an equal force upon her. She is propelled forwards with a force of 120 N.

9. A man stands in a lift, which is moving at a constant velocity upwards. The man has a mass of 75 kg.
   (a) Find the size of the force that the man exerts upon the floor of the lift.
   (b) Find the size of the normal reaction force upon the man from the floor of the lift.

(a) The man exerts a force of 75g upon the floor of the lift, which is 735 N.
(b) By Newton's third law, the reaction force upon the man is equal to the force exerted by the man upon the lift floor, i.e., 735 N.

Exercise 5F

Use $g = 10$ m s^{-2} as an approximation for the acceleration due to gravity throughout this exercise.

1. State whether each of the following statements is true or false.
   (a) A man stands on a cockroach. The cockroach dies, but the man does not. Therefore, the force exerted by the man on the cockroach must be larger than the force exerted by the cockroach on the man.
   (b) A truck pushes a small car. The force exerted by the truck on the small car is equal in size to the force exerted by the car on the truck.
   (c) Standing on one leg the reaction force I experience is half the reaction force I experience if I stand on two legs.
   (d) A rocket expels gases from its combustion chamber with a force of 2000 N. The gases push the rocket forwards with the same force.

Exercise 5F...

(e) A helicopter hovers over a field. Its rotor blades produce a downwards force on the air. The air provides an equal and opposite reaction force, the lift.

2. A cat of mass 4.5 kg sits on a table. Find the reaction force exerted by the table on the cat.

3. A boy takes his dog for a walk, with the dog in front. The tension in the lead is 200 N.
   (a) What is the size of the force pulling the dog back?
   (b) What is the size of the force pulling the boy forwards?

4. A hospital porter and a trolley are inside a lift. The porter has a mass of 80 kg and the trolley has a mass of 25 kg. The lift is moving downwards at a constant velocity.
   (a) Find the size of the force that the porter exerts upon the floor of the lift.
   (b) Find the size of the normal reaction force upon the porter from the floor of the lift.
   (c) Find the size of the force that the trolley exerts upon the floor of the lift.
   (d) Find the size of the normal reaction force upon the trolley from the floor of the lift.

5. A planet of mass $1000M$ kg and a comet of mass $0.001M$ kg are close enough to exert a gravitational force on each other.
   (a) Which force is greater: the force exerted by the planet on the comet, or the force exerted by the comet on the planet? Or are both forces equal in magnitude?
   (b) These gravitational forces cause both objects to accelerate. Which acceleration is greater in magnitude: the acceleration of the comet, or the acceleration of the planet? Or are both equal in magnitude?

6. A sprinter of mass 90 kg is starting a 100 metre race.
   (a) Explain how the sprinter can accelerate away from the blocks, referring to at least one of Newton's laws.
   (b) The initial forwards force acting on the sprinter is 630 N. Find the sprinter's initial acceleration.
   (c) The sprinter's foot is in contact with the blocks for 0.2 seconds. Find the distance he propels himself forwards in this time.

Exercise 5F...

7. An astronaut is performing a spacewalk to repair her module. She forgets to tether herself to the module and finds herself drifting away from the module into space. As well as her space suit, she is wearing a tool belt containing various heavy tools. What could she do to propel herself back towards the module?

5.7 Summary

Newton's three laws of motion are:

First law: If a body is stationary or moving with a constant velocity, it will continue in that state unless it is acted upon by some external force.

Second law: If there is a resultant force acting upon a body, then this resultant force, the body's mass and the body's acceleration are linked by the formula $F = ma$.

Third law: For every action there is a reaction; i.e. when two bodies interact they exert equal but opposite forces upon each other.

Chapter 6
Friction

6.1 Introduction

Newton's Laws were introduced in the previous chapter. Newton's second law was frequently combined with the constant acceleration formulae.

This chapter builds upon this work. Newton's second law is used frequently again, but now involving a friction force.

Sections 6.2 and 6.3 deal with **motion of a body on a rough surface**.

Section 6.4 looks at **limiting equilibrium**, where an object on a rough surface is **on the point of moving**.

Key words

- **Rough**: On a rough surface, friction impedes the movement of an object. Different surfaces impede this movement to different extents.
- **Smooth**: On a smooth surface there is no friction. An object on a smooth surface is not impeded by a friction force.
- **Coefficient of friction**: The coefficient of friction is a dimensionless quantity that measures the roughness of a surface when in contact with an object.
- **Limiting equilibrium**: A state in which an object is on the point of moving on a rough surface.

Before you start
You should know about:

- Kinematics.
- Resolving forces.
- Equilibrium.
- Newton's three laws of motion.

In this chapter you will learn about
- Motion of an object on a rough plane.
- The $F_r = \mu R$ model of friction.
- Limiting equilibrium.

In the real world...
A swimmer pushes against the side of the pool with her legs to propel herself forwards. Newton's third law states that the force she uses against the pool side will be equal in size to the force she is given, propelling her through the water.

Have you ever tried running on an icy surface? It's hard to get started.

On a rough surface the friction force between your shoes and the surface allows you to push back against the surface. Just like the swimmer, you receive a force of equal size pushing you forwards. In icy conditions there is less friction and therefore you experience a smaller forwards force.

Exercise 6A (Revision)

1. A car accelerates from rest to 10 m s^{-1} in 5 seconds. Find the distance travelled in 10 seconds, assuming that the acceleration is constant.

2. A force of 34 N acts on the particle P, as shown in the diagram. Resolve this force into its horizontal and vertical components.

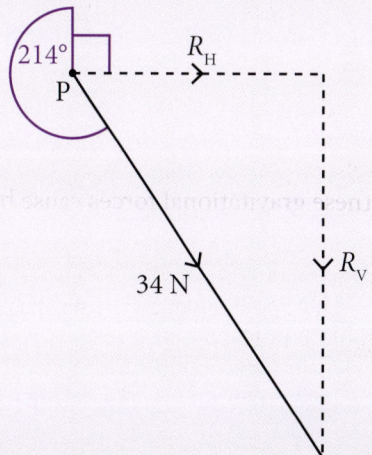

3. The particle P, shown in the next diagram, is in equilibrium under the action of the three forces shown. Find the value of F and the value of G.

AS 2: APPLIED MATHEMATICS

Exercise 6A...

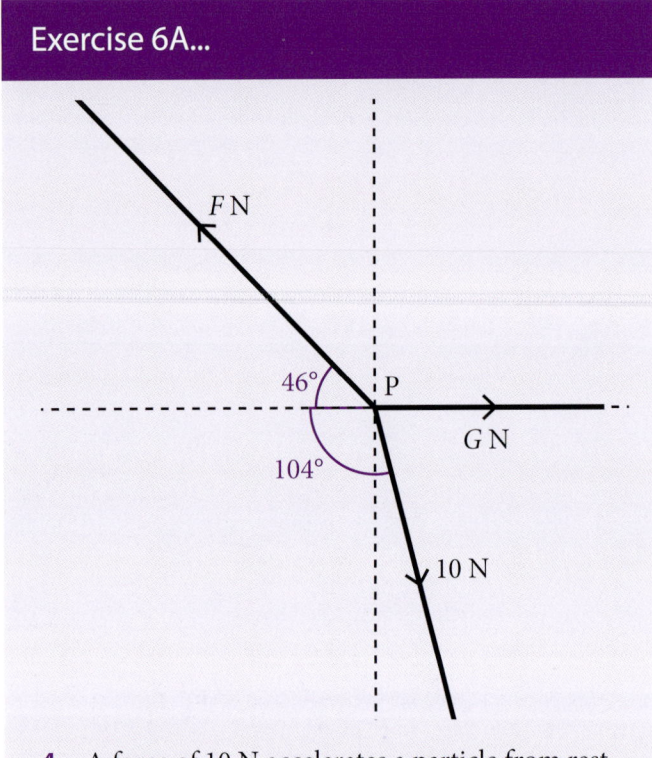

4. A force of 10 N accelerates a particle from rest to 5 m s^{-1} in 2 seconds. Find the mass of the particle.

6.2 The Friction Force

Friction is the resistance to motion when one object moves while in contact with another.

When an object moves while in contact with another object or surface, a frictional force may result. In previous chapters the frictional force has been neglected. The word 'smooth' indicates that the friction force is small enough to be neglected. The word 'rough' indicates that the friction force needs to be considered.

Friction works against the motion and therefore the friction force acts in the opposite direction to the motion. When an object slides across a rough surface, for example, a friction force will be acting, which opposes the motion and acts to slow the object down.

Worked Example

1. A crate of mass 8 kg is pulled by a rope across a rough surface at a constant speed, as shown in the diagram. The tension in the rope is 30 N.

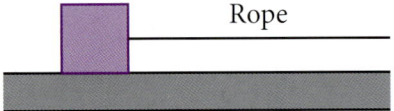

(a) Copy the diagram, showing all the horizontal forces acting on the crate.
(b) Find the size of the friction force opposing the motion.

(a) In the horizontal there is a 30 N force pulling the crate forwards and the friction force opposing the motion. Note: forces acting on the crate in the vertical are not considered.

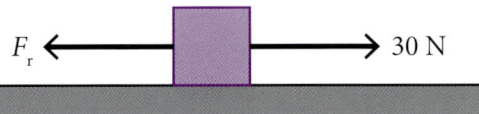

(b) Since the crate is moving at a constant speed in the horizontal, the force acting to the right is equal to the force acting to the left.
∴ $F_r = 30$ N

The next example demonstrates how Newton's second law can be used to calculate the acceleration of an object when frictional forces are acting. The acceleration can then be used in the constant acceleration formulae.

Worked Example

2. A chair of mass 35 kg is being moved from rest across a horizontal carpet with a force of 100 N. There is a frictional force of 30 N opposing the motion.
   (a) Find the acceleration of the chair.
   (b) Find how long it takes before the chair reaches a speed of 20 cm s^{-1}.

(a) Use $F = ma$ in the horizontal. Remember that F is the resultant force acting on the object, i.e. all forces acting in the direction of motion minus all those acting in the opposite direction:
$$100 - 30 = 35a$$
$$70 = 35a$$
$$a = \frac{70}{35} = 2 \text{ m s}^{-2}$$

(b) The constant acceleration formulae can be used here. The final velocity is 0.2 m s^{-1}.
$u = 0; v = 0.2; a = 2; t = ?$
$$v = u + at$$
$$0.2 = 0 + 2t$$
$$2t = 0.2$$
$$t = 0.1 \text{ s}$$

Similar calculations can be applied to an object on an inclined plane.

Worked Example

3. A box of mass 5 kg is placed onto a rough plane inclined at an angle of 15° to the horizontal, as shown in the diagram. It begins to slide down the plane. The size of the frictional force opposing the motion of the box is 10 N.

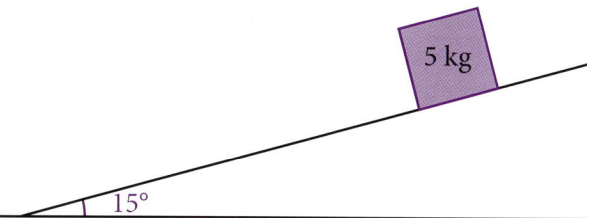

(a) Copy the diagram, adding all external forces acting upon the box.
(b) Calculate the acceleration of the box, giving your answer to a suitable level of accuracy.
(c) Find the distance moved by the box down the slope during the first second of its motion.

(a)

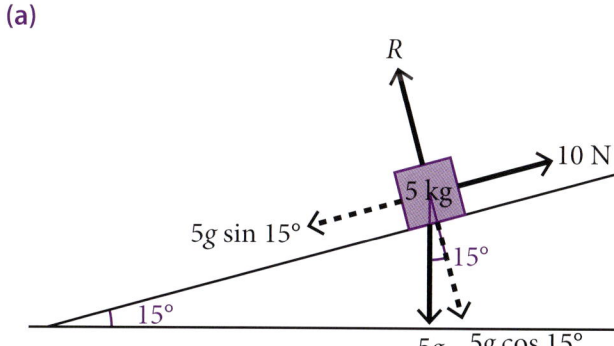

(b) Use $F = ma$ parallel to the slope:
$5g \sin 15 - 10 = 5a$
$5a = 2.682\ldots$
$a = 0.54 \text{ m s}^{-2}$ (2 s.f.)

Note: The acceleration is given to 2 significant figures since the value of g used in the calculation was 9.8 m s^{-2}. Alternatively, $g = 9.81$ m s^{-2} could be used as a more accurate approximation for g, in which case it would be appropriate to give the acceleration to 3 significant figures.

(c) Apply the constant acceleration formulae:
$s = ?\,;\, u = 0\,;\, a = 0.54\,;\, t = 1$
$s = ut + \frac{1}{2}at^2$
$s = 0 + \frac{1}{2} \times 0.54 \times 1^2$
$s = 0.27$ m (2 s.f.)

Exercise 6B

1. (a) Sean is doing an experiment in a science class. He sets an object sliding along a horizontal track. The object slows and stops before it reaches the end of the track. He says 'In the horizontal direction there are no forces acting upon the object.' Comment on Sean's statement, correcting any mistakes he has made.
   (b) Amy is talking about space missions to the moon. She says 'The rocket engines need to keep working the whole way, otherwise the capsule will slow down and stop.' Comment on Amy's statement, correcting any mistakes she has made.

2. An ice hockey player hits the puck of mass 150 grams from one end of the rink towards the other. He hits the puck with a velocity of 3 m s^{-1} and there is a constant friction force of 0.03 N opposing the puck's motion. The rink is 70 m long.
   (a) Find the deceleration of the puck as it moves, assuming this is constant.
   (b) Determine whether the puck reaches the other end of the rink. Show all your working.

3. A firefighter of mass 80 kg slides down a fireman's pole, moving from rest and descending vertically through a distance of 3 metres to the ground. As he slides, he grips the pole, generating a friction force of 200 N, which opposes his motion. Find the time taken for the fireman to reach the ground, giving your answer to a suitable level of accuracy.

4. A box of mass 3.5 kg is placed onto a rough plane inclined at an angle of 20° to the horizontal, as shown in the next diagram. It begins to slide down the plane. The size of the frictional force opposing the motion of the box is 10 N.

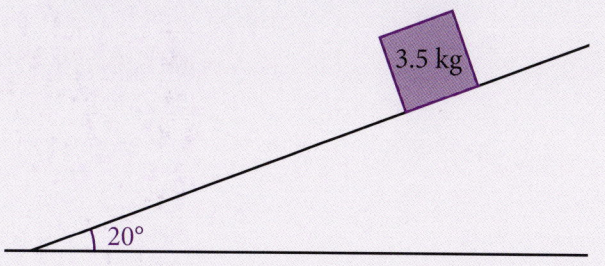

Exercise 6B...

(a) Copy the diagram, adding all external forces acting upon the box.
(b) Calculate the acceleration of the box down the slope, giving your answer to a suitable level of accuracy.
(c) Assuming the box continues with this acceleration, find the time taken for the box to reach a speed of 3 m s^{-1}.

5. A parcel of mass 2 kg is projected up a rough inclined plane with an initial velocity of 4 m s^{-1}. The plane is inclined at an angle of 30° to the horizontal, as shown in the diagram.

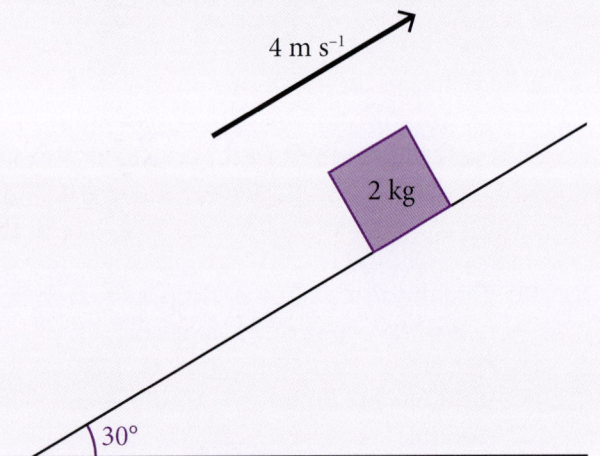

While the parcel is in motion, there is a constant friction force of 5 N opposing the motion.
(a) Copy the diagram, adding all the forces acting on the parcel while it moves up the plane.
(b) Show that the parcel moves up the plane $\dfrac{16}{g+5}$ metres.

The parcel then begins to slide down the plane. The magnitude of the friction force remains the same.
(c) Draw a second diagram, showing the forces acting on the parcel as it moves down the plane.
(d) Find how long the parcel takes to return from its highest point to its starting position, giving your answer in terms of g.

6.3 The $F_r \leq \mu R$ Model of Friction

The coefficient of friction μ

For an object upon a surface, the **coefficient of friction** μ is a measure of the resistance to the object sliding across the surface.

For a rougher surface, the value of μ is generally higher, but the value of μ depends on both the nature of the surface and the nature of the object itself.

Typically, the value of μ lies between 0.3 and 0.6 for a dry object on a dry surface. For very smooth surfaces or wet surfaces, such as ice or an oiled surface, the value of μ is closer to zero. For very rough or sticky surfaces, such as rubber, the coefficient of friction can be as high as 1, or even higher. The coefficient of friction has no units.

Consider this block on a rough surface.

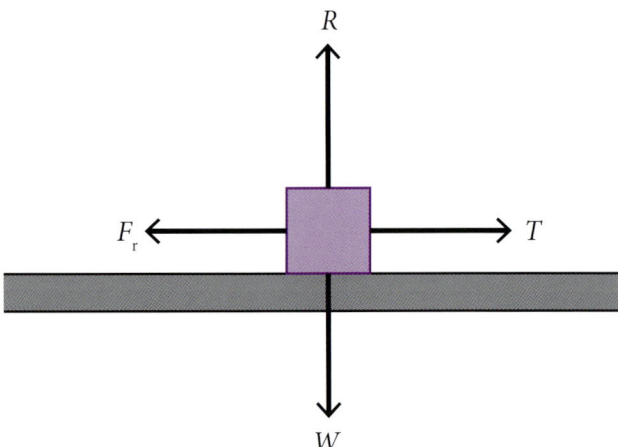

There is a force of size T N (for example tension in a rope) acting to the right.

Suppose $T = 10$ N and suppose that this force is not large enough to set the block in motion. Since the block is in equilibrium, there is a force of 10 N acting to the left, the friction force.

Now consider T being increased to 15 N. Suppose the block is now on the point of moving across the surface, but remains in equilibrium. This situation is called **limiting equilibrium**. Since the block is still in equilibrium, the size of the friction force has increased to 15 N. This is the maximum size of the friction force. Limiting equilibrium is discussed further in section 6.4.

T is increased again to 20 N. The friction force remains at 15 N, its maximum value. There is now a resultant force of 5 N acting to the right and the block begins to accelerate across the surface.

The friction force between an object and a surface reaches its maximum when the object is on the point of sliding across the surface. It remains at this maximum size if the object is then set in motion. The size of the maximum friction force is calculated using the formula:

$F_r = \mu R$

where μ is the coefficient of friction and R is the normal reaction. F_r is used for the friction force to distinguish it from F used elsewhere, for example in $F = ma$.

In summary:

- The friction force acts in the opposite direction to the motion, or to the motion that is about to take place.
- The friction force takes its maximum value when an object is in motion, or on the point of moving (in limiting equilibrium).
- This maximum value is calculated using the formula $F_r = \mu R$.
- So $F_r \leq \mu R$: equal when the object is in motion or on the point of moving; less than otherwise.

In the following example, a crate is in motion on a **rough** surface. If a surface is described as rough, friction forces must be considered. If a surface is **smooth**, friction can be ignored.

Worked Example

4. In Example 1 above, a crate of mass 8 kg is pulled by a rope across a rough surface at a constant speed. The tension in the rope is 30 N.

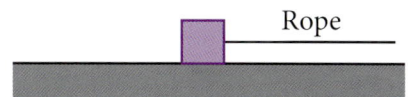

(a) Copy the diagram, showing all the forces acting on the crate.
(b) Find the size of the normal reaction force.
(c) Find the coefficient of friction μ.

(a)

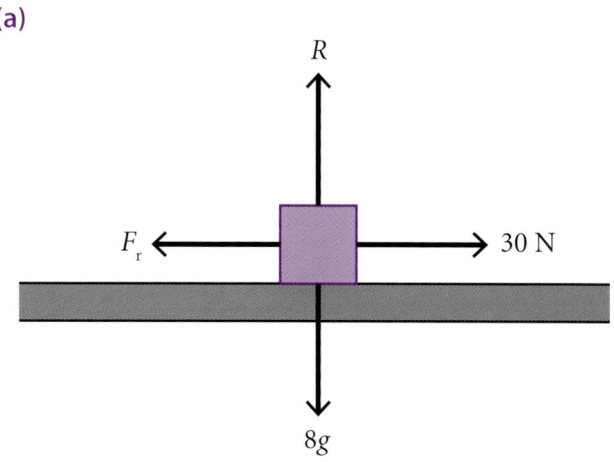

(b) Since the crate is in equilibrium in the vertical,
$R = 8g = 78.4$ N

(c) Since the crate is moving at a constant speed in the horizontal, the force acting to the right is equal to the force acting to the left.
$\therefore F_r = 30$ N

Now use $F_r = \mu R$ to calculate the coefficient of friction:
$F_r = \mu R$
$30 = \mu \times 78.4$
$\mu = \dfrac{30}{78.4} = 0.383$ (3 s.f.)

The following examples demonstrate situations involving rough inclined planes.

Worked Example

5. An object of mass 2 kg slides at a constant speed down a rough, inclined plane. Given that the plane is inclined at an angle of 20° to the horizontal, find the coefficient of friction between the object and the slope.

Draw a diagram, showing all the external forces acting on the object.

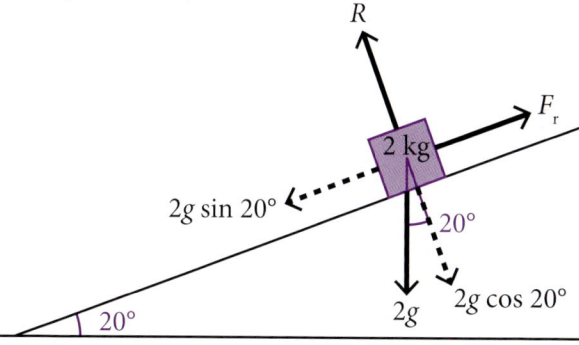

There is a friction force opposing the motion because the plane is rough.

The weight force of $2g$ N has been split into its components parallel and perpendicular to the plane.

Since the object is in equilibrium in the direction perpendicular to the slope,
$R = 2g \cos 20° = 18.418 ...$ N.

Since the object is moving at a constant speed in the direction parallel to the slope, forces acting up the slope are equal to the forces acting down the slope:
$F_r = 2g \sin 20° = 6.703 ...$

Using: $F_r = \mu R$

$\mu = \dfrac{F_r}{R} = \dfrac{6.703}{18.418} = 0.364$ N

AS 2: APPLIED MATHEMATICS

To determine whether an object slides down an inclined plane, resolve the object's weight force into its components parallel and perpendicular to the plane. Also calculate the maximum size of the friction force.

Assuming no other forces are at work, if the forces acting down the plane are greater than the maximum size of the friction force, there is a resultant force acting down the plane and the object begins to accelerate in this direction.

Otherwise, the friction force equals the parallel component of the weight force and the object remains in equilibrium.

Worked Example

6. A box of mass 8 kg is placed on an inclined plane. The plane is angled at 10° to the horizontal. The coefficient of friction between the box and the plane is 0.2. Determine whether the box slides down the plane, showing all your working.

Draw a force diagram:

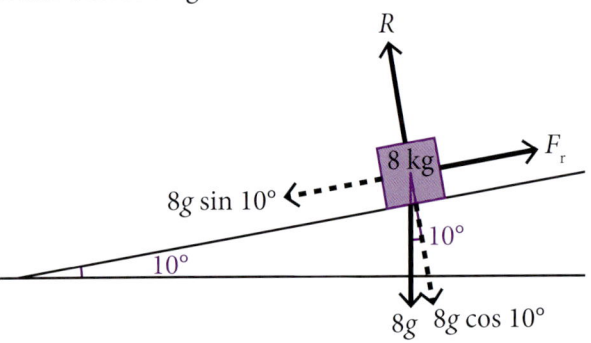

The component of the weight force parallel to the slope is:
$8g \sin 10° = 13.6$ N (3 s.f.)

The maximum friction force is μR. We know that $\mu = 0.2$ and $R = 8g \cos 10° = 77.2089 ...$ N
So the maximum friction force is:
$\mu R = 0.2 \times 77.2089 ... = 15.4$ N (3 s.f.)

The component of the weight force acting parallel to the plane is smaller than the maximum friction force. In this case, the friction force equals this component of the weight force. The object remains in equilibrium.

In Example 7, friction calculations are combined with the use of the constant acceleration formulae.

Worked Example

7. The diagram shows a box of mass M kg on a rough plane inclined at 30° to the horizontal. The coefficient of friction between the box and the plane is $\frac{\sqrt{3}}{6}$.

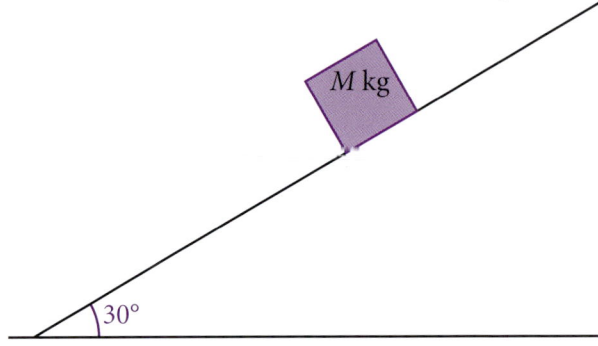

(a) Copy the diagram, showing all the external forces acting on the box.
(b) The box is released from rest and slides down the plane. Find how far the box travels down the plane in the first second of its motion. Give your answer in terms of g.
(c) The box is returned to its starting position and a rope is attached. The rope is parallel to the slope and is used to pull the box up the slope. Given that there is a constant tension of $\frac{5}{4}Mg$ N in the rope, find how far up the plane the box travels in the following one second. Give your answer in terms of g.
(d) What modelling assumptions have been made?

(a)

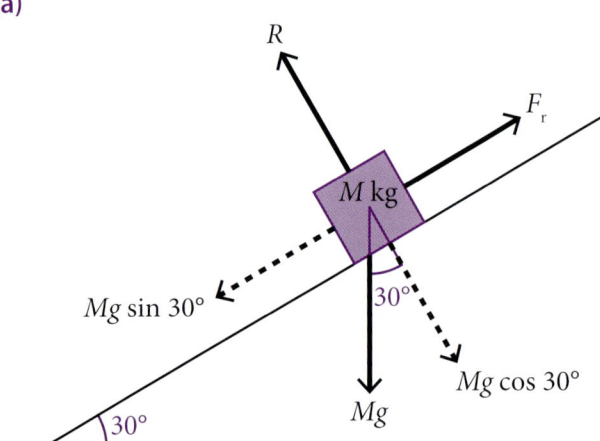

(b) In the perpendicular direction the box is in equilibrium:
$R = Mg \cos 30°$
$R = \frac{\sqrt{3}}{2} Mg$

Since the box is in motion, the friction force will be at its maximum and can be calculated using $F_r = \mu R$:

$$F_r = \frac{\sqrt{3}}{6} \times \frac{\sqrt{3}}{2} Mg$$

$$\Rightarrow F_r = \frac{Mg}{4}$$

In the direction parallel to the slope, the box is in motion. Calculate the acceleration down the slope using $F = ma$:

$$Mg \sin 30° - F_r = Ma$$

$$\frac{Mg}{2} - \frac{Mg}{4} = Ma$$

$$a = \frac{g}{4} \text{ m s}^{-2}$$

Next use the constant acceleration formulae to find the distance travelled in 1 second.
$u = 0$ since the box starts from rest. So we have:
$s = ?$
$u = 0$
$v =$
$a = \frac{g}{4}$
$t = 1$

The formula linking s, u, a and t is required:

$$s = ut + \frac{1}{2} at^2$$

$$s = 0 + \frac{1}{2} \times \frac{g}{4} \times 1^2$$

$$s = \frac{g}{8} \text{ m}$$

(c) Use a second diagram as the forces acting on the box have changed.

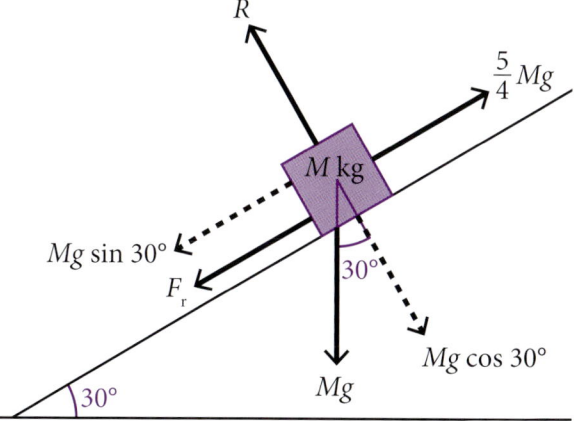

Note that the friction force now acts down the slope as the box is moving up the slope.
Use $F = ma$ in the direction of the motion:

$$\frac{5}{4} Mg - Mg \sin 30° - F_r = Ma$$

The size of the friction forces remains as $\frac{Mg}{4}$ because μ and R both remain unchanged.

$$\frac{5}{4} Mg - \frac{Mg}{2} - \frac{Mg}{4} = Ma$$

$$a = \frac{g}{2} \text{ m s}^{-2}$$

Calculate the distance travelled.

$$s = ut + \frac{1}{2} at^2$$

$$s = 0 + \frac{1}{2} \left(\frac{g}{2}\right)(1)^2$$

$$s = \frac{g}{4} \text{ m}$$

(d) The box is modelled as a particle; the rope is light and inextensible.

The next example shows an object on an inclined plane being acted upon by a force that is not parallel or perpendicular to the plane. In this case, the force in question should be resolved into components parallel and perpendicular to the plane, in the same way as the weight force is resolved.

Worked Example

8. The next diagram shows an object of mass 5 kg on a rough plane inclined at 24° to the horizontal. The object is pulled up the plane by a rope, which is at an angle of 30° to the plane. The tension in the rope is 60 N.

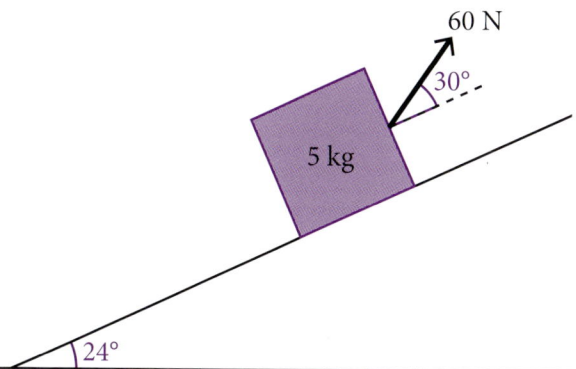

If the coefficient of friction between the object and the plane is 0.55, find the resultant force acting upon the object in the direction parallel to the plane.

Begin by drawing a force diagram, showing all the external forces acting on the object.

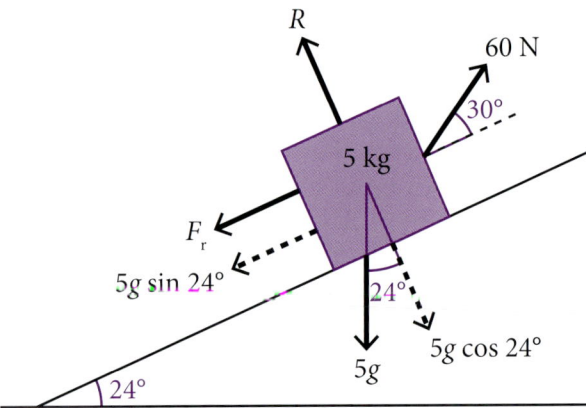

In the perpendicular direction forces are in equilibrium. Remember to consider the component of the tension acting in the perpendicular direction:

$R + 60 \sin 30° = 5g \cos 24°$
$R = 5g \cos 24° - 60 \sin 30°$
$R = 14.764 \text{ N}$

$F_r = \mu R$
$F_r = 0.55 \times 14.764 = 8.120 \text{ N}$

In the parallel direction, the resultant force F is given by:
$F = 60 \cos 30° - F_r - 5g \sin 24°$
$F = 23.911 \text{ N}$

Exercise 6C

1. Estimate the value of μ given the object and the surface with which it is in contact.
   (a) An ice hockey puck sliding across ice.
   (b) A tomato in a Teflon frying pan.
   (c) A book sliding across a polished wooden table
   (d) A chair being dragged across a bedroom carpet.
   (e) Bare feet on a tiled bathroom floor.
   (f) Feet wearing socks on a tiled bathroom floor.

2. A book of mass 300 grams is pushed at a constant speed across a rough horizontal table using a force of 1 N. Find the coefficient of friction between the table and book.

3. Consider the objects listed in the table below. Each object is on an inclined plane, with the angle of the slope to the horizontal given. Also given is the object's mass and the coefficient of friction. For each object, determine whether it will slide down the inclined plane. It may help to draw a force diagram in each case.

Exercise 6C...

Object	Mass	Angle of slope	Coefficient of friction
Matchbox	10 g	30°	0.45
Overturned car resting on its roof on a steep hill	1200 kg	15°	0.35
Skier	75 kg	35°	0.03
Brick	2.5 kg	20°	0.5

4. In an airport, luggage slides down a ramp onto a conveyor belt. The diagram shows a piece of luggage with a mass of 10 kg on the ramp. The ramp is inclined at an angle of 70° to the vertical and the coefficient of friction between the luggage and the ramp is 0.25.

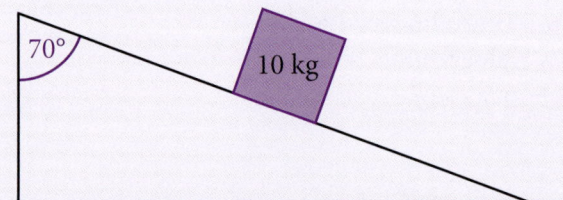

   (a) Copy the diagram and add all the external forces acting on this piece of luggage as it slides down the ramp.
   (b) Calculate the acceleration of the luggage as it slides down the ramp.
   (c) The airport manager says too many pieces of luggage are being damaged. To avoid this, she wants each piece of luggage to reach a maximum safe speed of 2 m s^{-1} as it reaches the end of the ramp. Given that the ramp is 3 m long, determine whether the piece of luggage shown in the diagram will reach the end of the ramp safely. You may assume the piece of luggage has an initial speed of 0 m s^{-1} at the top of the ramp. Show all your working.

5. The diagram shows a package being pulled by a rope along rough horizontal ground. The package has a mass of 18 kg and is accelerating at 0.15 m s^{-2}. The tension in the rope is 30 N. The angle between the rope and the ground is 25°.

Exercise 6C...

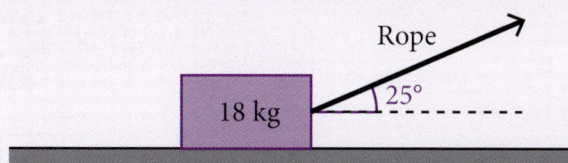

(a) Draw a diagram showing all the external forces acting on the package.
(b) Find the coefficient of friction between the package and the ground.
(c) State one modelling assumption you have made about the package and one you have made about the rope.
(d) State what would happen to the size of the frictional force if the angle between the rope and the ground was increased. Give a reason for your answer.

6. The following diagram shows a box of mass M kilograms moving up the line of greatest slope of a rough plane AB.

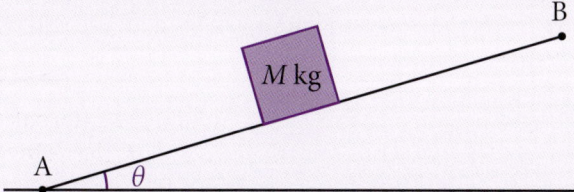

The plane is inclined at an angle θ to the horizontal, where $\cos \theta = \dfrac{24}{25}$.

The coefficient of friction between the box and the plane is μ.
(a) Draw a diagram showing all the external forces acting on the box.
(b) At A the box has a velocity of U m s^{-1}. T seconds later the box comes to rest at B. Show that the acceleration of the box up the plane is given by:
$$a = -\left(\dfrac{24\mu + 7}{25}\right)g$$
(c) Find U in terms of μ, g and T.
(d) What modelling assumptions have been made?

Exercise 6C...

7. A suitcase slides from rest down a rough ramp in an airport terminal. The ramp is 4 metres long and angled at 20° to the horizontal. The suitcase takes 2.5 seconds to slide from the top to the bottom of the ramp. If the size of the frictional force between the suitcase and the ramp is 31 N, find the mass of the suitcase, giving your answer to an appropriate degree of accuracy.

8. **Use $g = 9.81$ m s^{-2} in this question.**
The diagram shows a trunk of mass 8 kg on a rough plane inclined at 18° to the horizontal. The trunk is pulled up the plane by a rope, which is at an angle of 25° to the plane. The tension in the rope is 48 N.

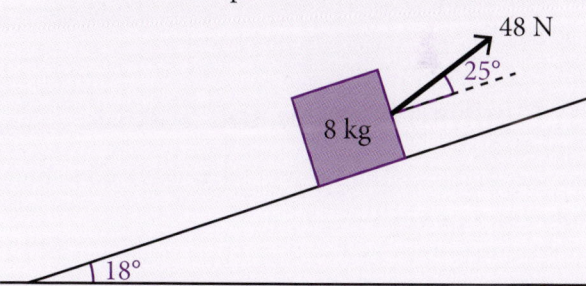

The coefficient of friction between the trunk and the plane is 0.31.
(a) Copy the diagram, adding to it all the external forces acting upon the trunk. Resolve forces into two components parallel and perpendicular to the plane where appropriate.
(b) Find the acceleration of the trunk up the plane.
(c) With the trunk starting from rest, find how far it moves up the plane in a time of 2 seconds.

9. A suitcase of mass 13 kg is placed on a rough slope inclined at an angle of 25° to the horizontal. The coefficient of friction between the suitcase and the slope is 0.22. A horizontal force of size P newtons acts on the suitcase, causing it to accelerate up the slope at 1.2 m s^{-2}.
(a) Draw a diagram showing all the external forces acting on the suitcase.
(b) Find P.

AS 2: APPLIED MATHEMATICS

Exercise 6C...

10. When a car driver applies the brakes, the car slows down, but the car slows more quickly in dry weather than in wet weather. This occurs because the coefficient of friction between the wheels and the road is higher when the road is dry. In this question, assume that the coefficient of friction in dry weather is 0.45 and in wet conditions is 0.35.
    (a) A car of mass 1000 kg is travelling at 30 m s^{-1} in dry weather conditions. The driver sees a stationary car 200 metres ahead and applies the brakes. Does the car stop in time? Show all your working.
    (b) What is the minimum safe distance between two cars travelling at 30 m s^{-1} in wet conditions?

6.4 Limiting Friction and Statics

An object that is on the point of moving is said to be in **limiting equilibrium**. In this state, forces are in equilibrium **and** $F_r = \mu R$ applies because the friction force is at its maximum.

Worked Examples

9. The coefficient of friction between a 20 kg crate and the horizontal floor is 0.52. A horizontal force of size P N is applied to the crate. If the crate is on the point of sliding across the floor, find the value of P, giving your answer to 3 significant figures.

A force diagram helps to visualise the problem.

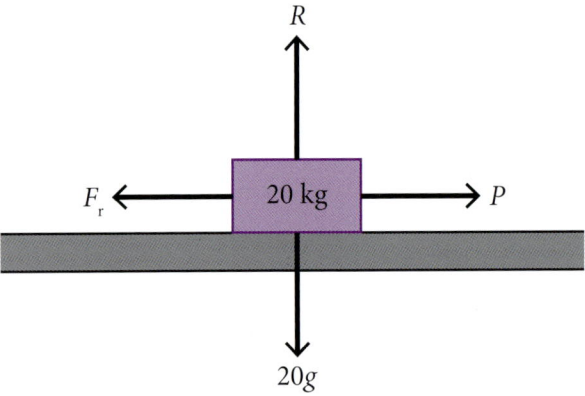

In the vertical, the crate is in equilibrium, therefore:
$R = 20g = 196$ N.

In the horizontal, since the crate is on the point of sliding, it is in limiting equilibrium. In this situation, friction is at its maximum, therefore:
$F_r = \mu R$
$ = 0.52 \times 196$
$ = 101.92$ N

And since the crate is in equilibrium, $P = F_r$, therefore:
$P = 101.92 = 102$ N (3 s.f.)

10. A parcel of mass M kg rests in equilibrium on a rough plane inclined at 30° to the horizontal, as shown. When a horizontal force, Q N, is applied to the parcel it is on the point of moving up the plane. The coefficient of friction between the parcel and the plane is μ.

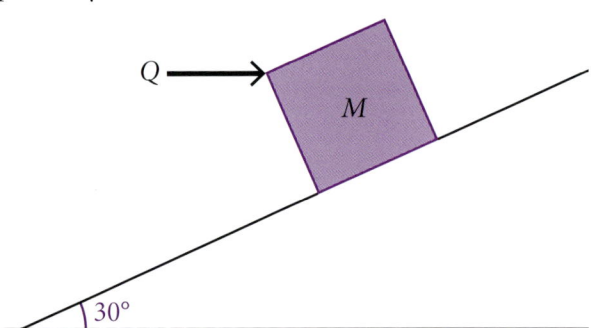

(a) Copy the diagram and mark on it all the external forces acting on the parcel.
(b) Show that the size of the normal reaction force
$$R = \frac{2\sqrt{3}Mg}{3 - \sqrt{3}\mu}$$
(c) Find Q in terms of g, M and μ.

(a) When drawing a force diagram, remember to show all forces as arrows **coming out of the object**. So, the force of Q N should appear as a horizontal arrow from the right-hand side of the box. This force should then be resolved into parallel and perpendicular components.

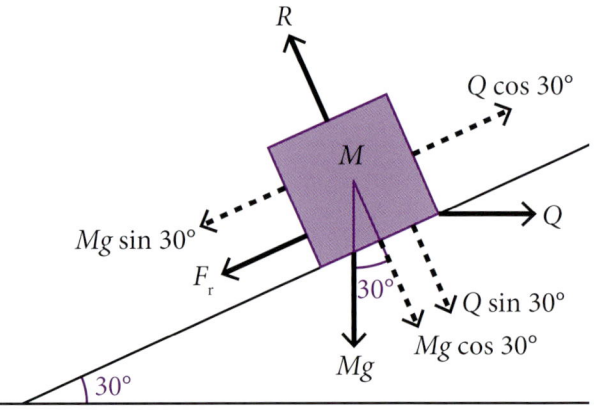

(b) Parallel to the slope:
$$Q \cos 30 = Mg \sin 30 + F_r$$
$$\frac{\sqrt{3}}{2}Q = \frac{Mg}{2} + F_r$$
$$\sqrt{3}Q = Mg + 2F_r$$
But $F_r = \mu R$, so:
$$\sqrt{3}Q = Mg + 2\mu R$$
$$Q = \frac{\sqrt{3}Mg}{3} + \frac{2\sqrt{3}\mu R}{3} \quad (1)$$

Perpendicular to the slope:
$$R = Q \sin 30 + Mg \cos 30$$
$$R = \frac{Q}{2} + \frac{\sqrt{3}Mg}{2}$$

Substitute for Q from (1):
$$R = \frac{\sqrt{3}Mg}{6} + \frac{\sqrt{3}\mu R}{3} + \frac{\sqrt{3}Mg}{2}$$
$$R - \frac{\sqrt{3}\mu R}{3} = \frac{2\sqrt{3}Mg}{3}$$
$$R\left(1 - \frac{\sqrt{3}\mu}{3}\right) = \frac{2\sqrt{3}Mg}{3}$$
$$R = \frac{\frac{2\sqrt{3}}{3}Mg}{1 - \frac{\sqrt{3}\mu}{3}}$$
$$R = \frac{2\sqrt{3}Mg}{3 - \sqrt{3}\mu}$$

(c) Using (1):
$$Q = \frac{\sqrt{3}Mg}{3} + \frac{2\sqrt{3}\mu R}{3}$$

Substituting for R:
$$Q = \frac{\sqrt{3}Mg}{3} + \frac{2\sqrt{3}\mu}{3}\left(\frac{2\sqrt{3}Mg}{3 - \sqrt{3}\mu}\right)$$
$$Q = \frac{\sqrt{3}Mg}{3} + \frac{12Mg\mu}{9 - 3\sqrt{3}\mu}$$
$$Q = \frac{\sqrt{3}Mg(3 - \sqrt{3}\mu)}{9 - 3\sqrt{3}\mu} + \frac{12Mg\mu}{9 - 3\sqrt{3}\mu}$$
$$Q = \frac{3\sqrt{3}Mg - 3Mg\mu}{9 - 3\sqrt{3}\mu} + \frac{12Mg\mu}{9 - 3\sqrt{3}\mu}$$
$$Q = \frac{3\sqrt{3}Mg + 9Mg\mu}{9 - 3\sqrt{3}\mu}$$
$$Q = \frac{\sqrt{3}Mg + 3Mg\mu}{3 - \sqrt{3}\mu}$$
$$Q = \left(\frac{\sqrt{3} + 3\mu}{3 - \sqrt{3}\mu}\right)Mg$$

Exercise 6D

1. In this question use $g = 10$ m s^{-2}. A book of mass 200 grams rests on a rough table.
   (a) A horizontal force of size 0.4 N is applied to the book, but the book does not move. Write down the size of the friction force.
   (b) The horizontal force is increased to 0.6 N. The book is now on the point of moving.
      (i) Write down the size of the friction force now.
      (ii) Find the coefficient of friction between the book and the table.
   (c) The force is increased again to 0.8 N.
      (i) Write down the size of the friction force now.
      (ii) With the help of a force diagram, find the acceleration of the book across the table.

2. A 6 kg parcel is on a rough horizontal table. The coefficient of friction between the parcel and the table is 0.36. A horizontal force of size T N is applied to the parcel. If the parcel is on the point of sliding across the table, find the value of T.

3. In a strong man competition, two competitors are attempting to pull a metal block across a rough horizontal concrete driveway using a rope, which is also horizontal. The mass of the block is 80 kg. Geoff is the first competitor. When he applies a force of 490 newtons the block is in limiting equilibrium.
   (a) Show that the coefficient of friction between the block and the driveway is 0.625.
   (b) Henry is the second competitor. When he pulls the block, it begins to accelerate at 0.1 m s^{-2}. Find the tension in the rope.
   (c) Find the speed at which the block is moving after 1 second.
   (d) After 1 second, Henry continues to pull the block at this speed for a further 5 seconds. Find how far he pulls the block **in total**.

Exercise 6D...

4. The diagram shows a box of mass 15 kg being pulled across a rough horizontal surface by a light inextensible rope inclined at 23° to the horizontal.

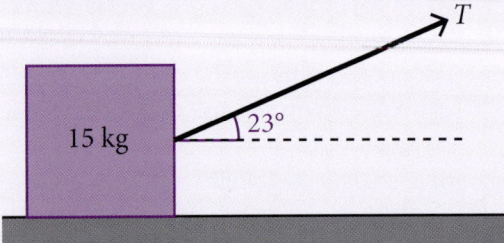

The tension in the rope is T newtons and the coefficient of friction between the box and the surface is μ.
   (a) Draw a diagram showing all the external forces acting on the box.
   (b) When $T = 12$ N, the box is in limiting equilibrium. Find μ.
   (c) T is now increased so that the box accelerates across the surface at 1.5 m s^{-2}. Find the new value of T.

5. A parcel of mass 25 kg rests in equilibrium on a rough plane inclined at 45° to the horizontal, as shown. The coefficient of friction is μ. When a horizontal force, Q N, is applied to the parcel it is on the point of moving up the plane.

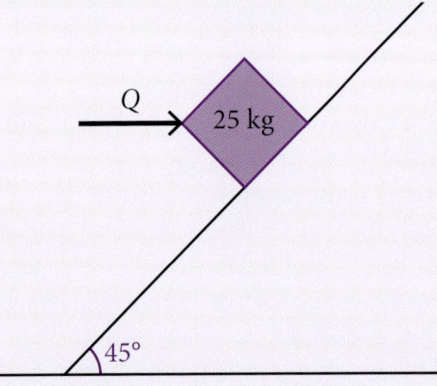

   (a) Copy the diagram and mark on it all the external forces acting on the parcel.
   (b) Show that the size of the normal reaction force R is given by $R = \dfrac{25g\sqrt{2}}{1-\mu}$.
   (c) Find Q in terms of g and μ.

Exercise 6D...

6. Katharine and Gary are moving to a new house. They are trying to move a crate of mass 50 kg up a rough slope. The slope is inclined at an angle of 20° to the horizontal. The coefficient of friction between the crate and the surface is 0.15. The diagram shows a rope attached to the crate.

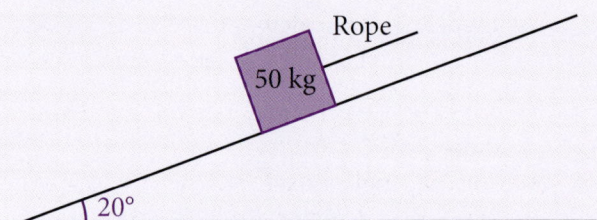

Gary tries to pull the crate up the slope using the rope, keeping the rope parallel to the plane. He exerts a force of G newtons. Despite Gary's efforts, the crate is on the point of sliding down the plane.
   (a) Draw a diagram, showing all the forces acting on the crate.
   (b) Find G, giving your answer to 3 significant figures.
   (c) Katharine now joins Gary to help him move the crate up the slope. She pushes with a force of K newtons parallel to the slope, as shown in the diagram below. The force Gary uses to pull the crate remains unchanged.

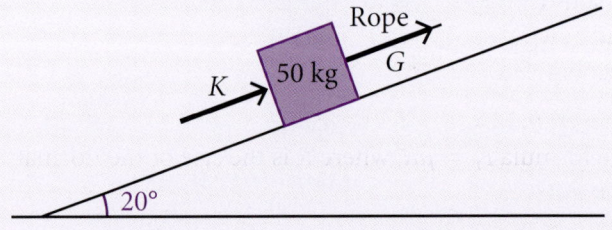

The crate is now on the point of sliding up the plane. With the help of a force diagram, find K. Give your answer to 3 significant figures.
   (d) State two modelling assumptions that have been made.

Exercise 6D...

7. David places his school bag and calculator on his school desk. The bag and the calculator have masses of 12 kg and 95 grams, respectively. The coefficient of friction between the bag and desk is 0.55. The calculator has rubber pads attached. This means the coefficient of friction is higher, at 0.85. David slowly tilts his desk, wondering whether the bag or calculator will slide off first.

(a) Show that, if David tilts his desk to an angle of $\tan^{-1}\left(\frac{11}{20}\right)$, the bag is on the point of sliding.

(b) To what angle must David tilt his desk to make the calculator slide? Give an exact answer in the form $\tan^{-1}\left(\frac{p}{q}\right)$, where $\frac{p}{q}$ is a fraction in its simplest form.

8. An object is in a state of limiting equilibrium on a rough inclined plane. Given that the plane is at an angle of $\theta°$ to the horizontal, show that the coefficient of friction μ is given by $\mu = \tan \theta$.

6.5 Summary

A friction force acts whenever an object is in contact with a rough surface.

The coefficient of friction μ is a measure of the resistance to sliding. It has no units and is usually between 0 and 1, with a value of 0 indicating a completely smooth (frictionless) surface.

The maximum size of the friction force is modelled using the formula $F_r = \mu R$, where R is the size of the normal reaction force.

The friction force only reaches its maximum value when the object is in motion, or is on the point of moving. It acts in the opposite direction to the motion.

An object that is on the point of moving is said to be in **limiting equilibrium**. In this state forces are in equilibrium **and** $F_r = \mu R$ applies.

If an object is accelerating, then its acceleration can be calculated using $F = ma$, where F is the resultant force, including any friction force.

It may be necessary to use the constant acceleration formulae in conjunction with $F = ma$ to calculate the distance travelled, the time taken, etc.

Chapter 7
Connected Bodies

7.1 Introduction

Many situations in mechanics involve connected bodies. In this chapter the word 'connected' means connected by a string or a towbar, or simply in contact.

Newton's third law is important in these studies. It states that when two bodies interact, the force exerted by the first body upon the second is equal in magnitude, but opposite in direction, to the force exerted by the second body upon the first. For example, when a car and trailer are connected by a towbar, there is a tension force in the towbar pulling the trailer forwards. There is an equal tension in the towbar acting to pull the car in the opposite direction.

Key words
- **Pulley**: A pulley is a wheel on an axle.
- **Lift**: A lift is a large vessel for transporting goods or people up or down.

Before you start
You should know:
- The constant acceleration formulae.
- Newton's three laws of motion.
- The $F_r \leq \mu R$ model of friction.

What you will learn
In this chapter you will learn about:
- Connected bodies in equilibrium.
- Connected bodies moving together.
- Pulleys.
- Connected bodies and friction.

In the real world...
A train leaving a busy train station in a rush hour may have eight or more carriages and hundreds of passengers.

The engine must provide a very large tractive force to accelerate such a huge mass to the speed required for the journey.

Engineers need to calculate the power required to accelerate the coaches. To do this they need to know the masses of the coaches, the passengers and the engine itself. They also need to know, or estimate, the size of any resistance forces and the top speed required.

In addition, they need to consider the tension in the couplings connecting the engine to the carriages, and between the carriages – are they strong enough to survive these huge forces?

Exercise 7A (Revision)

1. (a) An unmanned spacecraft is travelling at a constant velocity through space on a mission to study another planet.
    (i) What external forces are acting on the spacecraft?
    (ii) As the spacecraft approaches its destination, what may cause its velocity to change?
   (b) The spacecraft has a mass of 750 kg. It slows from 20 000 m s^{-1} to 5000 m s^{-1} over a distance of half a million kilometres. Find the force required for this deceleration.
   (c) A man walking on the surface of the moon has a mass of 150 kg, including his spacesuit. The acceleration due to gravity on the moon is 1.6 m s^{-2}. Find the size of the normal reaction force the man experiences from the moon's surface.

2. A boy pushes a stone of mass 100 grams across a rough table. He uses a force of 0.24 newtons for 1.4 seconds. The stone starts from rest and the coefficient of friction between the stone and the table is $\frac{1}{7}$.
   (a) How far does the stone travel across the table while it accelerates?
   (b) He then stops pushing and the stone comes to rest. For how long was the stone in motion altogether?

7.2 Connected Bodies in Equilibrium

When two bodies are connected vertically, the weight force of one body may affect the other.

Worked Example

1. Three maths textbooks are in a pile on a table, as shown in the diagram.

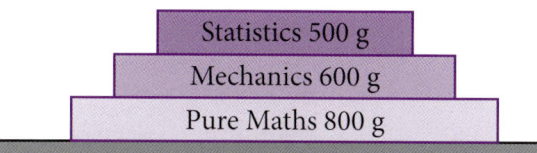

Draw three separate force diagrams, one for each book.

Calculate the size of each external force acting upon each book, giving your answers to 1 decimal place.

The **Pure Maths** book has three downwards forces acting upon it: its own weight force, and the downwards forces exerted by the Mechanics and Statistics books.

There is also an upwards reaction force R_1 from the table upon which it rests.

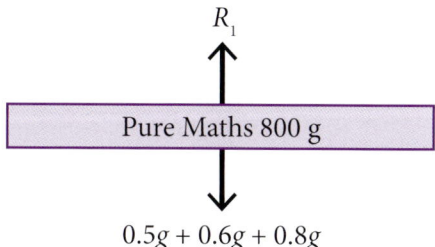

The total size of the downwards force is:
$0.5g + 0.6g + 0.8g = 1.9g = 18.6\,\text{N}$ (1 d.p.)

Since the Pure Mathematics book is in equilibrium:
$R_1 = 18.6\,\text{N}$ (1 d.p.)

The **Mechanics** book is affected by its own weight and the weight of the Statistics book on top.

There is also an upwards reaction force R_2 from the book upon which it rests.

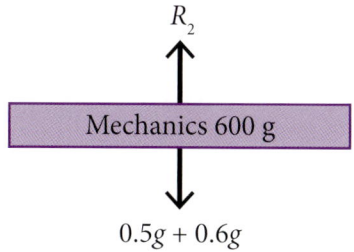

The total size of the downwards force is:
$0.5g + 0.6g = 1.1g = 10.8\,\text{N}$ (1 d.p.)

Since the Mechanics book is in equilibrium:
$R_2 = 10.8\,\text{N}$ (1 d.p.)

The only downwards force acting on the **Statistics** book is its own weight force.

There is also an upwards reaction force R_3 from the book upon which it rests.

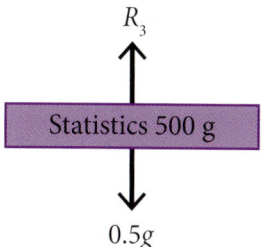

The downwards force is:
$0.5g = 4.9\,\text{N}$ (1 d.p.)

Since the Statistics book is in equilibrium:
$R_3 = 4.9\,\text{N}$ (1 d.p.)

Exercise 7B

1. A light scale pan is attached to a light, inextensible, vertical string. The scale pan carries two parcels A and B. The mass of Parcel A is 300 grams and the mass of Parcel B is 500 grams. Parcel A rests on top of Parcel B, as shown in the diagram.

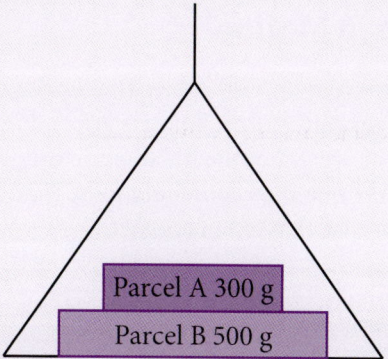

The scale pan is in equilibrium.
(a) Copy the diagram, showing all the forces acting on **Parcel B**.
(b) Find the force exerted by Parcel A on Parcel B.
(c) Find the reaction force exerted by the scale pan on Parcel B.
(d) Find the tension in the string.

AS 2: APPLIED MATHEMATICS

Exercise 7B...

2. Use $g = 10$ m s^{-2} as an approximation for the acceleration due to gravity in this question.
   The diagram shows a block of mass 0.1 kg hanging at rest at the end of a light inextensible vertical string. The other end of the string is attached to a mass of 0.2 kg which is attached to a second light inextensible vertical string.

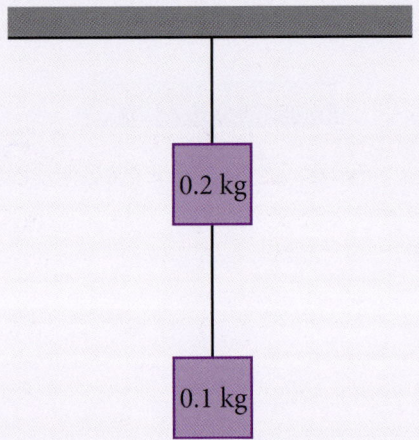

   Find the tension in each string.

3. A cube of mass 4 kg rests on a horizontal table. A smaller cube of mass 3 kg is placed on top of the 4 kg cube. Find:
   (a) The reaction force between the two cubes.
   (b) The reaction force between the 4 kg cube and the table.

7.3 Lifts in Motion

Lifts are supported by a lift cable. The tension in the cable provides an upwards force. There is also a downwards force from the weight of the lift and anything it is carrying. The lift and its contents can be modelled as connected bodies that are in contact.

Worked Example

2. A man of mass 70 kg is standing in a lift of mass 2000 kg. Use $g = 9.81$ m s^{-2} as an approximation for the acceleration due to gravity. Give all answers to 3 significant figures.
   (a) (i) Draw two separate diagrams to show all the external forces acting on the lift and on the man.
       (ii) Now consider the lift and man as a single system. Draw a third diagram showing the external forces acting on the system.
   (b) Find the normal reaction force between the lift floor and the man when the lift is:
       (i) stationary,
       (ii) accelerating upwards at 1 m s^{-2},
       (iii) accelerating downwards at 1 m s^{-2}.
   (c) Find the tension T in the lift cable when the lift is accelerating upwards at 1 m s^{-2}.

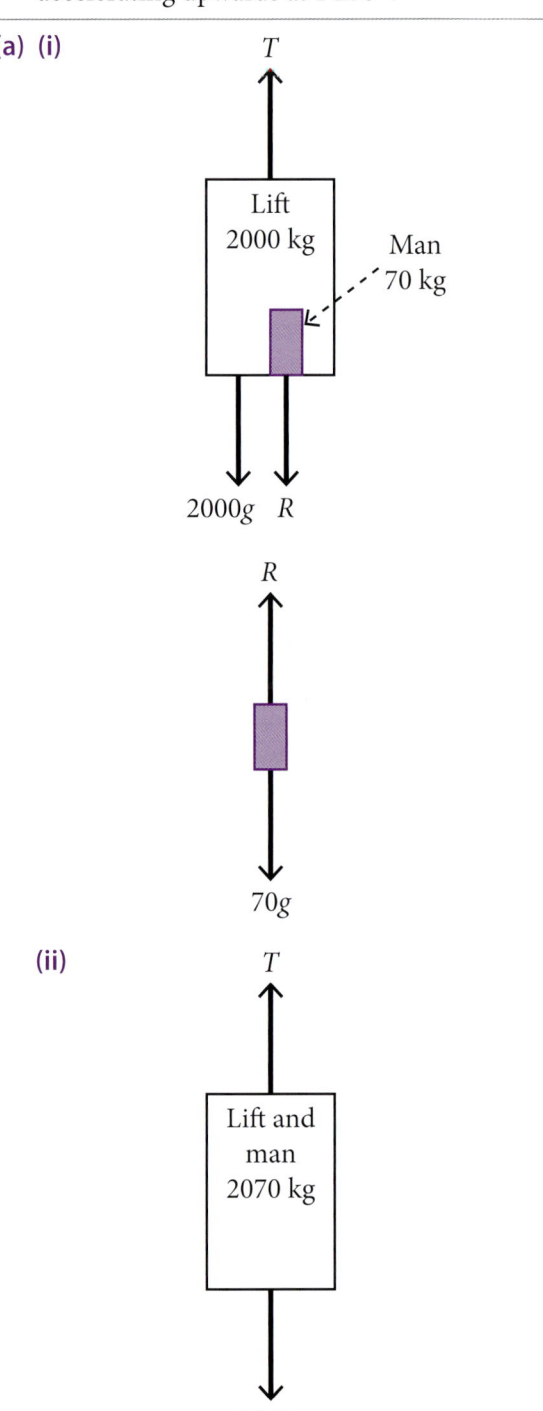

(b) Consider the forces acting on the man.
  (i) If the lift is stationary, forces upwards are equal to forces downwards.
  $R = 70g = 687$ N (3 s.f.)

  (ii) Use $F = ma$ for the man. Upwards is the positive direction.
  $R - 70g = 70 \times 1$
  $R = 70g + 70 = 756.7$
  $R = 757$ N (3 s.f.)

  (iii) Use $F = ma$ for the man. Downwards is the positive direction.
  $70g - R = 70 \times 1$
  $R = 70g - 70 = 616.7$
  $R = 617$ N (3 s.f.)

> **Note:** Do not assume the reaction force is always equal to the weight force! This is not true in an accelerating system.

> **Note:** The answers to parts (b) (i), (ii) and (iii) show that the reaction force is greater when the lift accelerates upwards and smaller when the lift accelerates downwards. You may have experienced this as a sense of heaviness in a rising lift and weightlessness in a descending lift.

> **Note:** The reaction force that the lift floor exerts upon the man is equal to the reaction force exerted by the man upon the floor. Newton's third law – for every force there is an equal and opposite reaction – applies whether the system is stationary or accelerating.

(c) **Method 1.** Consider the entire system. Its mass is 2070 kg, so its downwards weight force is $2070g$ newtons.
$F = ma$
$T - 2070g = 2070 \times 1$
$T = 2070 + 20306.7$
$T = 22376.7 = 22400$ N (3 s.f.)

> **Note:** Newton's second law can be used for the whole system because the lift and man are moving together with the same velocity and acceleration. Using this approach, the reaction force R does not appear in the equation.

Method 2. Consider just the lift. Its mass is 2000 kg, so its weight force is $2000g$ N.
$F = ma$
$T - 2000g - R = 2000 \times 1$
$T - 19620 - 756.7 = 2000$
$T = 22376.7 = 22400$ N (3 s.f.)

> **Note:** This method relies upon the value of R calculated in part (b) (ii). Method 2 becomes the longer method if R has not been previously calculated.

Exercise 7C

1. A woman of mass 65 kg stands in a lift. Find the reaction force between the floor of the lift and the woman when the lift:
   (a) descends with constant acceleration of 0.9 m s^{-2},
   (b) ascends with constant acceleration of 0.9 m s^{-2}.
   Give your answers to a suitable level of accuracy.

2. A lift carries a single passenger of mass 70 kg. The lift ascends from the ground floor to the first floor of a building. The journey is in three stages:
   - Firstly it accelerates upwards at 0.8 m s^{-2} until it reaches a certain velocity.
   - It then maintains this velocity for a period of time.
   - Finally it slows, with a retardation of 1 m s^{-2} until it comes to rest.

   Find the reaction between the floor of the lift and the passenger during each of the three stages, giving your answers to a suitable level of accuracy.

3. **Use $g = 9.81$ m s^{-2} as an approximation for the acceleration due to gravity in this question. Give your answers to an appropriate level of accuracy.**
   A woman of mass 55 kg is travelling in a lift of mass 750 kg.
   (a) Treating the lift and woman as a single system, draw a diagram showing the external forces acting on the system.
   (b) Find the tension in the cable when the lift is accelerating upwards at 0.3 m s^{-2}.

Exercise 7C...

(c) Draw a second diagram showing the external forces acting on the woman only.

(d) Find the reaction force between the floor of the lift and the woman when the lift is accelerating downwards at 0.25 m s^{-2}.

4. A lift attached to a vertical cable is descending into a basement with constant acceleration. The mass of the lift is 500 kg. Resting in the lift is a box of mass 50 kg. There is a constant upwards resistance force of 100 N on the lift. The box experiences a normal reaction of magnitude 340 N from the floor of the lift.
   (a) Draw a diagram showing all the forces acting on the lift.
   (b) Draw a diagram showing all the forces acting on the box.
   (c) Find the acceleration of the lift.
   (d) Find the tension in the lift cable.

5. **Use $g = 9.8$ m s^{-2} as an approximation for the acceleration due to gravity in this question. Give your answers correct to one decimal place.**
   A light scale pan is attached to a light, inextensible, vertical string. The scale pan carries two parcels A and B. The mass of parcel A is 300 grams and the mass of parcel B is 500 grams. Parcel A rests on top of parcel B, as shown in the diagram.

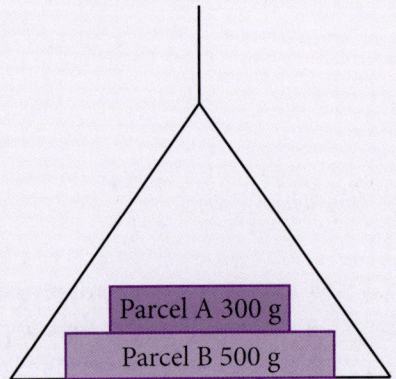

   The scale pan is raised vertically, using the string, with acceleration 0.4 m s^{-2}.
   (a) Find the tension in the string.
   (b) Find the force exerted on parcel B by parcel A.
   (c) Find the force exerted on parcel B by the scale pan.

Exercise 7C...

6. **Use $g = 9.8$ m s^{-2} as an approximation for the acceleration due to gravity in this question. Give your answers to a suitable level of accuracy.** A lift has mass 650 kg.
   (a) Find the tension in the lift cable when the lift is accelerating downwards at 0.5 m s^{-2} without any passengers.
   (b) The greatest tension the lift cable can support is 13 000 N. The average mass of a person travelling in the lift is 75 kg. Find the largest number of passengers that the lift can safely carry when it is accelerating upwards at 0.75 m s^{-2}.
   (c) State two modelling assumptions you have used in answering parts (a) and (b).

7.4 Bodies Moving Together Horizontally

When two bodies are connected, Newton's second law can be applied to each body.

The two equations arising often involve two unknowns; therefore problems involving connected bodies often require the solution of simultaneous equations.

In a situation where the two bodies move in the same direction with the same velocity and acceleration, the second law can be applied to the entire system. In this way, it is possible in some cases to avoid using simultaneous equations. Such situations include a car and trailer connected by a rigid towbar, and a lift carrying a passenger.

.....

Worked Examples

3. A car of mass 1500 kg pulls a trailer of mass 2700 kg along a straight, horizontal road, as shown in the diagram. The car and trailer are attached to each other by a light, rigid towbar.

   The car's engine exerts a constant tractive force of 4200 N. The car and trailer both experience constant resistive forces when they are in motion due to air resistance and the effects of the road. The resistance to the motion of the car is 400 N and the resistance to the motion of the trailer is 800 N.

   The car and trailer start from rest.

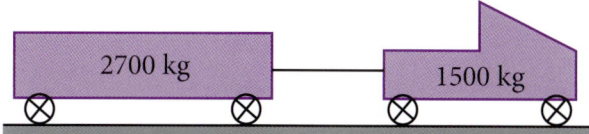

(a) Copy the diagram, showing all the horizontal forces acting on both the car and the trailer.
(b) Find the acceleration of the system.
(c) Find the tension in the towbar while the car and trailer are in motion.
(d) When the car and trailer reach a speed of 20 m s^{-1}, the towbar breaks. Assuming the resistive force on the trailer remains a constant 800 N, find how much further the trailer travels along the road.
(e) What is the meaning of these words in the question?
 (i) Light
 (ii) Rigid
(f) Are the car and trailer modelled as particles?

(a)

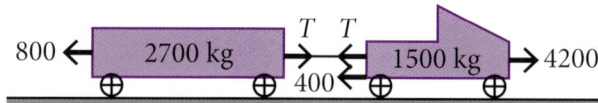

(b) **Method 1: use simultaneous equations**
Use $F = ma$ for the car:
$$4200 - T - 400 = 1500a$$
$$3800 - T = 1500a \text{ (1)}$$

Use $F = ma$ for the trailer:
$$T - 800 = 2700a \text{ (2)}$$

Add equations (1) and (2) to eliminate T:
$$3000 = 4200a$$
$$a = \frac{5}{7} = 0.714 \text{ m s}^{-2} \text{ (3 s.f.)}$$

Method 2: use $F = ma$ for the entire system
The combined mass of the car and trailer is 4200 kg. Use $F = ma$ for the entire system:
$$\Rightarrow 4200 - 400 - 800 = 4200a$$
$$3000 = 4200a$$
$$a = \frac{5}{7} = 0.714 \text{ m s}^{-2} \text{ (3 s.f.)}$$

(c) Substitute a into (2):
$$T - 800 = 2700\left(\frac{5}{7}\right)$$
$$T = 2700\left(\frac{5}{7}\right) + 800 = 2728.57 \ldots$$
$$T = 2730 \text{ N (3 s.f.)}$$

(d) After the towbar breaks, the only force acting on the trailer is the constant resistive force of 800 N.

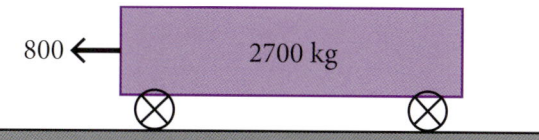

Use $F = ma$ for the trailer:
$$-800 = 2700a$$
$$a = -\frac{8}{27} = -0.296 \ldots \text{ m s}^{-2}$$

The acceleration is negative since the trailer is slowing down (decelerating).

Use the constant acceleration formulae (suvat formulae) to calculate the extra distance travelled during this phase of the motion. During this phase, the initial velocity is the velocity of the trailer when the towbar breaks, 20 m s^{-1}. The final velocity is zero, since the trailer comes to rest.

For the trailer, from the instant the towbar breaks until the trailer comes to rest:
$s = ?$
$u = 20$
$v = 0$
$a = -\dfrac{8}{27}$
$t =$

The formula linking s, u, v and a is required:
$$v^2 = u^2 + 2as$$
$$0 = 20^2 + 2\left(-\frac{8}{27}\right)s$$
$$0 = 400 - \frac{16}{27}s$$
$$s = 400 \div \frac{16}{27}$$
$$s = 675 \text{ m}$$

The trailer travels an extra 675 m after the towbar breaks.

(e) The towbar is described as light and rigid. Light means that its mass can be neglected. Rigid means that it does not bend or extend.
(f) The objects are not treated as particles since air resistance is included. Instead, a very simple model for the resistance forces is used, whereby the total size of the resistance forces is assumed to be constant.

4. A light aeroplane pulls an advertising banner through the sky. The plane has a mass of 3920 kg and the banner has a mass of 10 kg. They are attached by an inextensible cable, which remains taut. The plane flies horizontally through the sky at a constant velocity. There is a constant air resistance force of 0.15 newtons per kilogram acting on the plane and a constant air resistance force of 0.2 newtons per kilogram acting on the banner. The plane's engine produces a forward thrust of P newtons.
   (a) Calculate the magnitudes of the resistance forces acting on the plane and the banner.
   (b) Draw a diagram showing all the horizontal forces acting on both the plane and the banner.
   (c) Show that the magnitude of the forward thrust P is 590 N.
   (d) Find the tension in the cable while the plane is moving at this constant velocity.

   The plane's engine now increases the forward thrust to 668.6 N. Assume that the resistance forces remain the same.
   (e) Find the acceleration of the plane and banner and the new tension in the cable.

(a) For the plane, there is a constant resistance of 0.15 newtons per kilogram. The total size of the resistance force is $0.15 \times 3920 = 588$ N.
For the banner, there is a constant resistance of 0.2 newtons per kilogram. The total size of the resistance force is $0.2 \times 10 = 2$ N.

(b)

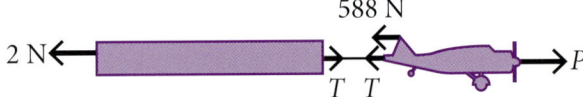

(c) The plane and banner are travelling at a constant velocity, so the forward forces are equal to the backward forces. When considering the whole system, the tension in the cable is not considered:
$P = 588 + 2 = 590$ N

(d) Considering just the banner, forwards forces are equal to backwards, because it is moving at a constant velocity. Therefore $T = 2$ N.

Note: The same answer could be obtained by considering the forces on the plane, although this is slightly more complicated because three forces are involved.

(e) Using $F = ma$ for the entire system:
$668.6 - 590 = 3930a$
$a = \dfrac{78.6}{3930} = 0.02$ m s^{-2}

Using $F = ma$ for the banner only:
$T - 2 = 10 \times 0.02$
$T = 2.2$ N

Note: The alternative method is to use $F = ma$ for each object separately. The acceleration and the new tension are then found using simultaneous equations.

Exercise 7D

1. A horse of mass 800 kg is pulling a cart of mass 200 kg. The horse is attached to the cart using a tether. The horse provides a forward thrust of 1000 N to get the cart in motion and the cart experiences a constant resistance force of 100 N.

   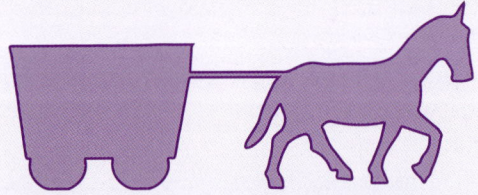

   (a) Draw a diagram, showing all the horizontal forces acting on the horse and the cart.
   (b) Show that the acceleration of the horse and cart is 0.9 m s^{-2}.
   (c) Find the size of the tension force in the tether pulling the cart forwards.

2. A car of mass 1200 kg pulls a trailer of mass 800 kg, as shown in the diagram.

   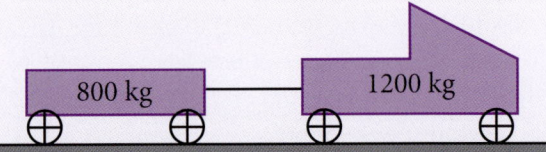

   The car and trailer are connected by a light, rigid towbar. The car experiences a constant resistance force of 400 N while it is in motion and the trailer experiences a constant resistance force of 600 N. The car and trailer start from rest, with the car's engine providing a forwards tractive force of 2600 N.
   (a) Copy the diagram, showing all the forces acting on both the car and the trailer.
   (b) Find the acceleration of the car and trailer.

Exercise 7D...

When the car and trailer have a velocity of 24 m s^{-1}, the towbar breaks.
(c) Find how far the car and trailer travel before the towbar breaks.
(d) Find the deceleration of the trailer after the towbar breaks.
(e) Find the extra distance travelled by the trailer.
(f) Find the **total** time that elapses between the car and trailer starting from rest and the trailer coming to rest again.

3. A tow truck of mass $2M$ kg is towing a car of mass M kg along a straight, horizontal road. The truck and the car are connected by a light, inextensible towbar. The tow truck exerts a driving force of 40 000 N causing the truck and car to accelerate at 5 m s^{-2}. The truck and car experience resistance forces of 5000 N and 8000 N respectively.
   (a) Find the mass of the tow truck and the mass of the car.
   (b) Find the tension in the towbar.
   (c) The towbar is assumed to be light and inextensible. Explain how these modelling assumptions affect the calculations.

4. A boy of mass 20 kg sits in a toy car of mass 16 kg. The boy moves the car by pedalling it, propelling it forwards with a force of 89 N. It pulls a trailer of mass 10 kg using a short, rigid towbar. The tension in the towbar is 25 N while the car and trailer are in motion. The car experiences a constant resistance force of 1.5 newtons per kilogram of the combined mass of the boy and car. The trailer experiences a resistance force of 1.2 newtons per kilogram.
   (a) Find the resistance forces experienced by the car and the trailer while they are in motion.
   (b) Draw a diagram, showing all the horizontal forces acting upon the car and the trailer.
   (c) Find the acceleration.
   (d) Find the tension in the towbar while the car and trailer are accelerating.
   (e) The car and trailer start from rest. Find how long it takes for the boy to accelerate to a velocity of 1.5 m s^{-1}.

Exercise 7D...

(f) When the car and trailer reach the velocity of 1.5 m s^{-1}, the boy begins to pedal more slowly to keep the car and trailer moving at this constant speed. Find the force with which his pedalling is now moving the car forward.
(g) Find the tension in the towbar while the car and trailer are moving at this constant velocity.

5. A train engine pulls a single carriage out of the sidings along a straight horizontal track. It takes the engine 40 s to pull the carriage 200 m from rest with a constant acceleration.
   (a) Find the acceleration of the engine and carriage as they leave the sidings.

   The mass of the engine is 10 000 kg and the mass of the carriage is 6600 kg. The engine and carriage both experience resistance forces of 0.8 N per kilogram during the motion. The engine exerts a tractive force of P N during the motion.
   (b) Draw a diagram showing all the forces acting upon the engine and carriage during the motion.
   (c) Find the tension in the couplings.
   (d) Find P.

6. A train engine pulls two carriages with constant acceleration along a straight horizontal track, as shown in the diagram.

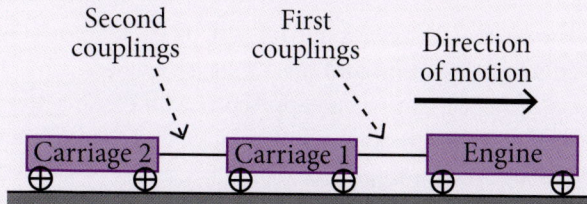

   The tension in the first set of couplings is T_1 newtons and the tension in the second set is T_2 newtons. Each carriage has the same mass and experiences the same constant resistance force while in motion.
   (a) Copy the diagram, adding all external forces acting on the engine and both carriages.
   (b) Show that $T_1 = 2T_2$.

AS 2: APPLIED MATHEMATICS

7.5 Pulleys

You will encounter systems in which two objects are connected by a string or rope passing over a pulley or a fixed peg.

In such a system, the two objects are generally moving in different directions, so they have different velocities and different acceleration vectors. Therefore, it is not possible to use Newton's second law for the entire system. Instead, each object must be considered separately.

The most common type of problem involves a pulley with the two objects hanging freely on either side.

Worked Example

5. The diagram shows two blocks of mass 1.5 kg and 2 kg connected on either side of a smooth pulley by a light inextensible cord. Both blocks start at a height of 2 metres above the ground.

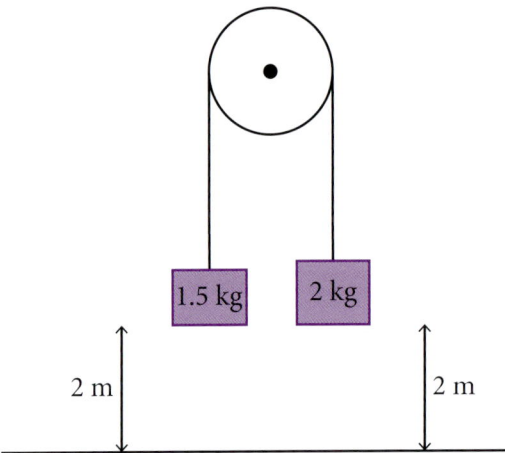

The blocks are released from rest. The 2 kg block falls to the ground and the 1.5 kg block rises. Use $g = 9.81$ m s^{-2} as an approximation for the acceleration due to gravity and give all your answers to 3 significant figures.
(a) Copy the diagram, showing all the external forces acting upon both blocks. Show also the forces acting on the pulley.
(b) Find the acceleration of the system and the tension in the cord while the blocks are in motion.
(c) What is the downwards force upon the pulley while the blocks are in motion?
(d) Find the time taken for the 2 kg block to hit the ground.
(e) Find the velocity of the 2 kg block as it hits the ground.
(f) When the 2 kg block hits the ground, the cord becomes slack and the 1.5 kg block continues to rise. Assuming the 1.5 kg block does not hit the pulley, find the extra distance through which it rises.
(g) Find the time taken, after the cord goes slack, for the cord to become taut again.
(h) What is the meaning of the following words in the question:
(i) smooth (ii) light (iii) inextensible
(i) State one other modelling assumption that has been made.

(a)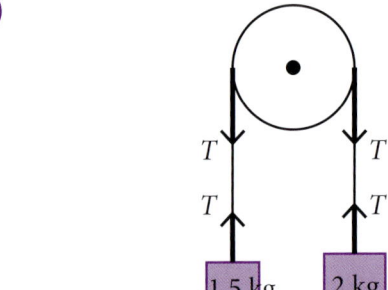

(b) Use $F = ma$ for the 1.5 kg block. Since this block is rising, upwards is the positive direction.
$T - 1.5g = 1.5a$ (1)

Use $F = ma$ for the 2 kg block. Since this block is descending, downwards is the positive direction.
$2g - T = 2a$ (2)

Add equations (1) and (2). The T in equation (1) and the $-T$ in equation (2) cancel out.
$0.5g = 3.5a$

$a = \frac{1}{7}g$

$a = 1.4014... = 1.40$ m s^{-2} (3 s.f.)

Substitute a into equation (1) to find T:
$T - 1.5g = 1.5(1.4014...)$
$T = 2.102... + 1.5g = 16.8$ N (3 s.f.)

(c) The cord exerts a total downwards force of $2T$ on the pulley.
$2T = 33.6$ N (3 s.f.)

(d) Use the constant acceleration formulae to find the time taken for the 2 kg block to reach the ground. For the 2 kg block, from its starting position to the ground:
$s = 2$
$u = 0$
$v = ?$
$a = 1.4014 \ldots$
$t = ?$

The formula linking s, u, a and t is required:
$s = ut + \frac{1}{2}at^2$
$2 = 0 + \frac{1}{2} \times 1.4014 \ldots \times t^2$
$t^2 = \frac{2}{0.7007 \ldots} = 2.854 \ldots$
$t = 1.689 \ldots$ s
$t = 1.69$ s (3 s.f.)

(e) Use the same suvat table, since the same journey, from the starting point to the ground, is being considered. Choose the formula linking s, u, v and a:
$v^2 = u^2 + 2as$
$v^2 = 0^2 + 2 \times 1.4014 \ldots \times 2$
$v^2 = 5.6057 \ldots$
$v = 2.367 \ldots = 2.37$ m s^{-1} (3 s.f.)

Note: Other formulae could be used to find v, but it is safest to use the one that doesn't involve t.

(f) Now consider the journey for the 1.5 kg block from the moment the cord becomes slack to the moment the block reaches its highest point. It begins this journey with the velocity that both blocks had when the 2 kg block hit the ground. At its highest point its velocity is zero. The block is now free-falling under gravity, so its acceleration is $-g$:
$s = ?$
$u = 2.367$
$v = 0$
$a = -g$
$t = ?$

So we use:
$v^2 = u^2 + 2as$
$0 = 5.6057 \ldots + 2 \times -9.81 \times s$
$19.62s = 5.6057 \ldots$
$s = \frac{5.6057 \ldots}{19.62} = 0.286$
$s = 0.286$ m (3 s.f.)

(g) To find the time taken for this journey:
$v = u + at$
$0 = 2.367 \ldots - 9.81t$
$t = \frac{2.367 \ldots}{9.81} = 0.241 \ldots$ s

The cord becomes taut when the block falls again. The time taken to fall is the same as the time taken to rise, so the total time to rise and fall is:
$2 \times 0.241 \ldots = 0.4826 \ldots$
$= 0.483$ s

(h) (i) The pulley is smooth. This means that frictional forces can be ignored as the cord passes over the pulley.
(ii) The cord is light means that the mass of the cord can be ignored.
(iii) The cord is inextensible means that the cord does not stretch.

(i) The two blocks are treated as particles, i.e. they have no size and therefore air resistance can be ignored.

Another common configuration features a pulley attached to the corner of a table.

One object is hanging freely; the second is on the tabletop.

Worked Example

6. **Use $g = 9.8$ m s^{-2} in this question and give all answers to a suitable degree of accuracy.** The diagram shows two blocks A and B with masses 0.1 kg and 0.2 kg respectively. They are connected by a light, inextensible string, which passes smoothly over a pulley.

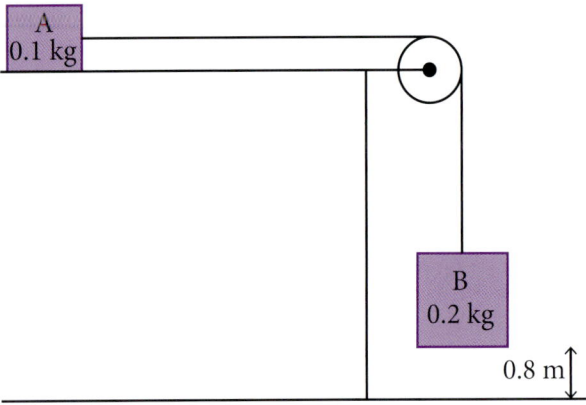

At time $t = 0$, object A starts at rest on a smooth horizontal table while block B hangs freely from the string, as shown, 0.8 m above the ground. The system is released from rest.

(a) Copy the diagram, showing the forces acting on the two blocks. Show also the forces acting upon the pulley.
(b) Find the acceleration of the blocks during the motion.
(c) Find the tension in the string while the blocks are in motion.
(d) Find the time taken for block B to fall to the ground.
(e) Find the velocity of B as it hits the ground.
(f) Find the resultant force acting on the pulley while the blocks are in motion.
(g) State any assumptions that have been made in modelling the motion of the two blocks.

In this question $g = 9.8$ m s^{-2} is used as an approximation for the acceleration due to gravity. Therefore, it is appropriate to give all answers to 2 significant figures.

(a)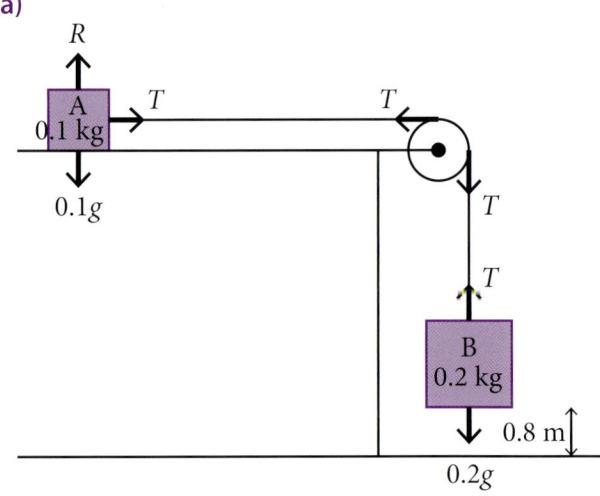

(b) Use $F = ma$ for both blocks.
For A, in the horizontal:
$$T = 0.1a \quad (1)$$

For B, in the vertical:
$$0.2g - T = 0.2a \quad (2)$$

Add equations (1) and (2):
$$0.2g = 0.3a$$
$$a = \frac{2}{3}g = \frac{98}{15}$$
$$a = 6.5 \text{ m s}^{-2} \text{ (2 s.f.)}$$

(c) Substitute a into equation (1):
$$T = 0.1 \left(\frac{98}{15}\right)$$
$$T = 0.65 \text{ N (2 s.f.)}$$

(d) Use the constant acceleration formulae to find the time taken for block B to reach the ground.
For block B, from its starting position to the ground:
$s = 0.8$
$u = 0$
$v = ?$
$a = \dfrac{98}{15}$
$t = ?$

The formula linking s, u, a and t is required.
$$s = ut + \frac{1}{2}at^2$$
$$0.8 = 0 + \frac{1}{2} \times \frac{98}{15} \times t^2$$
$$0.8 = \frac{49}{15}t^2$$

$$t^2 = \frac{0.8}{49/15} = \frac{12}{49}$$
$t = 0.49487\ldots$ s
$t = 0.49$ s (2 s.f.)

Note: Using an exact value for a in the working ensures no loss of accuracy. If working with a decimal value for a, use more than 2 significant figures in the working, to ensure the final answer is accurate to 2 significant figures.

(e) Use the same suvat table, since the same journey is being considered. Choose the formula linking s, u, v and a:
$v^2 = u^2 + 2as$
$v^2 = 0^2 + 2 \times \frac{98}{15} \times 0.8$
$v^2 = \frac{784}{75}$
$v = 3.233\ldots$
$v = 3.2$ m s^{-1} (2 s.f.)

Note: Other formulae could be used to find v, but it is safest to use the one that doesn't involve t.

(f) The tension in the string is pulling the pulley to the left and downwards. Since forces are vectors, add them using a vector diagram:

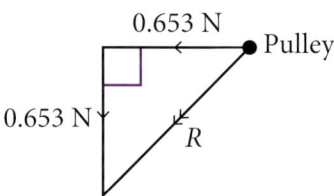

The resultant force is marked R. Using Pythagoras' Theorem:
$R^2 = 0.653^2 + 0.653^2$
$R^2 = 0.8528$
$R = 0.92$ N (2 s.f.)

(g) Apart from the modelling assumptions stated in the question (the string is light and inextensible; the pulley and the table are smooth), both blocks are being treated as particles.

Exercise 7E

Throughout this exercise use $g = 9.8$ m s^{-2} as an approximation for the acceleration due to gravity, unless the question specifies a different value. Choose an appropriate level of accuracy for your answers.

1. Two blocks of mass 2.6 kg and 1.4 kg are connected by a light, inextensible string, which passes smoothly over a pulley, as shown in the diagram.

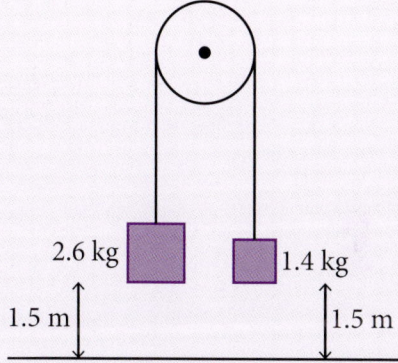

The blocks are released from rest at a height of 1.5 m above the ground.
   (a) Copy the diagram, adding all the forces acting upon both blocks.
   (b) Find the acceleration of the 2.6 kg block as it falls. Also find the tension in the string.
   (c) Find the time taken by the 2.6 kg block to reach the ground.
   (d) Find the maximum height reached by the 1.4 kg block.

2. Use $g = 10$ m s^{-2} as an approximation for the acceleration due to gravity in this question. Give your answers to 1 decimal place where appropriate.
The diagram shows two particles, of mass M kg and 5 kg, connected by a light inextensible string passing over a smooth fixed pulley.

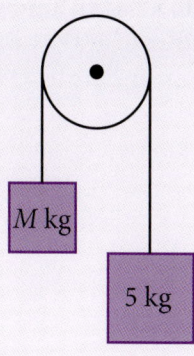

Exercise 7E...

The system is released from rest. Given that $M < 5$ and that the acceleration of each particle is 2.5 m s^{-2}, calculate:
(a) The tension in the string.
(b) The value of M.

3. The diagram shows two objects A and B attached to the ends of a light inextensible rope, which passes over a smooth fixed pulley.

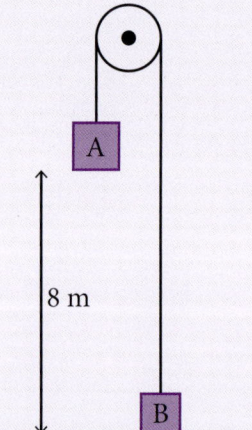

Object A has mass M_A kg and B has mass M_B kg, where $M_A > M_B$. Initially A is suspended 8 m above a horizontal surface and B is on the surface. The objects are released from rest.
(a) Given that object A hits the ground with a speed of 5 m s^{-1}, find the acceleration of the system, giving your answer as a simplified fraction.
(b) Copy the diagram, adding all the external forces acting upon objects A and B.
(c) Given that the tension in the rope is 32.95 N, find M_A.
(d) Find M_B.
(e) State a modelling assumption you have used relating to the objects.

4. Use $g = 9.81$ m s^{-2} as an approximation for the acceleration due to gravity in this question. Choose an appropriate level of accuracy for your answers.
Two blocks of mass 2 kg and 3 kg are connected by a light, inextensible string, which passes smoothly over a pulley, as shown in the diagram. The 2 kg object is at rest on a smooth horizontal table and the 3 kg object hangs 2 metres above the ground.

Exercise 7E...

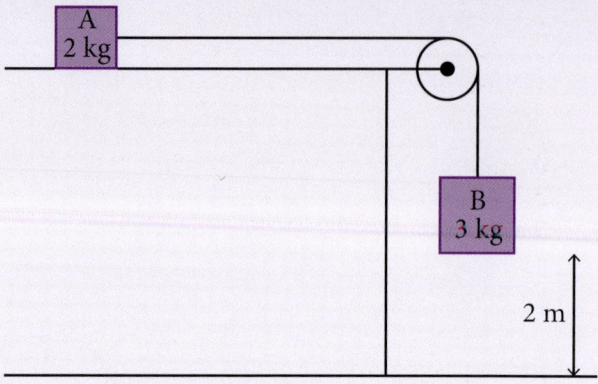

The blocks are released from rest. The 3 kg object falls to the ground.
(a) Copy the diagram, showing all the forces acting on the two blocks while they are in motion. Also show the forces acting on the pulley.
(b) Find the acceleration of the objects and the tension in the string while the objects are in motion.
(c) Find the magnitude of the resultant force acting on the pulley while the objects are in motion.
(d) Find how long the 3 kg object takes to fall to the ground.
(e) What modelling assumptions have been made?

5. Use $g = 9.8$ m s^{-2} as an approximation for the acceleration due to gravity in this question. Give your answers correct to 1 decimal place.
The diagram shows two blocks A and B connected by a light inextensible string, which passes smoothly over a pulley.

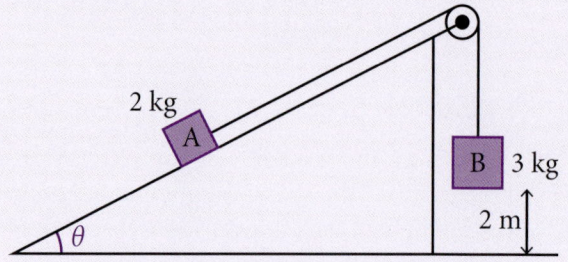

The block A has a mass of 2 kg and is held on a smooth fixed plane inclined at an angle $\theta°$ to the horizontal, where $\theta = 27°$. The block B has a mass of 3 kg. It hangs freely, 2 metres above the horizontal ground. The system is released from rest.

Exercise 7E...

(a) Copy the diagram, adding all the forces acting on both blocks while the system is in motion. Also include any forces acting upon the pulley.

Assuming block A does not reach the pulley, calculate:

(b) The acceleration of the blocks while they are in motion and the tension in the string.

(c) The velocity of block B as it reaches the ground.

(d) The magnitude and direction of the resultant force acting upon the pulley.

6. Two objects A and B, with masses m_1 and m_2 kilograms respectively (where $m_1 > m_2$), are connected by a light, inextensible cord, which passes smoothly over a pulley as shown in the diagram.

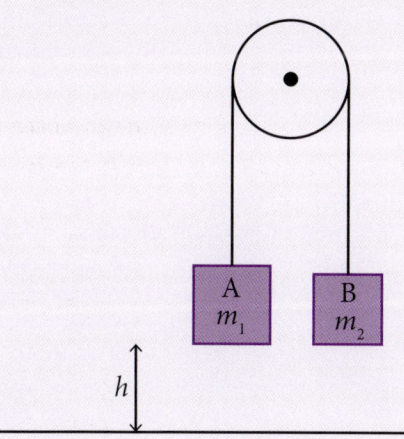

The objects are released from rest.

(a) Show that the acceleration a of the system is given by:
$$a = \left(\frac{m_1 - m_2}{m_1 + m_2}\right)g$$

(b) Show that the tension T in the string, while the objects are in motion, is given by:
$$T = \frac{2m_1 m_2 g}{m_1 + m_2}$$

(c) If object A starts h metres above the ground, show that the velocity with which it hits the ground is:
$$v = \sqrt{2\left(\frac{m_1 - m_2}{m_1 + m_2}\right)gh}$$

Exercise 7E...

(d) Show that the time taken for object A to hit the ground is given by:
$$t = \sqrt{\frac{2(m_1 + m_2)h}{(m_1 - m_2)g}}$$

7.6 Connected Bodies on Rough Surfaces

There are cases where one or more of the connected bodies may be in contact with a rough surface. In these cases, friction forces must be considered.

Worked Example

7. The two boxes A and B shown in the diagram are connected by a light, inextensible string passing over a smooth pulley.

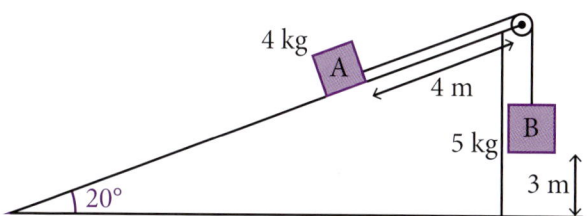

Box A is on a rough plane inclined at 20° to the horizontal. The coefficient of friction between Box A and the plane is 0.1. Box A begins 4 m from the pulley. Box B is hanging freely and begins 3 m above the ground.

The system is released from rest and the boxes both start to move. Determine whether Box A reaches the pulley, showing all required working.

Construct a force diagram.

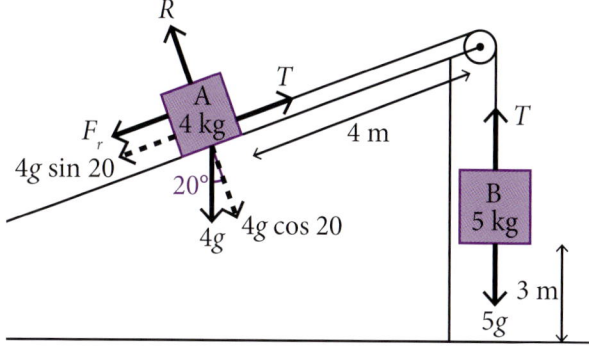

Step 1: Use Newton's second law for both boxes to find the acceleration:
$F = ma$ for box A, parallel to the plane:
$$T - F_r - 4g \sin 20 = 4a \quad (1)$$
$F = ma$ for box B, vertically:
$$5g - T = 5a \quad (2)$$
Add (1) and (2)
$$\Rightarrow 5g - F_r - 4g \sin 20 = 9a \quad (3)$$
Resolving perpendicular to the plane for Box A:
$$R = 4g \cos 20$$
Since:
$$F_r = \mu R$$
$$F_r = 0.1 \times 4g \cos 20 = 3.6836 \ldots \text{ N}$$
Substitute for F_r in (3):
$$5g - 3.6836 \ldots - 4g \sin 20 = 9a$$
$$\Rightarrow a = \frac{1}{9}(5g - 3.6836 \ldots - 4g \sin 20)$$
$$a = 3.545 \ldots \text{ m s}^{-2}$$

Step 2: Use the constant acceleration formulae to find the speed of the boxes as B hits the ground.
For B:
$$v^2 = u^2 + 2as$$
$$v^2 = 0 + 2(3.545 \ldots) \times 3$$
$$v^2 = 21.27 \ldots$$
$$v = 4.61 \ldots \text{ m s}^{-1}$$

Box A is also moving at this speed as B hits the ground. Box B has descended 3 metres, so Box A has moved 3 metres up the plane.

Step 3: When Box B hits the ground, the string goes slack. Box A continues to slide up the plane for a short time. Use Newton's second law for Box A for this second phase of the motion. The string is slack, so there is no tension force:

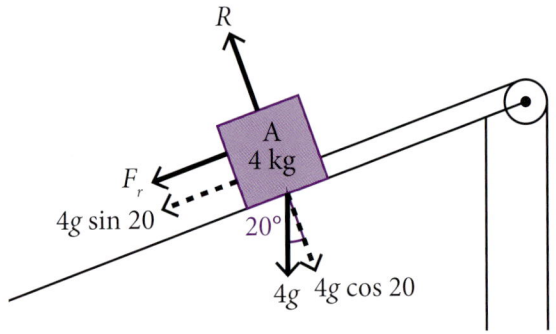

$F = ma$ for Box A parallel to the plane. There is no tension force in the string.
$$-F_r - 4g \sin 20 = 4a \quad (1)$$
$$a = \frac{1}{4}(-3.6836 \ldots - 4g \sin 20)$$
$$a = -4.27 \ldots \text{ m s}^{-2}$$

Step 4: Use the constant acceleration formulae again to determine the additional distance that Box A travels. When it comes to rest at the top of its trajectory, $v = 0$.
$$s = ?$$
$$u = 4.61 \ldots$$
$$v = 0$$
$$a = -4.27 \ldots$$
$$v^2 = u^2 + 2as$$
$$0 = 21.27 \ldots - 2 \times 4.27 \ldots \times s$$
$$s = 2.49 \ldots \text{ m}$$

The total distance travelled by Box A would be 5.49 m. Therefore it does reach the pulley.

In the following example three objects are connected.

Worked Example

8. The diagram shows three particles, X, Y and Z, connected by two light, inextensible strings. Both strings pass smoothly over the two pulleys shown.

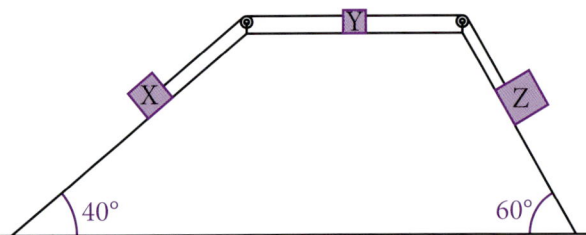

The three particles X, Y and Z have masses 20 kg, 10 kg and 30 kg respectively. Particle Y is on a smooth horizontal surface. Particle X is in contact with a rough surface inclined at 40° to the horizontal. Particle Z is in contact with a smooth surface inclined at 60° to the horizontal. The coefficient of friction between X and the plane is 0.3. The other two surfaces are smooth.

The system is released from rest and Z moves down the plane. Find the acceleration of the particles and the tension in each string while the particles are in motion, giving all answers to a suitable level of accuracy.

Begin by copying the diagram and showing all the external forces acting on the three particles.

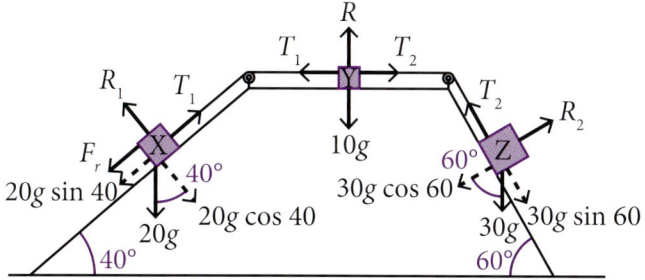

Note: It is important to understand that the tension is not necessarily the same in the two strings. Therefore, the two tension forces have been labelled T_1 and T_2.

For X, $F = ma$:
$T_1 - F_r - 20g \sin 40 = 20a$
$T_1 - F_r - 125.9864... = 20a$ (1)

For Y, $F = ma$:
$$T_2 - T_1 = 10a \quad (2)$$

For Z, $F = ma$:
$30g \sin 60 - T_2 = 30a$
$254.6115... - T_2 = 30a$ (3)

Add equations (1), (2) and (3):
$128.6251... - F_r = 60a$

To find the size of the friction force, use $F_r = \mu R$ for A:
$F_r = 0.3 \times 20g \cos 40 = 45.0434...$ N
$\therefore 60a = 128.6251 - 45.0434$
$a = 1.393... = 1.4$ (2 s.f.)

Substitute a into (1):
$T_1 - F_r - 125.9864... = 20(1.393...)$
$T_1 = 27.8606 + 45.0434 + 125.9864$
$T_1 = 198.89...$
$T_1 = 200$ N (2 s.f.)

Substitute a and T_1 into (2):
$T_2 - 198.89... = 10(1.393...)$ (2)
$T_2 = 212.82...$
$T_2 = 210$ N (2 s.f.)

Note: A suitable level of accuracy is 2 significant figures, since the value of g used is 9.8, accurate to 2 significant figures.

Exercise 7F

1. The diagram shows two blocks A and B with masses 2 kg and 3 kg respectively. The blocks are connected by a light, inextensible string, which passes smoothly over a pulley. Block A starts at rest on a rough table. The coefficient of friction between block A and the table is 0.2. Block B hangs from the string, 2 m above the ground, as shown.

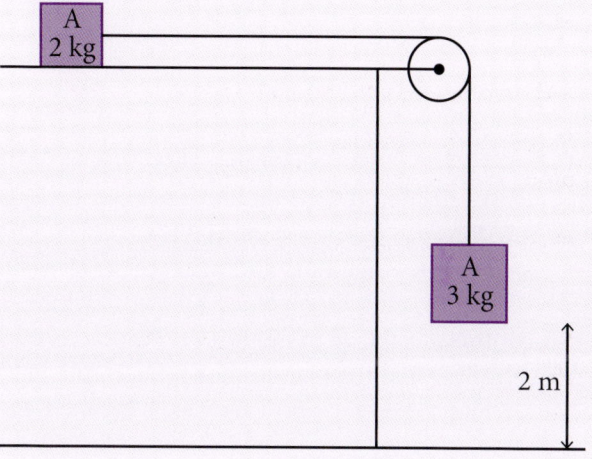

   The blocks are released from rest.
   (a) Copy the diagram, showing all the forces acting on the two blocks while they are in motion. Also show the forces acting on the pulley.
   (b) Find the time taken for block B to fall to the ground.
   (c) When block B hits the ground, block A continues to move across the table for a time. Find:
     (i) How far block A moves **in total**.
     (ii) The **total time** that block A is in motion.

2. Two objects A and B have masses 5 kg and 4 kg respectively. Object A is on a rough plane inclined at an angle of 120° to the vertical. Object B hangs freely. They are attached by a light inextensible cord, which passes smoothly over a pulley, as shown in the diagram.

Exercise 7F...

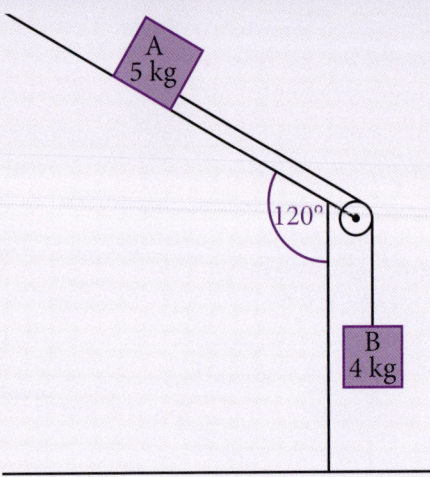

The coefficient of friction between object A and the slope is 0.6. The system is released from rest.
(a) Copy the diagram, adding all the external forces acting upon objects A and B while the objects are in motion.
(b) Show that the magnitude of the frictional force between object A and the inclined plane is $\frac{3\sqrt{3}}{2}g$ N.
(c) Find the acceleration of object A while the string is taut, giving an exact answer in terms of g.
(d) When calculating the acceleration in part (c), the two objects are treated as particles. Explain briefly how this modelling assumption affects the calculations.

3. The two parcels P and Q shown in the diagram have masses 2.5 kg and 3 kg respectively. They are connected by a light, inextensible string passing over a smooth pulley.

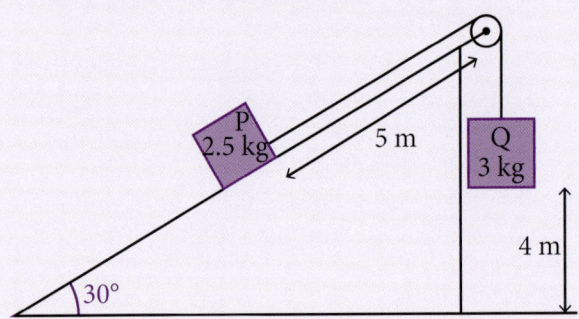

Exercise 7F...

Parcel P is on a rough plane inclined at 30° to the horizontal. The coefficient of friction between Parcel P and the plane is 0.25. Parcel P begins 5 m from the pulley. Parcel Q is hanging freely and begins 4 m above the ground. The system is released from rest and the parcels both start to move.
(a) Find the acceleration of the two parcels as they move.
(b) Find the speed of Parcel Q as it hits the ground.
(c) As soon as Parcel Q hits the ground, the string goes slack and Parcel P decelerates. Find the deceleration of Parcel P.
(d) Determine whether Parcel P reaches the pulley, showing all required working.

4. **In this question, take $g = 10$ m s^{-2} as an approximation to the acceleration due to gravity.**
The diagram shows two boxes, A and B, of masses 3 kg and 4 kg respectively. The boxes are connected by a light, inextensible string which passes over a light, smooth fixed pulley.

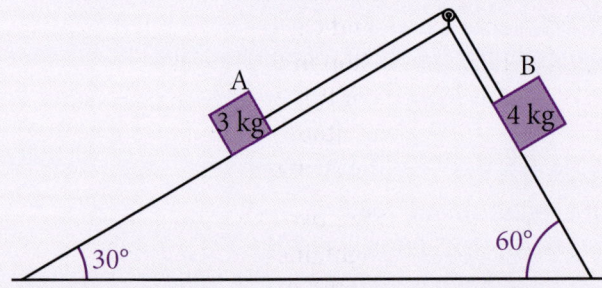

Box A lies on a rough plane inclined at 30° to the horizontal. The coefficient of friction between Box A and the plane is 0.6. Box B lies on a smooth plane inclined at 60° to the horizontal.
(a) Copy the diagram showing the external forces acting on the boxes.
(b) The system is released from rest and Box B starts to move down the plane.
(c) Show that the acceleration a of Box B is given by:
$$a = \frac{1}{7}(p\sqrt{3} - q)$$
where p and q are integers to be found.

Exercise 7F...

5. **Use $g = 9.8$ m s^{-2} as an approximation for the acceleration due to gravity in this question. Give your answers to a suitable level of accuracy.**

   Two boxes, P and Q, are connected by a light, inextensible rope which passes smoothly over a fixed pulley. Box P, with a mass of 24 kg, rests on a rough plane that is inclined at 20° to the horizontal. The coefficient of friction between P and the plane is 0.3. Box Q, of mass 30 kg, rests on the rough plane that is inclined at 70° to the horizontal. The coefficient of friction between Q and the plane is 0.15.

   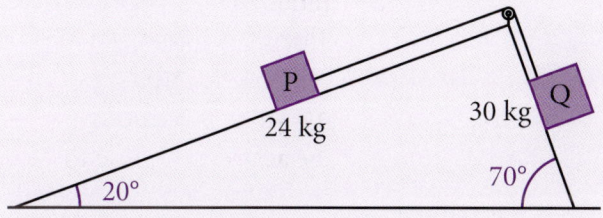

   The boxes are released from rest. Q slides down the plane it is in contact with. P slides up the opposite plane, towards the pulley.
   (a) Draw a diagram showing the external forces acting on P and Q.
   (b) Find the tension in the rope and the acceleration of P.
   (c) Find the magnitude of the resultant force exerted by the string on the pulley.

6. The diagram shows two boxes, A and B, connected by a light, inextensible rope passing smoothly over a light, fixed pulley. Box A is held at rest on a rough horizontal table 6 m from the pulley. Box B hangs 2.4 m vertically above the floor. A and B have masses 2.8 kg and 4.2 kg respectively.

Exercise 7F...

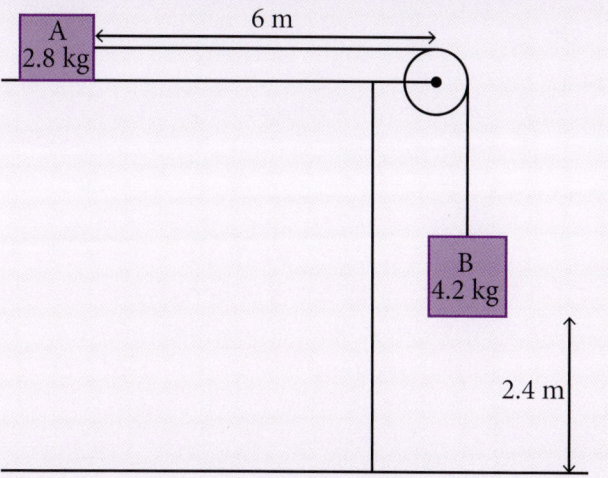

The system is released from rest.
(a) Copy the diagram showing all the external forces acting on A and B.
(b) Box B takes 1.08 seconds to reach the floor. Find the value of the coefficient of friction between A and the table.
(c) Determine whether box A will collide with the pulley.
(d) State one modelling assumption you have made about the rope.

7. **In this question, take $g = 10$ m s^{-2} as an approximation to the acceleration due to gravity.**

   The diagram below shows two boxes A and B connected by a light inextensible string which passes smoothly over a fixed pulley.

   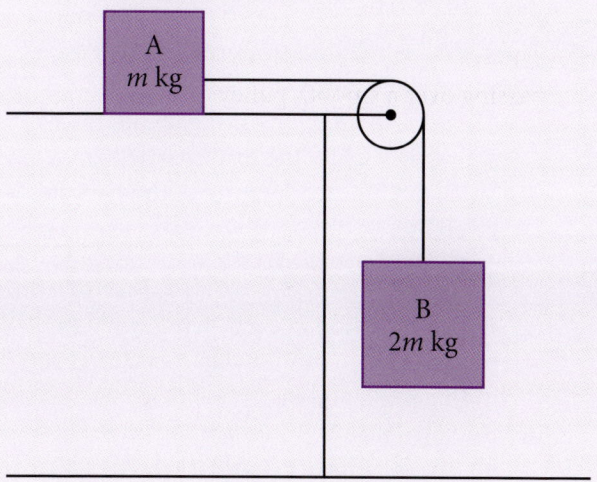

   Box A has mass m kg and is on a rough horizontal surface. The coefficient of friction between Box A and the surface is 0.5.

Exercise 7F...

Box B has mass $2m$ kg and hangs freely. Box A is being pulled to the left along the surface by a constant horizontal force P newtons. Box B rises vertically.

(a) Copy the diagram, adding the external forces acting on both boxes. Also show any forces acting upon the pulley.

(b) Find, **in terms of P and m**:
   (i) The acceleration of the system.
   (ii) The tension in the string.
   (iii) The magnitude of the resultant force acting on the pulley.

8. The diagram shows three particles, A, B and C, connected by two light, inextensible strings. The strings pass smoothly over the two pulleys shown.

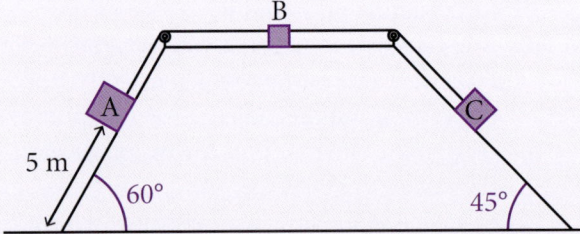

The three particles A, B and C have masses $3m$ kg, m kg and $2m$ kg respectively. Particle B is on a horizontal surface. Particle A and particle C are in contact with planes inclined at 60° and 45° to the horizontal respectively. The coefficient of friction between C and the plane is $\frac{1}{\sqrt{2}}$. The other two surfaces are smooth.

The system is released from rest and A moves down the plane.

(a) Copy the diagram, marking all the forces acting on each of the three particles.

(b) Show that the acceleration of the particles is given by:
$$a = \frac{g}{12}(3\sqrt{3} - 2\sqrt{2} - 2) \text{ m s}^{-2}$$

(c) Find the tension in each string while the particles are in motion. Give exact answers in terms of g and m.

(d) Initially Particle A is positioned 5 metres up the slope from ground level. Assuming neither particle B nor particle C collide with the pulleys, find the speed of Particle A when it reaches the ground, giving your answer to 2 significant figures.

Exercise 7F...

9. The diagram shows two bodies A and B connected by a light inextensible string, which passes smoothly over a pulley attached to a rough table. Body A has a mass of m_1 kg and body B has a mass of m_2 kg, where $m_2 > m_1$. The coefficient of friction between body A and the table is μ. The bodies are released from rest, with block B h metres above the ground.

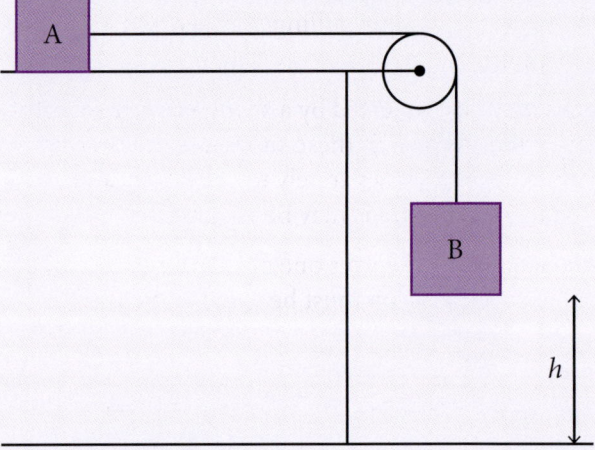

(a) Show that the acceleration a of the blocks is given by:
$$a = \left(\frac{m_2 - m_1\mu}{m_2 + m_1}\right)g$$

(b) Show that the tension in the string while the blocks are in motion is given by:
$$T = \frac{m_1 m_2 g(1 + \mu)}{m_1 + m_2}$$

(c) Find the time taken for block B to hit the ground. Give your answer in terms of m_1, m_2, μ and h.

(d) When B hits the ground, A continues to move across the table for a time, decelerating because of the friction force. Find the deceleration of body A in terms of μ and g.

7.7 Summary

Newton's three laws of motion can be applied to connected bodies.

The second law, $F = ma$, can be applied to each body separately. In a problem involving two bodies, this approach often leads to simultaneous equations.

In cases where both bodies are moving in the same direction with the same acceleration, the second law can be applied to the whole system. This approach usually avoids the need for simultaneous equations. Such situations include a car pulling a trailer, and a lift carrying passengers.

Bodies may be connected by a string passing smoothly over a pulley or peg. In these situations, the second law must be applied separately to each body and simultaneous equations may be required.

When a question involves one object in motion on a rough surface, friction must be considered.

Chapter 8
Statistical Sampling

8.1 Introduction

Sampling is an important tool used in statistical research. It allows for the collection of a subset of all the relevant data, therefore saving time and resources. Various techniques are used to ensure the sample is representative of the entire set of relevant data.

Key words
- **Population**: The entire set of items of interest.
- **Sample**: A selection of observations taken from a subset of the population.
- **Sampling unit**: Individual members or items in a population.
- **Sampling frame**: A list of all the units in the population.
- **Census**: A survey of the entire population.
- **Discrete data**: Data that can take certain values within a range, often integer values.
- **Continuous data**: Data that can take any numerical value.
- **Qualitative data**: Data that have numerical values.
- **Quantitative data**: Non-numerical data.

Before you start
You should:
- Know how to find the mean, median, mode and range from a set of data.
- Be familiar with surveys and the best way to structure them.

What you will learn
In this chapter you will learn:
- The definitions of statistical terms, including **population** and **sample**.
- How to use a variety of sampling techniques, including simple random sampling and stratified sampling.
- How to use samples to make informal inferences about the population.
- How to select or critique different sampling techniques in the context of solving a statistical problem.

In the real world...
A census is held in the United Kingdom once every ten years. Every member of the population is required by law to complete and return the census form by a certain date. This census records a wide variety of information, such as the number of people living in a house, their ages, genders, occupations, religions, etc.

Exercise 8A (Revision)

1. Find the mean, median, mode and range for this data set:

   39, 35, 33, 39, 27, 45, 37, 31

2. Gráinne is conducting a survey on the number of mobile phones in households. Below is one question from her survey.

How many people have mobile phones in your family?
☐ 1 person
☐ 1–2 people
☐ 3–4 people

   Give two criticisms of the question and write an improved question.

8.2 Definitions

Data
You will need to know the difference between **quantitative** and **qualitative** data.

Quantitative data are data that have numerical values, for example temperature data, or the number of people passing a checkpoint.

Qualitative data are data that have no numerical value, for example the colours of the cars passing a checkpoint, or the surnames of the pupils in a school.

You are also expected to know the difference between **discrete** and **continuous** numerical data.

Discrete data are data that can only take certain values (usually, but not always, integer values), for example the number of people travelling in the cars going into Belfast.

Continuous data can take any numerical value (in some cases within a range), such as heights or weights. For example, the height of a pupil could be 1.715 metres.

Worked Example

1. State whether the data in the following surveys are quantitative or qualitative. If quantitative, state whether the data are discrete or continuous.
   (a) Shoe sizes of 50 boys in Year 8.
   (b) The prices of 20 randomly chosen items in a shop.
   (c) The names of the world's 20 largest deserts.
   (d) The sea temperature at a buoy in Belfast Lough every day for a year.

   (a) Shoe sizes are quantitative and discrete. They usually go up in half sizes, but it is not possible to get any value in between.
   (b) Price data are quantitative and discrete. It is not possible, for example, to have a price of £1.4261
   (c) Names of deserts are qualitative data.
   (d) Temperature data are quantitative and continuous. A temperature could, in theory, take any value, e.g. 9.6223 °C.

   > **Note:** If we were discussing temperatures **reported** and/or **recorded**, these would probably be discrete data, since they would have to be rounded, e.g. to the nearest 0.1 °C. A similar situation arises for other examples of continuous data, for example the heights of all the boys in a class. Height is a continuous variable, but when recorded usually it would be rounded to the nearest centimetre. Heights rounded to the nearest centimetre are discrete data.

Population, census and sample

A **population** is the entire set of items that are of interest. For example, if a researcher wishes to find the mean height of the women in Northern Ireland, the women in Northern Ireland would form the population.

The letter N is usually used for the size of the population.

Information can be obtained from a population by taking a **census** or a **sample**.

A **census** records information from *every* member of the population. In the example above, to find the mean height of the women in Northern Ireland, the researcher would measure the height of every woman.

A **sample** is a selection of observations taken from a subset of the population, which is used to find out information about the population.

If a sample is taken, the letter n is used to represent the sample size.

Worked Example

2. Define the following terms and give an example for each:
   (a) a census,
   (b) a sample survey.

   (a) Every member of the population is observed. **Example:** the United Kingdom census carried out once every 10 years.
   (b) A small portion of the population is observed. **Example:** asking one in ten people leaving a supermarket how much they spent on their shopping.

Other terms

Individual members or items in a population are known as **sampling units**.

To carry out some types of sampling, we require a **sampling frame**. This is a list of all the units in the population.

The information obtained from the census or sample is called the **raw data**.

Exercise 8B

1. (a) Define the term census.
   (b) Give an example of a census.

2. Explain what is meant by the term sample.

3. Is the situation described below a census or a sample?

   The manager of a supermarket wishes to know her staff's attitudes towards customer service. The manager questions every member of staff and records the results.

4. (a) Explain what is meant by a sampling frame.
   (b) What effect does an increase in a population have on the sampling frame?
   (c) What effect would an increase in the variability of a population have on the sampling frame?

Exercise 8B...

5. State whether the following are qualitative or quantitative data. Also state, for quantitative data, whether they are discrete or continuous.
   (a) The price of a pint of milk in 10 different shops.
   (b) The ages of all the people in Belfast.
   (c) The religions of all the people in Belfast.
   (d) The times for each of a group of skiers to complete a downhill slalom course.
   (e) The first names of all the pupils in a school.
   (f) The shoe sizes of all the pupils in a school.

8.3 The Size of a Survey

Census or sample?

When collecting data in a survey, one of the first decisions for the researcher is whether to conduct a census or to take a sample.

The advantages of a sample are:

- A sample requires fewer resources than a census, i.e. less time and money. A census can be very time-consuming and costly.
- Results are obtained more quickly since fewer units of the population are observed (fewer people have to respond to a survey, or be questioned, etc).
- Overall, the data processing is easier because a sample generates a smaller volume of data.
- A census cannot be used when observation of a unit would result in destruction of that unit. For example, if testing the expected lifetime of the light bulbs produced in a factory, any light bulbs tested would be rendered useless. As small a sample as possible should be used in this case.

The advantages of a census are:

- A census is appropriate if a high degree of accuracy, or a completely accurate result, is required for the entire population.
- With a census there is always enough data to gain information about sub-sections of the population. Certain sampling techniques also ensure this is the case.
- If the population is small, a census is often practical, since the resources required are smaller than they would be for a census of a large population.

The size of a sample

If the researcher chooses to survey a sample of the population, their next decision is about the sample size.

A larger sample is usually required for a larger population, but population size is not the only factor; the size of a sample also depends upon the accuracy required and the resources (time, money, etc.) available for collection of the data.

A larger sample will usually be more accurate than a smaller one but will need greater resources. In addition, it may be important to consider the nature of the population: a very varied population may require a larger sample than one that is more uniform.

Worked Examples

3. A researcher is interested in finding the mean height of women in Northern Ireland. She surveys 50 women from a total of 500 000.
   (a) What is the population size?
   (b) What is the sample size?
   (c) Why is a sample more suitable than a census in this situation?
   (d) How could the researcher improve her estimate of the mean height?

   (a) The population size $N = 500\,000$
   (b) The sample size $n = 50$
   (c) A census would be too time-consuming and costly for a single researcher, because the population is large.
   (d) She could increase the sample size, for example from 50 to 1000.

4. (a) Which factors determine the size of a sample?
   (b) For each of the scenarios below, discuss whether it is appropriate to take a large sample of the population, or a relatively small one.
      (i) A government survey to obtain the average height of the people of Northern Ireland correct to the nearest millimetre.
      (ii) Conducting a survey using a single researcher, investigating the reason for people travelling to Belfast Docks throughout one day in January.

   (a) Sample size depends upon:
   - population size;
   - accuracy required;
   - resources available (time, money, etc);
   - the nature of the population (very varied, or fairly uniform).

(b) (i) This may need a large sample, since the population is large and a high level of accuracy is required. The fact that this is a government survey may mean adequate resources are available.
(ii) There will be relatively few people travelling to the docks – possibly a few hundred – on this day. The fact that this survey is being carried out by a single person indicates there are restrictions on the resources available. Additionally, there is no specified accuracy. All of these factors indicate that a smaller sample may be appropriate.

5. A greengrocer wants to test the quality of the apples being delivered to his shop by slicing them open.
(a) For what reason should the greengrocer not perform a census of the apples when conducting this research?

The greengrocer tests 10 apples and finds that one of them is bad. He concludes that 90% of the apples are fine.
(b) Suggest one way in which the greengrocer could improve his estimate.

(a) A census would involve testing every apple in the delivery by slicing it open. This would leave no apples left to sell. This is known as a **destructive test**.
(b) He could use a larger sample of apples to gain a better estimate. In general, a larger sample leads to a more accurate conclusion about the population.

Exercise 8C

1. (a) Give three advantages of taking a sample compared with taking a census.
   (b) Give two reasons for taking a census rather than a sample.
2. A factory produces pallets for the construction industry. Each pallet is designed to take a load of 5 tonnes in weight. Explain why a census would not be appropriate when testing the pallets.
3. In each of the following cases, state whether a census or a sample would be appropriate. If a sample, suggest a rough size. Briefly give a reason for your choice in each case.

Exercise 8C...

(a) A study to gain a complete list of all citizens of a country and other information about them. In this case accurate data about the population is required for government records.
(b) A study of the bees in a hive to determine how many have a disease.
(c) A study to determine the number of households in Belfast that have milk delivered to the door. A team of 10 university students are working on the project.
(d) Research to determine whether giraffes in Belfast Zoo have longer necks than those in London Zoo.
(e) Research to determine an accurate figure for the average daily temperature during July at Belfast International Airport.
(f) A student wishes to research the average height of men in Banbridge.
(g) A survey in a school to determine the precise number of pupils getting free school meals.
(h) A factory manager wishes to know the average lifetime of the light bulbs coming off the production line.

4. A supermarket produces carrier bags for their customers to use. The supermarket carries out testing of the bags to make sure they are strong enough.
(a) Explain briefly why the supermarket should conduct a sample rather than a census to test the bags.
(b) The supermarket tests five carrier bags, which break when loaded with the following masses:

19 kg
17.5 kg
21 kg
20 kg
23 kg

The supermarket then makes the claim that their bags are strong enough to carry 20 kg of shopping.
(i) Comment on this claim.
(ii) How could the test be improved?

8.4 Sampling Methodologies

When selecting units from the population for a sample, the researcher tries to collect a sample that is representative of the entire population.

The sample should be free from **bias**. To achieve this, the sample should be **randomised**.

Three sampling methodologies are discussed below: **simple random sampling**, **systematic sampling** and **stratified sampling**. Each one has its own advantages and disadvantages.

Simple random sampling

In **simple random sampling**, units are chosen from the population at random. Each unit has an equal and known chance of being included in the sample.

This technique uses sampling without replacement. A unit cannot be chosen more than once.

Worked Example

6. Give an example of a simple random sampling process.

   The selection of the numbers for the National Lottery is a simple random sample of 6 balls from a population of 59.

Two methods are widely used to implement simple random sampling:

- Random number sampling
- Lottery, or ticket, sampling

These are discussed in more detail below. However, both types of simple random sampling have the following two advantages:

- For small samples and small population sizes, they are relatively cheap and easy to implement.
- Each sampling unit has a known and equal chance of being selected.

A disadvantage is that a sampling frame (a list of all the units in the population) is always required.

Random number sampling

In random number sampling, each unit within the sampling frame is assigned a number. Random numbers are generated within the required range and the corresponding items in the population are selected for the sample.

The advantages of random number sampling are that:

- The process is truly random and free from bias.
- It is easy to use.
- Each unit has an equal chance of being chosen.

One disadvantage is that the process may not be suitable for a large population and/or large sample size because it can become time-consuming to allocate a number to each of the sampling units and to select the random numbers.

In the past random number tables were used to generate random numbers. No random number table is included in the A-Level Mathematics formula sheet, so you must generate random numbers using your calculator. There are two random number functions on Casio calculators:

- **Ran#** gives a random decimal value between 0 and 1 to 3 decimal places;
- **RanInt#**(a,b) gives an integer random number between a and b.

Worked Example

7. Describe carefully how simple random sampling could be used in the following situation:

   From a population of 500 school pupils select a random sample of 20 pupils.

   Step 1: Each pupil could be given a number from 1 to 500 and these numbers recorded next to the names on a list or a spreadsheet.

   Step 2: On a calculator use the RanInt function:

   RanInt(1,500)=

   This will give an integer between 1 and 500 inclusive. Record the number and repeat another 19 times. If a number appears more than once, ignore the second occurrence and generate a different number. (Simple random sampling is 'without replacement', so a pupil cannot be included in our sample more than once.)

   Step 3: When 20 unique integers have been generated, convert these numbers to the names of pupils using the list of pupils with their assigned numbers.

Lottery sampling

The second method widely used in simple random sampling is called lottery sampling. In this method, each unit within the population is represented by a ticket, making sure all the tickets are the same size and shape.

All the tickets are placed into a container and they are drawn one at a time until the required number of tickets has been chosen.

The advantages of lottery sampling are that:
- It is easy to use.
- Each unit has an equal, known chance of being chosen.
- The process is completely random.

Like random number sampling, the main disadvantage is that it is not suitable for large populations.

Worked Examples

8. The 40 members of a youth club are listed in a spreadsheet. The organisers want to choose 12 members at random to complete a survey. Describe how they could use simple random sampling to choose these 12 members.

   ### Method 1: Random number sampling

   Step 1: Each member is assigned a unique number from 1 to 40.

   Step 2: On a calculator the random number function is used to generate a random integer between 1 and 40. On a Casio calculator this can be done using:

   RanInt(1,40)=

   Step 3: Step 2 is repeated until 12 unique integers have been generated. If a number recurs, ignore it and generate another one. (This is sampling without replacement.)

   Step 4: The names of the 12 members are looked up in the spreadsheet by their numbers.

   ### Method 2: Lottery sampling

   Step 1: Write down the names of the 40 members on identical tickets.

   Step 2: Place all 40 tickets into a container.

   Step 3: Draw 12 of the tickets to obtain the 12 names.

9. Explain the benefits of taking a sample survey over taking a census.

   A sample survey requires fewer resources: time, money, etc.

   Results can be obtained more quickly, because fewer people have to respond.

Exercise 8D

1. Newry, Mourne and Down District Council wishes to carry out a survey on attitudes to bin collections. The population of the council area is 200 000 and the council wishes to survey 15% of the population. The council sends letters to the people in the survey.
   (a) Find the number of letters the council sends.
   (b) Describe briefly the process used to select the households involved in the survey.

2. Thirty marbles are to be chosen out of a large pot containing 600 marbles. Describe the simplest way to obtain a simple random sample.

Systematic sampling

In systematic sampling, the units for the sample are chosen at regular intervals from an ordered list of the entire population. The process is as follows:

Step 1. Find the **sampling interval** k. This can be found using the formula:

$$k = \frac{\text{population size } (N)}{\text{sample size } (n)}$$

Step 2. Choose a number m between 1 and k inclusive, at random. This is the position of the first item to be selected from the ordered list.

Step 3. When finding the units for the sample, select items $m, m + k, m + 2k, \ldots$

Systematic sampling is appropriate when the population and/or the sample size are large, because it does not rely on choosing a large set of random numbers.

Worked Examples

10. The principal of a school in Ballymoney wants to select 50 out of the 500 pupils, chosen at random, to be members of the audience at a concert. She decides to use a systematic sample.
    (a) Find the sampling interval k.

    The principal then chooses a random number m between 1 and k and gets $m = 2$.
    (b) Describe fully how the 50 students can be selected using systematic sampling with these values of k and m.

    (a) The sampling interval $k = \dfrac{500}{50} = 10$.

(b) The principal has selected a random value of $m = 2$, so she begins with the 2nd name on the list. The names of all 500 pupils are put in alphabetical order. The 2nd, 12th, 22nd etc names are selected from the list to be a part of the audience.

11. One hundred and fifty passengers are booked on a flight from Dublin to London Heathrow. The airline decides to give 10 of them a complimentary free drink. Describe how this can be achieved using systematic sampling. Show your working to determine the sampling interval k.

Firstly, the population (the names of the passengers on the flight) is listed alphabetically.

$N = 150$ (the population size)
$n = 10$ (the sample size)

Find the sampling interval:
$$k = \frac{N}{n} = \frac{150}{10} = 15$$
The sampling interval is 15.

The airline then chooses a number at random from 1 to 15. If we assume the number chosen was 6, then the following ten names would be chosen from the list of passengers:

6, 21, 36, 51, 66, 81, 96, 111, 126, 141.

The advantages of systematic sampling are:
- It is simple to use.
- It can be used for large populations and large samples.

The disadvantages are:
- It is only random if the ordered list is truly random.
- It can introduce bias. An example of how a systematic survey can introduce bias is given in Example 14 in section 8.5.

Exercise 8E

1. A factory manager wants to get information on the workers' opinions on the factory canteen. There are 200 workers in the factory and each one has a clock-in number from 1 to 200. Explain how the manager could take a systematic sample of size 25 from these workers.

2. Roadworks are currently taking place in 15 different locations in Northern Ireland. The roads authority would like to inspect a sample of 20% of these work sites to make sure

Exercise 8E...

contractors are providing a good service and value for money.
   (a) How many sites should be inspected?
   (b) Explain how the sites for inspection should be selected using systematic sampling.
   (c) Do you think systematic sampling is the best approach in this case? Explain your answer.

Stratified sampling

Stratified sampling is a form of random sampling in which the population is divided into groups or **strata** (singular **stratum**), which are non-overlapping, so that no unit can appear in more than one group. The groups would be decided using one or more criteria, for example age or gender.

Within each stratum we use simple random sampling, with the same proportion of each stratum being selected for the sample. This ensures each group is fairly represented in the sample.

This formula can be used to decide how many units should be taken from each group:

   Number sampled from group
   = sampling fraction × size of group

Worked Examples

12. The sixth form at a school in Dungannon had 180 pupils in the lower sixth and 120 in the upper sixth. A senior teacher wishes to take a stratified sample of size 15 from the sixth form for a survey on school uniforms. How many pupils should be chosen from the lower sixth and how many from the upper sixth?

The total number of pupils in the sixth form is 300.

The sampling fraction is $\frac{15}{300}$.

From the lower sixth the teacher must take:
$$\frac{15}{300} \times 180 = 9 \text{ pupils.}$$

From the upper sixth the teacher must take:
$$\frac{15}{300} \times 120 = 6 \text{ pupils.}$$

13. In a church choir there are 19 women and 11 men. One fifth of the choir are to represent the church in a special service at Belfast Cathedral. The choristers chosen are selected using a stratified sample.

(a) How many men and how many women will be chosen to go?
(b) What method could be used to decide which men and which women will be chosen?

(a) The sampling fraction is $\frac{1}{5}$.

The number of women chosen is $\frac{1}{5} \times 19 = 3.8$

The number of men chosen is $\frac{1}{5} \times 11 = 2.2$

Rounding to the nearest integers gives 4 women and 2 men.

(b) Within each stratum (group) we use simple random sampling. The 4 women could be chosen either by random number sampling from the population of 20, or by lottery sampling. The 2 men should be chosen from the population of 10 using the same method.

Stratified sampling is used when:
- the population is large; and/or
- the population divides naturally into non-overlapping groups.

The advantages of stratified sampling are:
- It can give a more accurate result than simple random sampling.
- Each group within the population is fairly represented within the sample.

The disadvantages of stratified sampling are:
- Within each stratum any of the disadvantages associated with simple random sampling (see earlier in this chapter) may occur.
- Care must be taken to divide the population into well-defined, non-overlapping groups.
- There is slightly more work involved than with a simple random sample.

Exercise 8F

1. The mayor is visiting North Street Primary School. The school principal wishes to take a stratified sample of 10% of the pupils to meet the mayor. The school has the following numbers of pupils.

P1	P2	P3	P4	P5	P6	P7
26	30	24	28	32	35	32

(a) Work out the number of pupils from each year group to be included in the sample that will meet the mayor.

Exercise 8F…

(b) Describe one benefit to the principal of using a stratified sample.

2. A market researcher finds that, out of 100 people surveyed, 38 said they usually shop in Stainberry's supermarket. 27 said they shop in Priceland, 23 shop in Superstuff and 12 shop in Scrounders. The market researcher wants a panel of 15 shoppers to answer some more detailed questions about their shopping habits. She wishes to take a stratified sample of these 100 people.
(a) Find the number of shoppers who would usually shop in each of the four supermarkets to be chosen for the panel.
(b) A different market researcher allows the population of 100 shoppers to name **one or more** supermarkets they shop in. Is it possible for him to use a stratified sample to find a panel of 15? Explain your answer.

3. In a choir there are 4 soprano singers, 11 alto singers, 8 tenors and 5 baritones. For a special Christmas concert the choirmaster wishes to have a choir of size 7. He takes a sample of these singers stratified by voice type.
(a) How many of each type should he select?
(b) What are the advantages of using a stratified sample in this case?

8.5 Selecting and Critiquing a Sampling Technique in Context

In section 8.1 we discussed when it may be more appropriate to take a census and when a sample.

It is also important to understand that different sampling methods can lead to different conclusions about the population. In extreme cases, a poorly chosen sample can lead to very wrong conclusions.

Be careful when selecting a sample using systematic sampling. If the units in the ordered list show some cyclical trend, you may introduce a bias in the sample.

When describing advantages or disadvantages of a particular sampling method, always give your answer in the context of the question.

AS 2: APPLIED MATHEMATICS

Worked Examples

14. The PSNI has data for the number of road traffic accidents on the M1 motorway every day for the 147 days between 1st January and 27th May. They need to take a sample of 21 of these days for analysis and compilation of a report. Explain why systematic sampling may not be the best sampling method and suggest an alternative.

Find the sampling interval:
$$k = \frac{N}{n} = \frac{147}{21} = 7$$

If we take every 7th day, we will get 20 of the same day of the week, e.g. 20 Fridays or 20 Sundays.

There would be a bias in the sample, since Sundays are quieter on the roads (possibly leading to fewer accidents) and Fridays busier. Ideally, we require a mixture of different days of the week in our sample.

Alternative sampling strategies would be:

- Simple random sampling. Using this approach, 21 days would be picked at random from the 147 days.
- Stratified sampling. Using this approach, the population of 147 days would be split into 7 strata – one for each day of the week. Then 3 days could be selected at random from each of the 7 strata to obtain the 21 days required for the sample.

> **Note:** Systematic sampling has these problems in this case because the dataset is not truly randomised; it is cyclical with a period of 7 days.

15. A school population comprises 200 boys and 400 girls. Following complaints about the school dinners, the headteacher decides that if **more than half** of the pupils rate the school dinners as poor, there should be a change in the menu.

A sixth form statistics class is asked to conduct the research. The statistics class is split into three teams (Teams A, B and C), each carrying out their research independently.

Each team conducts a survey of the pupils in the school, to learn about pupils' opinions on school dinners. All three teams use a questionnaire in which pupils are asked to rate school dinners as poor, satisfactory, good or very good. All three teams take a sample of 10% of the pupils in the school.

It is known that the boys in the school are more likely than the girls to rate the school dinners as good or very good.

(a) **Team A** uses a simple random sample. They choose names at random from the school registers. Discuss, using numbers to help you, how this sampling methodology could be flawed.

(b) **Team B** uses a systematic sample. They list all 600 pupils in alphabetical order and pick the 5th, 10th, 15th, etc, until they have the names of 60 pupils. From the 60 pupils in their sample, there are 36 girls and 24 boys. Give two reasons why this methodology could be flawed.

(c) **Team C** uses a stratified sample.
  (i) Calculate the number of boys and the number of girls in the sample.
  (ii) Discuss the advantages and disadvantages of this approach.

(d) What is the best sampling methodology? Justify your choice. Can you think of any further improvements that could be made?

There are 600 pupils in total, so each team takes a sample of size 60.

(a) **Team A** uses a simple random sample. Suppose they choose names at random, resulting in a sample with 30 boys and 30 girls. The boys are over-represented in the sample, since boys make up only one third of the school population, but one half of the sample.

Since boys are more likely to say that school dinners are good or very good, Team A would be more likely to conclude that the menu does not need to change.

Alternatively, the girls could be over-represented and the conclusion would be biased in the other direction.

(b) **Team B** uses a systematic sample. From the 60 pupils in their sample, there are 36 girls and 24 boys. This methodology could be flawed because:

1) Again, with 24 boys in a sample of 60, boys are over-represented.

2) Team B work through the alphabetical list, choosing every 5th pupil until they have 60 names. The 60th pupil chosen is at most the 300th person in the list. Therefore, pupils in the second half of the list cannot be chosen. In a fair sample, all pupils have an equal chance of being included in the sample.

(c) **Team C** uses a stratified sample. This ensures that the boys and girls are fairly represented in the sample. A sample of 10% is needed, so the sampling fraction is $\frac{1}{10}$.

(i) The number of boys chosen for the sample is:
$$\frac{1}{10} \times 200 = 20$$

The number of girls chosen for the sample is:
$$\frac{1}{10} \times 400 = 40$$

(ii) The advantage of stratified sampling is that boys and girls are fairly represented in the sample. One disadvantage is that the random selection of 20 boys from the 200 boys in the school may be affected by the usual problems associated with random sampling. Likewise, with the random sampling of 40 girls from the 400 in the school.

(d) The best sampling methodology in this case is the stratified sampling of Team C. This is because boys and girls are represented fairly in the sample. A better approach still would be to stratify the school population by its year groups, as well as by gender. In this way the team would ensure that each year group is represented fairly within the sample.

Exercise 8G

1. A market researcher is interested in the opinions of employees of the police service in Northern Ireland. To get a random sample he questions men that he meets entering a police station in Belfast on three consecutive Wednesday mornings after 9 am. Give two reasons to explain why this sampling strategy is not satisfactory.

2. The owner of a fitness centre decides to take a survey of its members. Each member has a unique membership number. There are 16 members who use the gym, 20 members who use the swimming pool and 14 members registered for the pilates class.
   (a) Explain how the owner could take a simple random sample of 10 members and state one disadvantage of this sampling method.
   (b) Suggest a more suitable sampling method and use it to work out the number of gym users, pool users and pilates class members to be included in the sample.

Exercise 8G...

3. A garage manager wishes to survey his customers to determine whether the service needs to improve. There are 300 customers in the database: 200 men and 100 women.

   The manager requires a sample of 10% of the entire customer base and he tasks three of his employees to conduct surveys. He says that if **more than half** of the customers say the service is poor, he will pay for retraining for all staff.

   The three employees work independently of each other, each choosing different sampling methodologies.

   Annie uses a simple random sample. Brad uses a systematic sample and Cormac takes a stratified sample.

   Annie uses a simple random sample. She chooses names at random from the database, resulting in a sample with 15 male customers and 15 female. Here are her findings.

   Annie's result

	Men	Women	Total
Poor		3	
Satisfactory	3		
Good	1	4	5
Very Good	1	5	
Total	15		30

   (a) Calculate the six missing values in the table.
   (b) Based on these figures, should the manager pay for retraining?
   (c) Comment on Annie's sampling methodology.

   Brad uses a systematic sample. He listed all 300 customers in alphabetical order and picked the 5th, 15th, 25th, etc. From the 30 customers in his sample, there were 18 men and 12 women, as shown in the table below.

   Brad's results

	Men	Women	Total
Poor	13	2	15
Satisfactory	4	2	6
Good	1	4	5
Very Good	0	4	4
Total	18	12	30

Exercise 8G...

(d) What is Brad's conclusion?
(e) Give two ways in which Brad's methodology could be flawed.

Cormac uses a stratified sample.

(f) Calculate the number of men and the number of women he will include in the sample.
(g) Copy and complete the table below.

Cormac's results

	Men	Women	Total
Poor	14	2	16
Satisfactory		2	6
Good			5
Very Good	0	3	3
Total			30

(h) What is Cormac's conclusion?
(i) Discuss the advantages and disadvantages of Cormac's approach.

4. Fionnuala has 600 photos of her holiday to Rwanda. They are on her computer with filenames 1.JPG to 600.JPG. She wishes to take a sample of 20 photos to put on her social media page.
   (a) Explain how Fionnuala could take a systematic sample.
   (b) Do you think this is a good way to choose the photos? Justify your answer.

5. A company employing 500 workers has data on the number of staff members taking their annual leave each day for the 250 working days between 2nd January and 24th December. The office is closed at weekends, so the dataset includes data for Mondays to Fridays.

   The PR department wishes to take a sample of 10 of these days for analysis and compilation of a report. Explain why systematic sampling may not be the best sampling method and suggest an alternative.

8.6 Summary

You will need to learn the definitions of terms such as population, sample, census, quantitative, qualitative, continuous and discrete.

It is important to remember why, in some situations, a census is more appropriate than a sample, and vice versa.

Three random sampling strategies are discussed in this chapter. You should know the methodologies, the advantages and disadvantages of each one.

Simple random sampling: A random selection of n units from the population. Every unit has an equal chance of being included in the sample.

Advantages:
- Free of bias.
- Easy and cheap to implement for small populations and small samples.
- Each sampling unit has a known and equal chance of selection.

Disadvantages:
- Not suitable for large populations or samples as it can be time-consuming and expensive.

Systematic sampling: The required elements are chosen at regular intervals from an ordered list.

Advantages:
- Simple and quick to use.
- Suitable for large populations and samples.

Disadvantages:
- A sampling frame is required.
- It can introduce bias if the sampling frame is not random, e.g. if there is a cyclical nature to the data.

Stratified sampling: The population is divided into non-overlapping strata (e.g. year groups or males and females) and a random sample is taken from each group.

Advantages:
- The sample should reflect the underlying population structure.
- Guarantees a fair representation of each group within the population.

Disadvantages:
- The population must be clearly classified into distinct strata or groups.
- Selection of units from each stratum can suffer from the same problems as simple random sampling.

Chapter 9
Data Presentation and Interpretation

9.1 Introduction

This chapter discusses various ways to present data, both in table form and graphically. Many of the data presentation techniques will already be familiar to you.

Key words

- **Frequency table**: A table that displays the frequency of various outcomes in a survey or dataset.
- **Histogram**: A representation of the distribution of numerical data using bars, in which the area of a bar is proportional to the related frequency.
- **Box plot** (or **box and whisker plot**): A method for graphically demonstrating the central tendency and spread of a numerical dataset using the quartiles of the data.
- **Stem and leaf diagram**: A method for displaying quantitative data, grouped in horizontal rows, useful for visualising the distribution of the data.
- **Vertical line graph:** A chart in which a single variable is displayed graphically, where it would not be appropriate to join the data points into a continuous line or curve.

Before you start
You should:
- Have come across frequency tables and grouped frequency tables.
- Be familiar with all the methods of data presentation discussed in this chapter:
  - Histograms,
  - Box plots,
  - Stem and leaf diagrams,
  - Vertical line graph.

What you will learn
In this chapter you will learn:

- How to interpret diagrams for single-variable data, including understanding that area in a histogram represents frequency and connections to probability distributions.
- Methods of presenting data, including frequency tables for ungrouped and grouped data, box plots and stem and leaf diagrams.

In the real world...

Visual representations can help to illustrate the key features of a dataset without the need for calculations. Graphs and charts appear on websites, in newspapers and in magazines, often stylised to suit the subject matter. For example the pie chart opposite shows the proportions of each food type thought to make up a balanced diet.

Approximate food types in a balanced diet

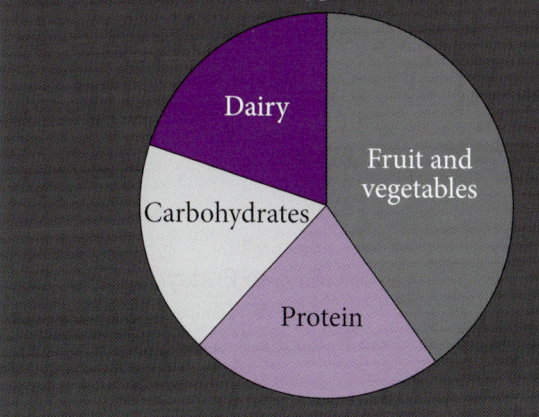

Exercise 9A (Revision)

1. Charlie records the amount of rainfall each day for a month for a school project. The results are shown in the table below.

Amount of rainfall x mm	Frequency
0	16
1	6
2	4
3	3
4	0
5	1

Find:
(a) The number of dry days during the month.
(b) The number of days in the month.
(c) The mean rainfall.

9.2 Frequency Tables

Definitions
You have come across grouped and ungrouped frequency tables in GCSE Mathematics. This section introduces some definitions of terms you will use at A Level.

Class interval
A large quantity of data may be collected into **groups**, also known as **classes** or **class intervals**. These are often displayed in grouped frequency tables, like the one shown below. In this example, the third class interval is 59 – 61 kg.

Mass, measured to nearest kg	Frequency
53 – 55	1
56 – 58	2
59 – 61	4
62 – 64	10
65 – 67	9

Class limits
In the example, the lower class limit of the third group is 59 kg and the upper class limit is 61 kg.

Class boundaries
For the third group, the lower class boundary is 58.5 kg and the upper class boundary is 61.5 kg.

Note: There are no gaps between class boundaries, so the lower boundary of the fourth class is also 61.5 kg, etc.

Note: If the class intervals were written: 53 – 56, 56 – 59, 59 – 62, etc, then the class limits and class boundaries are the same.

Class width
The class width (or class size) is the difference between the upper and lower class boundaries (**Note:** it is not the difference between the class limits). The class width for the third class in the table is 3 (61.5 – 58.5 = 3).

Class mark, midpoint or mid-mark
The **class mark**, **mid-mark** or **midpoint** of a class interval is the mean of the class limits. Hence, the mid-mark for the fifth class interval is 66 kg.

Median
The median value is the value halfway through the dataset. If there are n items in total, then:

- For grouped continuous data, or for data given in a cumulative frequency table, use $\frac{n}{2}$ for the position of the median item.

- For grouped or ungrouped discrete data use $\frac{n+1}{2}$

Worked Example
1. The frequency table given in the text on the left summarises the masses of 26 athletes in groups or classes. The masses are rounded to the nearest kilogram. Using the data in the table, find:
   (a) The modal class.
   (b) The median class.
   (c) For the fifth class, find:
       (i) The class boundaries.
       (ii) The class limits.
       (iii) The class width.
       (iv) The class midpoint.

(a) The 62 – 64 kg class has the highest frequency. This is the modal class.
(b) Note that this dataset contains discrete data, since the masses are rounded to the nearest kilogram. There are 26 items in the dataset. The position of the median is given by $\frac{n+1}{2} = \frac{26+1}{2} = 13.5$.
The 13th and 14th items both lie in the 62 – 64 kg class, so this is the median class.

(c) The fifth class is the 65 – 67 kg class.
  (i) The class boundaries are 64.5 kg and 67.5 kg.
  (ii) The class limits are 65 kg and 67 kg.
  (iii) The class width is 67.5 – 64.5 = 3 kg.
  (iv) The midpoint is the mean of the upper and lower class limits, i.e. $\frac{65 + 67}{2} = 66$ kg

Exercise 9B

1. (a) Copy and complete the following table.

Mass rounded to nearest kg	Midpoint	Class limits	Class boundaries
35 – 37			
	39		
41 – 43			
44 – 46			

(b) What is the size (class width) of the 35 – 37 kg class?
(c) Within which group would a mass of 43.5 kg lie in this table?

9.3 Diagrams Used to Represent Single-Variable Data

This section introduces methods of presenting data, including frequency tables for ungrouped and grouped data, box plots and stem and leaf diagrams.

You will not be asked to construct these tables and diagrams, but you may be expected to interpret them and draw inferences from them.

To compare two datasets, compare one average and one measure of spread. Either use the mean and standard deviation or the median and interquartile range. If the dataset contains extreme values, it is better to use the median and interquartile range, since these measures are not affected by extreme values. Extreme values, known as outliers, are discussed further in Chapter 12.

Stem and leaf diagrams

You should remember how to read a stem and leaf diagram.

Worked Example

2. The following stem and leaf diagram shows the ages of people in a hospital waiting room.

0	9					
1	1	5				
2	0	3	7			
3	2	3	3	7		
4	1	6	7	8	9	
5	0	0	4	8	9	
6	1	1	1	4	7	8
7	0	0	2	4	5	6	9
8	1	2	3	5	6	8

Key: 1|5 means 15

Note: When working with a stem and leaf diagram, there should always be a key.

(a) How many people are in the waiting room?
(b) What is the modal age of the people waiting?
(c) What is the median age of the people waiting?
(d) If a person under 16 years of age is considered a child, what fraction of those waiting are children?

(a) There are 39 people (ages 9, 11, 15, 20, … up to 88).
(b) The mode is 61. This is the age with the highest frequency; it appears 3 times in the diagram.
(c) There are 39 ages. The position of the median is $\frac{n+1}{2}$, i.e. the median is the 20th item in the ordered list of ages. The median is 59.
(d) There are three children (ages 9, 11 and 15) out of 39 people. The fraction of people that are children is $\frac{3}{39} = \frac{1}{13}$.

You may come across a back-to-back stem and leaf diagram, such as in the following example. Note that the lowest leaf values are closest to the stem on both sides of the diagram.

Worked Example

3. The following back-to-back stem and leaf diagram shows the heights of 24 pupils in a class: 12 girls and 12 boys.

Girls					Boys			
			4	12	1			
		7	5	2	13	3	7	
		6	5	3	14	1	3	5	7
	8	7	2	15	0	6	7	
			2	16	2	7		
				17	1			

Key for girls: 2|13 means 132 cm
Key for boys: 13|3 means 133 cm

AS 2: APPLIED MATHEMATICS

What is the difference in the median heights for the girls and the boys?

There are 12 girls and 12 boys.

Position of the median $= \dfrac{n+1}{2} = \dfrac{12+1}{2} = 6.5$

So the position of the median is 6.5 for both girls and boys, so take the mean of the 6th and 7th heights.

For the girls: median $= \dfrac{143 + 145}{2} = 144$ cm.

For the boys: median $= \dfrac{147 + 150}{2} = 148.5$ cm.

The difference is 148.5 − 144 = 4.5 cm.

Note: The shape of the diagram shows that the boys are generally taller. The girls' heights occupy the first 5 rows of the table from 120 cm to 160 cm. The boys' heights are in the 130 cm to 170 cm rows.

Exercise 9C

1. The stem and leaf diagram below shows the ages of the people in a TV talent show. Each person takes part on their own.

0	7	8				
1	0	2	5	5	8	9
2	0	1	3	7	9	
3	1	2	3	6	7	
4	0	5	7	9		
5	1	3				

   Key: 1|0 means 10

   (a) How many people take part in the talent show?
   (b) What is the modal age of the people taking part?
   (c) What is the median age of the people taking part?
   (d) What fraction of the competitors are under 20?

2. The following back-to-back stem and leaf diagram shows the cost of petrol per litre in 22 petrol stations: eleven in Belfast and eleven in Omagh.

Exercise 9C...

Belfast				Omagh		
	9	8	12			
7	4	3	13	4	7	
8	6	3	14	0	1	5	7
6	5	1	15	0	6	8
			16	2	3	

Key: 13|4 means £1.34

(a) Compare the distribution of prices in Belfast with the distribution in Omagh.
(b) What explanation may account for the difference in price between the two locations?

Box plots and cumulative frequency diagrams

Given the median, upper and lower quartiles, as well as the highest and lowest values in a dataset, a **box plot** allows you to visualise the data.

The left- and right-hand sides of the box are in the positions of the lower and upper quartiles respectively. The ends of the two whiskers are the lower and upper extreme values within the dataset. The central bar inside the box is in the position of the median value.

If two box plots are shown on the same diagram, this facilitates comparison of the two datasets.

Box plots are sometimes known as a box and whisker plots.

Worked Example

4. The box plots show the distribution of the speeds of cars in miles per hour on two motorways on the same day at 5pm.

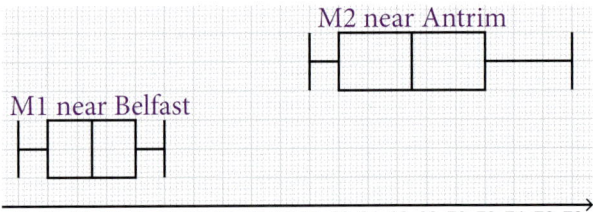

(a) Compare the two distributions.
(b) Comment on your findings.

(a) Using a box plot, it is not possible to compare the mean and standard deviation. Instead, consider the median and interquartile range for each plot.

The medians are: for the M1 near Belfast, 45 mph; for the M2 near Antrim, 67 mph.

The interquartile ranges are:
for the M1, 48 − 42 = 6 mph;
for the M2, 72 − 62 = 10 mph.

Note that the interquartile range is the width of the box and that the distance between the whiskers is the range. From the diagram, it is clear that the M2 dataset has a larger spread.

(b) Clearly there is a higher median speed and a bigger spread of speeds on the M2 near Antrim. This may be because of traffic jams near Belfast. The cars at this time are not able to move faster than 50 mph and all cars are forced into travelling at roughly the same speed, accounting for the lower spread.

Box plots are often used in conjunction with **cumulative frequency diagrams**. A cumulative frequency diagram can be used to estimate the median, lower and upper quartiles from grouped data.

Chapter 10 describes the method of **linear interpolation** for estimating the median from a grouped frequency table.

Worked Example

5. A new airport is being built. Eighty cranes are on the building site. The heights of these cranes are shown in the following frequency table.

Height h (m)	Frequency
$69 \leq h < 72$	4
$72 \leq h < 75$	8
$75 \leq h < 78$	14
$78 \leq h < 81$	33
$81 \leq h < 84$	17
$84 \leq h < 87$	4

(a) Calculate the cumulative frequency for each class.
(b) Use the cumulative frequency curve that follows to:
   (i) Estimate the median height of the cranes.
   (ii) Estimate the lower and upper quartiles.
(c) Draw a box plot to represent the data.

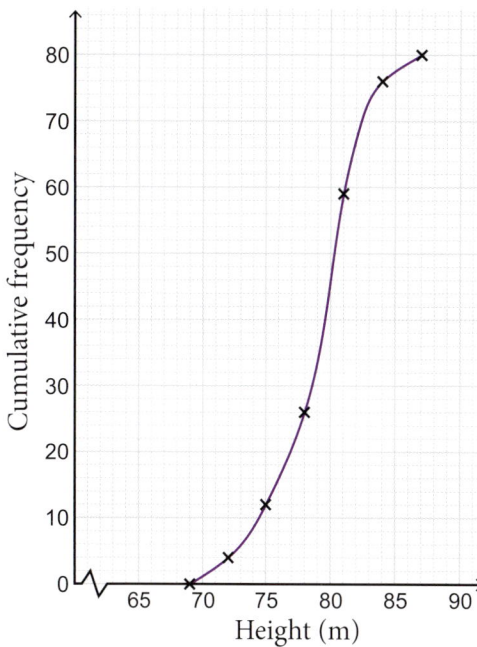

(a) Add a column to the table to show the cumulative frequency.

For the second class, the cumulative frequency = 4 + 8 = 12

For the third class, the cumulative frequency = 12 + 14 = 26

For the fourth class, the cumulative frequency = 26 + 33 = 59, and so on.

Height h (m)	Frequency	Cumulative frequency
$69 \leq h < 72$	4	4
$72 \leq h < 75$	8	12
$75 \leq h < 78$	14	26
$78 \leq h < 81$	33	59
$81 \leq h < 84$	17	76
$84 \leq h < 87$	4	80

(b) Note the following points about the cumulative frequency curve given in the question:
- The **upper bound** of each class is plotted against the cumulative frequency.
- The cumulative frequency tells you the number of cranes less than this upper bound in height. For example, the cumulative frequency for the fourth group is 59. This tells you there are 59 cranes less than 81 metres in height.
- The point (69, 0) can be used as the first point on the curve, since the number of cranes less than 69 m in height is zero.
- The points are joined with a smooth curve.

(i) The total number of cranes $n = 80$. Since height is a continuous variable, the position of the median is $\frac{n}{2} = 40$.

> **Note:** These are grouped continuous data, so use $\frac{n}{2}$ to find the position of the median item.

Draw a horizontal line segment at $y = 40$ (shown on the diagram below). Where this intersects the curve, draw a vertical segment to meet the x-axis. This gives an estimate of the median, roughly 79.5 m.

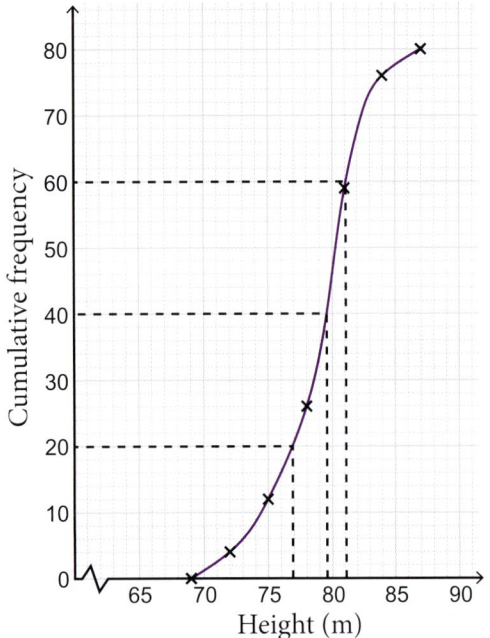

(ii) To find the lower and upper quartiles, find $\frac{n}{4} = 20$ and $\frac{3n}{4} = 60$.

Construct horizontal segments at $y = 20$ and $y = 60$. From where these lines meet the curve, drop to the x-axis, as shown in the previous diagram. The resulting estimates for the lower and upper quartiles are 77 m and 81 m.

(c) With grouped data, it is not always possible to state exact values for the lowest and highest values in the dataset. Without any additional information, we assume that the smallest crane has a height of 69 m and the highest has a height of 87 m.

Using the median value of 79.5, the lower and upper quartiles of 77 and 81 m respectively, the box plot is as follows.

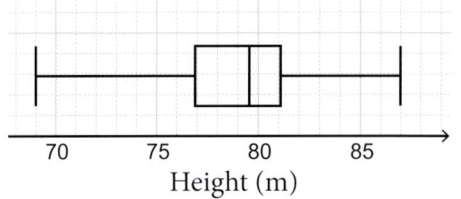

Exercise 9D

1. Runners in two running clubs took part in a 5 km timed trial. The distributions for the two clubs are shown in the diagram.

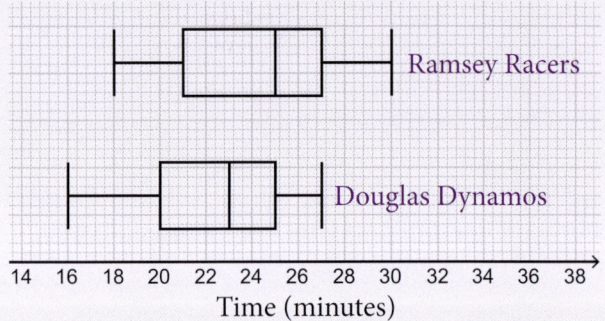

   (a) Write down the time by which 25% of the Ramsey Racers had completed the trial.
   (b) Write down the time by which 50% of the Douglas Dynamos had completed the trial.
   (c) Compare the times taken by the members of the two clubs and comment on your answers.

2. Forty truck drivers are stopped randomly and asked how far they drove the previous day. The results are shown in the following cumulative frequency diagram.

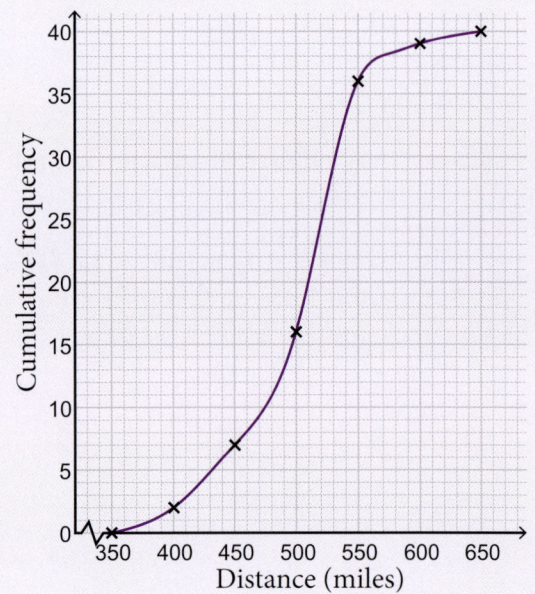

Exercise 9D...

(a) Copy and complete the table below.

Distance travelled d (miles)	Frequency	Cumulative frequency
$350 \leq d < 400$	2	2
$400 \leq d < 450$		7
$450 \leq d < 500$		
$500 \leq d < 550$		
$550 \leq d < 600$		
$600 \leq d < 650$		40

(b) Use the cumulative frequency curve to:
 (i) Estimate the median distance.
 (ii) Estimate the lower and upper quartiles.
(c) Draw a box plot to represent the data.

3. A wildlife group carries out a study into the size of red and grey squirrels in Northern Ireland. The group captures and weighs 60 of each type of squirrel. The following cumulative frequency diagram shows the masses of the two sets of squirrels.

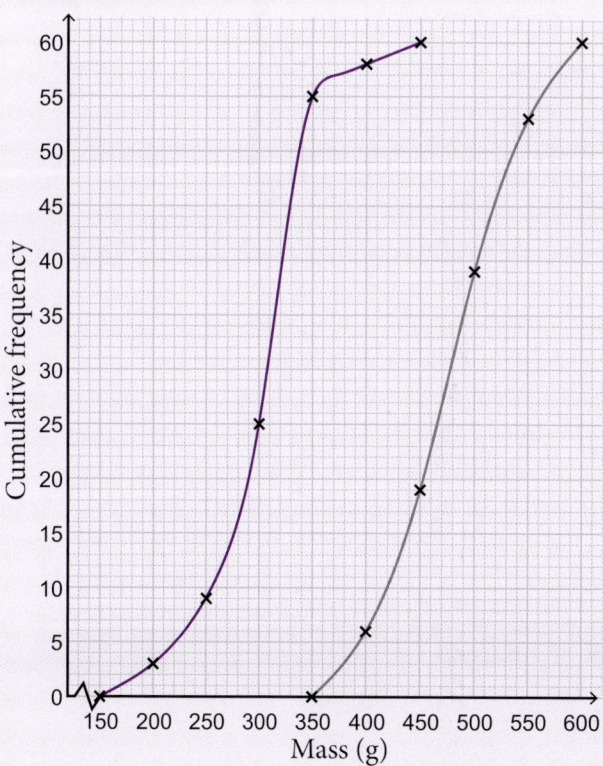

Compare the two sets of squirrels by answering the following questions.
(a) From the cumulative frequency graph, which type of squirrel is generally larger?

Exercise 9D...

(b) Which of the box plots below relates to the grey squirrels and which to the red?

Plot 1:

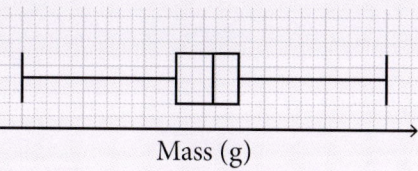

Mass (g)

Plot 2:

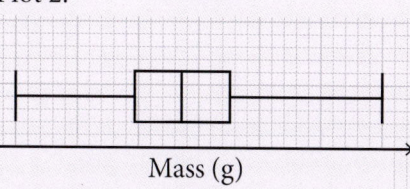

Mass (g)

(c) Copy both box plots and complete the horizontal scale for each one.
(d) Which set of squirrels has the larger spread of weights?

Histograms

A histogram allows you to visualise the data in a grouped frequency table.

Each class of the grouped frequency table is represented by a rectangular bar on the histogram.

The histogram is drawn so that the area of the bar is the frequency of the class.

The **frequency density** is plotted on the y-axis. This is the height of the bar.

Since:
$$\text{Area} = \text{width} \times \text{height}$$
so:
$$\text{Frequency} = \text{class width} \times \text{frequency density}$$

To calculate the frequency density, therefore:

$$\text{Frequency density} = \frac{\text{frequency}}{\text{class width}}$$

It is important to remember that:
- There are no gaps between the bars of a histogram.
- The area of a bar represents the frequency of the class. Therefore, the entire bar must be visible, meaning that the numbering on the y-axis must begin at zero. There can be an axis break on the x-axis, however, as in the next example.

Worked Example

6. The histogram shows the distribution of heights of 100 year 10 boys. One bar is missing.

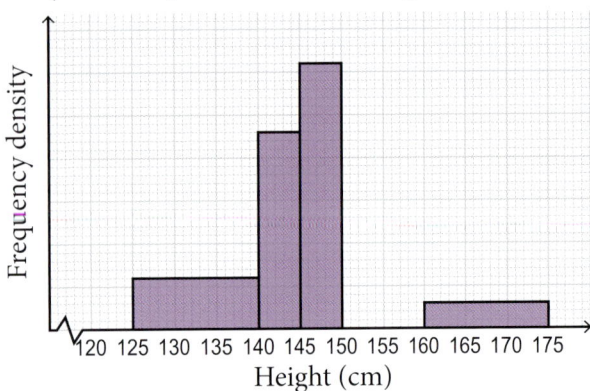

(a) There are 18 boys in the 125 – 140 cm group and 18 boys in the 150 – 160 cm group. Copy the histogram above and on your copy:
   (i) complete the scale on the frequency density axis;
   (ii) complete the histogram by adding the missing bar.
(b) Copy and complete the frequency table below.

Height (cm)	Frequency
$125 \leq h \leq 140$	18
$140 \leq h < 145$	
$145 \leq h < 150$	
$150 \leq h < 160$	18
$160 \leq h < 175$	

(c) A stratified sample of size 6 is to be taken from the boys with heights of 155 cm and over. Estimate the number of boys in the range 155 – 160 cm in the sample.

(a) (i) There are 18 boys in the 125 – 140 cm group. So the frequency is 18 and the class width is 15 cm. The frequency density is given by:

$$\frac{\text{frequency}}{\text{class width}} = \frac{18}{15} = 1.2$$

Therefore, the first bar has a height of 1.2 units and the scale on the frequency density axis can be completed (see above right).

(ii) The missing bar relates to the 150 – 160 cm group. The frequency is 18, so the area of the bar is 18. Its width is 10 cm. The frequency density is given by:

$$\frac{\text{frequency}}{\text{class width}} = \frac{18}{10} = 1.8$$

So the bar can be drawn with this height.

The completed graph is as follows:

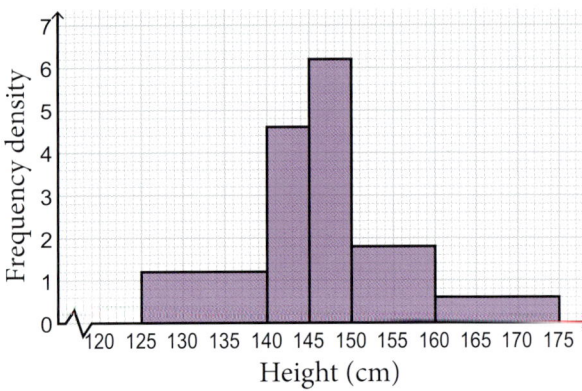

(b) For the 140 – 145 cm group:
frequency = class width × frequency density
= 5 × 4.6
= 23

For the 145 – 150 cm group:
frequency = 5 × 6.2
= 31

For the 160 – 175 cm group:
frequency = 15 × 0.6
= 9

The frequency table can be completed as follows.

Height (cm)	Frequency
$125 \leq h \leq 140$	18
$140 \leq h < 145$	23
$145 \leq h < 150$	31
$150 \leq h < 160$	18
$160 \leq h < 175$	9

(c) The boys whose heights are greater than 155 cm are to be sampled. Estimate there are 18 of these boys: half of the 150 – 160 cm group and all of the 160 – 175 cm group.

A sample of size 6 is taken, i.e. one third of the boys. Since this is a stratified sample, take one third of the boys in each height range.

If there are 9 boys altogether in the 155 – 160 cm range, then there are 3 in this range in the sample.

Special treatment is required for discrete variables, as demonstrated in the following example. This example also demonstrates scaling on both axes.

Worked Example

7. The variable w is measured to the nearest integer. 38 observations of w are made and are recorded in the following frequency table.

w	10 – 15	16 – 18	19 – 25
Frequency	15	9	14

A histogram is drawn. The bar representing the 10 – 15 group has a width of 2 cm and a height of 5 cm. For the bar representing the 16 – 18 group find, in centimetres:

(a) The width of the bar.
(b) The height of the bar.

(a) **In the horizontal**
Since histograms never have gaps between the bars, the class boundaries are used, rather than the class limits. The table has been copied and the boundaries added. The frequency density has also been calculated.

w	10 – 15	16 – 18	19 – 25
Frequency	15	9	14
Boundaries	9.5 and 15.5	15.5 and 18.5	18.5 and 25.5
Class width	6	3	7
Frequency density	$\frac{15}{6} = 2.5$	$\frac{9}{3} = 3$	$\frac{14}{7} = 2$

For the 10 – 15 group, the lower and upper boundaries are 9.5 and 15.5.

Therefore, the group has a class width of 6. We are told that the bar representing this group has a width of 2 cm. Therefore 1 cm represents 3 units in the horizontal direction.

For the 16 – 18 group, the lower and upper boundaries are 15.5 and 18.5. It has a class width of 3. Therefore, the bar representing this group has a width of 1 cm.

(b) **In the vertical**
For the 10 – 15 group, the height of the bar is 2.5 (the frequency density).

Since this is represented by a bar of height 5 cm, the vertical scale is 2 cm for every unit.

The 16 – 18 group has a frequency density of 3. Therefore, it is represented by a bar of height 6 cm.

The shape of a histogram can be inspected to give indications of the mean and spread within a dataset, without carrying out any calculations.

In addition, the size of each bar can be related to a probability for each class, in the case of a random selection being made.

9: DATA PRESENTATION AND INTERPRETATION

Worked Example

8. The histogram below shows the distribution of heights for the 1000 tallest buildings in central Toronto in Canada.

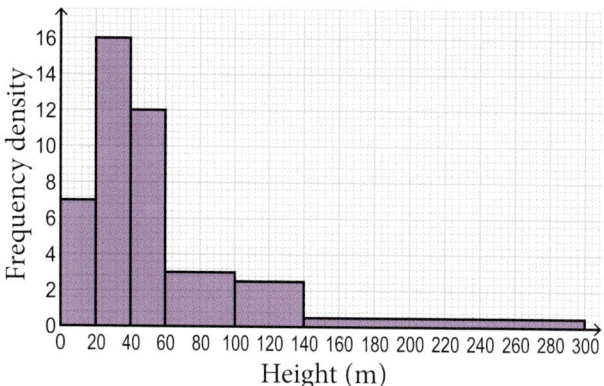

The following box plot shows equivalent data for the 1000 tallest buildings in central Ottawa, Canada's capital city.

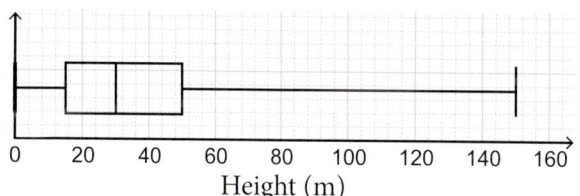

(a) Compare these two distributions.
(b) A building is taken at random from the 1000 buildings in central Toronto. Estimate the probability this building has a height between 10 and 20 metres.

(a) From the box plot, the median height for Ottawa is 30 m. Convert the information in the histogram for Toronto into a frequency table.

Height h (m)	Frequency
$0 \leq h < 20$	140
$20 \leq h < 40$	320
$40 \leq h < 60$	240
$60 \leq h < 100$	120
$100 \leq h < 140$	100
$140 \leq h < 300$	80

Note: Recall that the frequency of a group is the area of its rectangular bar.

The median is the 500th value, which lies in the 40 – 60 m group. Therefore, the 1000 tallest buildings in central Toronto are taller on average than those in Ottawa.

It is not possible to know the height of the tallest building in Toronto, but there are 80 buildings in the height range 140 – 300 m. We can assume the tallest building is close to 300 m. Therefore, the range of heights is greater in Toronto (roughly 300 m for Toronto, compared with 150 m for Ottawa).

A comparison of the interquartile ranges is left as an exercise.

(b) In Toronto, there are 140 buildings between 0 and 20 m in height. Assume that 70 are between 10 and 20 m. The probability of a randomly selected building being between 10 and 20 m is approximately $\frac{70}{1000} = 0.07$

The following example demonstrates how a histogram can be used to calculate relative frequencies and probabilities.

Worked Example

9. Forty boys are on a scout camp. The histogram shows the distribution of their ages.

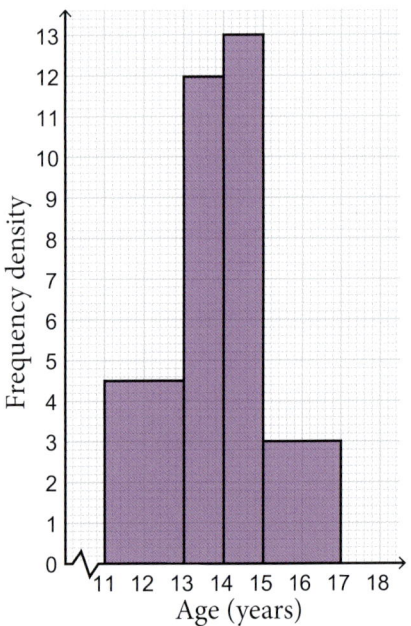

(a) Find the number of boys in the 11 – 13 age group.
(b) Find the relative frequency for the 11 – 13 age group.
(c) Calculate frequencies and relative frequencies for each of the age groups.

A boy is picked at random.
(d) Find the probability that he is over 15.
(e) Estimate the probability that he is over 16.
(f) Find the probability that he is over 14.

(a) This question part is asking for the frequency of the 11 – 13 group, which is the area of the bar. The class width is 2 and the frequency density is 4.5. So $2 \times 4.5 = 9$

(b) The relative frequency for this group is the frequency divided by the total number of boys, i.e. $\frac{9}{40}$

(c) Similar calculations for the other groups give the frequencies and relative frequencies in the table below.

Age range	Frequency	Relative frequency
11 – 13	9	$\frac{9}{40}$
13 – 14	12	$\frac{12}{40} = \frac{3}{10}$
14 – 15	13	$\frac{13}{40}$
15 – 17	6	$\frac{6}{40} = \frac{3}{20}$

(d) Relative frequencies can be used to estimate probabilities. Let X be the random variable 'the age of the boy chosen at random'.
$P(X > 15) = \frac{3}{20}$

(e) To estimate $P(X > 16)$, assume that the 6 people in the 15 – 17 group are distributed uniformly through the group. Then, 3 are younger than 16 and 3 older than 16.
Then: $P(X > 16) = \frac{3}{40}$

(f) $P(X > 14) = \frac{13}{40} + \frac{6}{40} = \frac{19}{40}$

Exercise 9E

1. The histogram below shows the weekly incomes of a random sample of households in a part of Belfast.

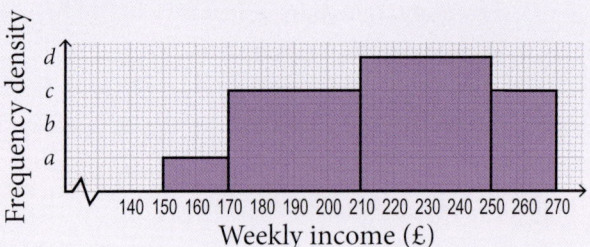

Exercise 9E...

(a) The number of households with a weekly income in the range £170 to £210 is 120. Use this information to find the values of a, b, c and d on the frequency density axis.

(b) Copy and complete the table below.

Weekly income (£)	Frequency
150 –	
170 –	120
210 –	
250 –	60

(c) How many households were surveyed?

(d) Estimate the number of households whose weekly income is less than £180.

(e) Explain why your answer to part (d) is an estimate, not an accurate value.

2. The histogram shows the distribution of heights for a group of P6 pupils at St Joseph's Primary School.

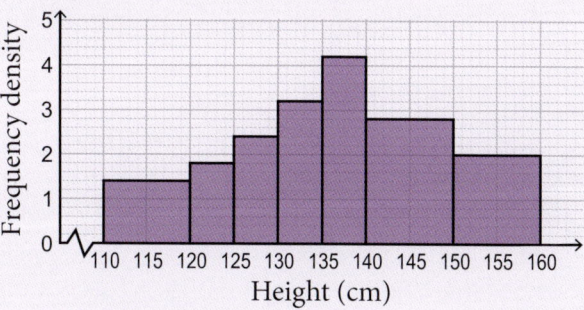

(a) How many P6 pupils are there at St Joseph's Primary School?

A P6 pupil is chosen at random.

(b) Find the probability this pupil is less than 120 cm tall.

(c) Find the probability the pupil is between 130 and 140 cm tall.

(d) Find the relative frequencies for:
   (i) the 140-150 cm group; and
   (ii) the 150-160 cm group.

(e) Estimate the probability the pupil is over 145 cm tall.

(f) Explain why your answer to part (e) is an estimate.

Exercise 9E...

3. A table of the lengths m of mobile phone calls to a company is shown.

Length of call (minutes)	Frequency
0 – 3	18
3 – 12	81
12 – 20	96
20 – 32	216
32 – 36	40

A histogram for these data is also shown below.

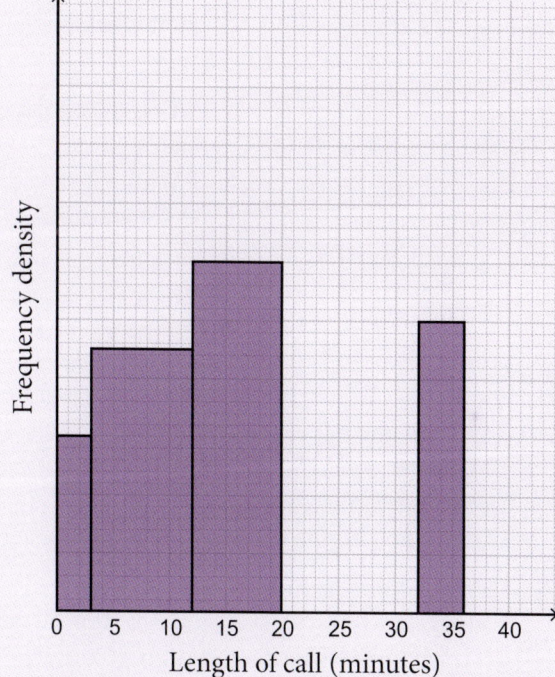

(a) Copy and complete the histogram by:
   (i) Completing the scale on the frequency density axis.
   (ii) Drawing the one missing bar.

(b) Explain what a stratified sample is.

(c) A stratified sample of size 18 is taken from those calls that are shorter than 9 minutes. Estimate how many of these calls are less than 2 minutes long.

Exercise 9E...

4. The number of letters per word is analysed in a page of text. The results are shown in the frequency table below.

Number of letters, n	Frequency
1 – 3	135
4 – 6	120
7 – 10	44
> 10	5

   A histogram is drawn. The bar representing the group 1 – 3 is 1.5 cm wide and 4.5 cm high. Find the width and height of the bar representing the 7 – 10 group.

5. The mass m grams of a collection of small objects is measured to the nearest gram. Twenty-eight mass measurements are taken and are recorded in the frequency table below.

m (grams)	1 – 6	7 – 9	10 –
Frequency	12	9	7

   A histogram is drawn. The bar representing the 1 – 6 class has a width of 3 cm and a height of 4 cm. For the bar representing the 7 – 9 class find, in centimetres:
   (a) The width of the bar.
   (b) The height of the bar.

6. Ian is training to run a marathon. The histogram displays the number of runs he takes, grouped by the length of the run in minutes.

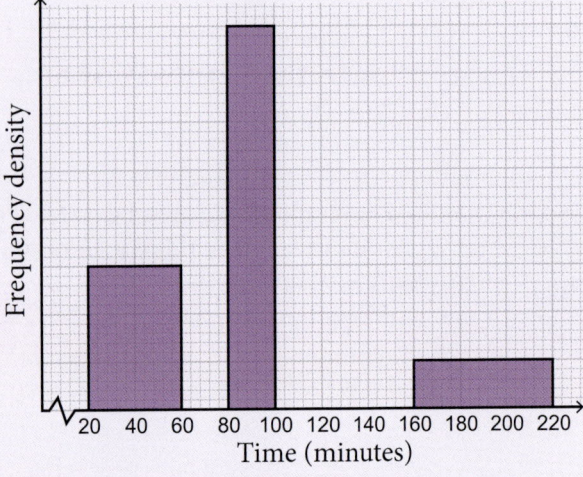

Exercise 9E...

During his training, Ian did 6 runs that took between 20 and 60 minutes and 7 runs between 60 and 80 minutes.
   (a) Copy the histogram and complete the scale on the y-axis.
   (b) Add the bar for the runs lasting between 60 and 80 minutes.

   The mean length of Ian's runs is 93 minutes 20 seconds.
   (c) Estimate the number of runs Ian took between 100 and 160 minutes.
   (d) Complete the histogram by adding the bar for the runs lasting between 100 and 160 minutes.

Vertical line graphs

A vertical line graph (or vertical line chart) is often used to plot real-life data in situations where it is not possible or appropriate to interpolate between values. In a vertical line graph the variable of interest is plotted on the y-axis, often – but not always – against some measure of time on the x-axis.

Worked Example

10. Jessica plants a sunflower. She measures its height on the first day of each month. She draws the following vertical line chart.

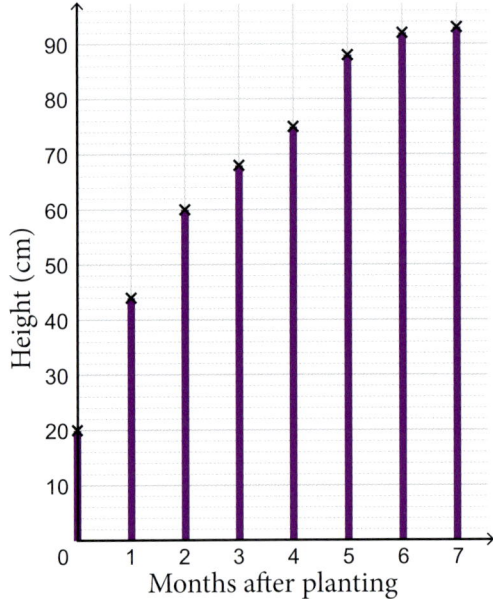

   (a) What is the height of the sunflower 3 months after planting?

(b) In which month does the sunflower grow the most?
(c) In which month does the sunflower grow the least?
(d) Do you think the sunflower will grow much more? Explain your answer.
(e) Jessica's mother says 'Instead of the vertical lines, you should join the points together with straight lines.' Do you think Jessica should follow her mother's advice? Explain your answer.

(a) 68 cm
(b) During the first month the sunflower grows the most. Its height increases by 24 cm from 20 cm to 44 cm.
(c) In the 7th year the shrub only grows 1 cm from 92 cm to 93 cm.
(d) The shrub probably won't grow much more. The increases from one year to the next have become very small.
(e) Jessica may be right to keep this as a vertical line graph. Joining the points with straight lines would indicate steady growth throughout each month, which is probably not the case.

Exercise 9F

1. Archie and Tilly are carrying out a traffic survey in Armagh. They record the number of people travelling in each car on a main road into the city. Unfortunately they lost their data after doing the survey. Before they lost it, however, they had written down some important information:
   - We surveyed 100 cars altogether.
   - The number of cars with 2 people was a square number.
   - The number of cars with 2 people was greater than the number with 3 people.
   - The number of cars with an even number of people was 31.
   - There were 17 cars with 3 people.
   - No cars had more than 5 people.
   - Three quarters of all the cars had only 1 or 2 people.

   (a) Use the clues to find the frequencies: how many cars had 1 person, how many had 2 people, etc.
   (b) Draw a vertical line graph to illustrate the data.

Exercise 9F...

2. The following table gives the number of guests at a small bed-and-breakfast in County Fermanagh in the months of one year.

Month	Number of Guests
Jan	12
Feb	20
Mar	26
Apr	60
May	40
Jun	45
Jul	54
Aug	75
Sep	20
Oct	15
Nov	4
Dec	42

(a) Show these data on a vertical line chart. Plot the months of the year on the horizontal axis and the number of guests on the vertical axis.
(b) Discuss briefly the shape of the graph, explaining why bookings are lower at some times of the year and higher at others. Explain any peaks in the data.
(c) Explain briefly why a vertical line chart is appropriate for these data.

9.4 Select or Critique Data Presentation Techniques in the Context of a Statistical Problem

You may be asked to select a data presentation technique appropriate to the context of a problem. Alternatively, you may be asked to critique a technique that has been chosen.

Exercise 9G

1. Siobhán is given summary data relating to the number of hours of sunshine during 2020 in her home town, Portadown. She is given the median, the upper and lower quartiles, the lowest value and highest value. What type of diagram could Siobhán draw to show these data?

Exercise 9G...

2. Richard has been given the grouped frequency table below relating to the number of hours of sunshine during 2020 in his home town, Bangor.

Number of hours of sunshine, rounded to the nearest half hour	Frequency
0.0 – 0.5	45
1.0 – 1.5	57
…	…

   He is asked to draw a diagram that allows him to estimate the median and interquartile range. What type of diagram could Richard draw to show these data?

3. The following frequency table is in a council report on the number of people using a skateboard park. The data were collected each day for the month of August 2021.

Number of people	Frequency
0 – 2	10
2 – 4	12
4 – 6	6
6 – 8	4

   State two problems with the way in which the data are presented in this frequency table.

4. Malachy has drawn a histogram, shown below, to show air pollution levels in Belfast throughout 2020. State four problems with Malachy's histogram.

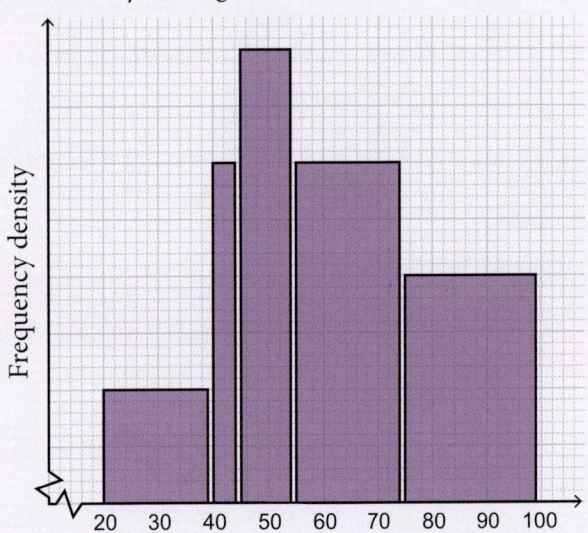

Exercise 9G...

5. Redstar Colt is a football team and has a club shop next to the stadium. The back-to-back stem and leaf diagram below shows the number of home and away shirts sold during each of the last 10 seasons.

Home		Away				
	29	1	4	7	7	9
	30	0	0	1	1	2
	31					
7	32					
6 8 9	33					
1 8	34					
4 7	35					
1	36					

   State three problems with this diagram.

6. A student is giving a presentation on the lengths of her stick insects. The raw data are below.

   Male stick insects (mm):
   9, 10, 12, 13, 15, 17, 20

   Female stick insects (mm):
   14, 18, 22, 23, 25, 26, 29, 30, 31, 32, 34

   She presents the data in three different ways: a frequency table, a back-to-back stem and leaf diagram and a pair of box plots.
   (a) Which of these diagrams best illustrates the difference in the spread of the male and female stick insects?
   (b) State one way in which the stem and leaf diagram is better than the box plots.

7. Look at the two histograms that follow. They show the heights of two groups of children at a holiday sports camp. Answer the following questions by inspecting the diagrams and without doing any calculations.
   (a) Without doing any calculations, compare the two height distributions.
   (b) State one possible reason for your answer to part (a).

Exercise 9G...

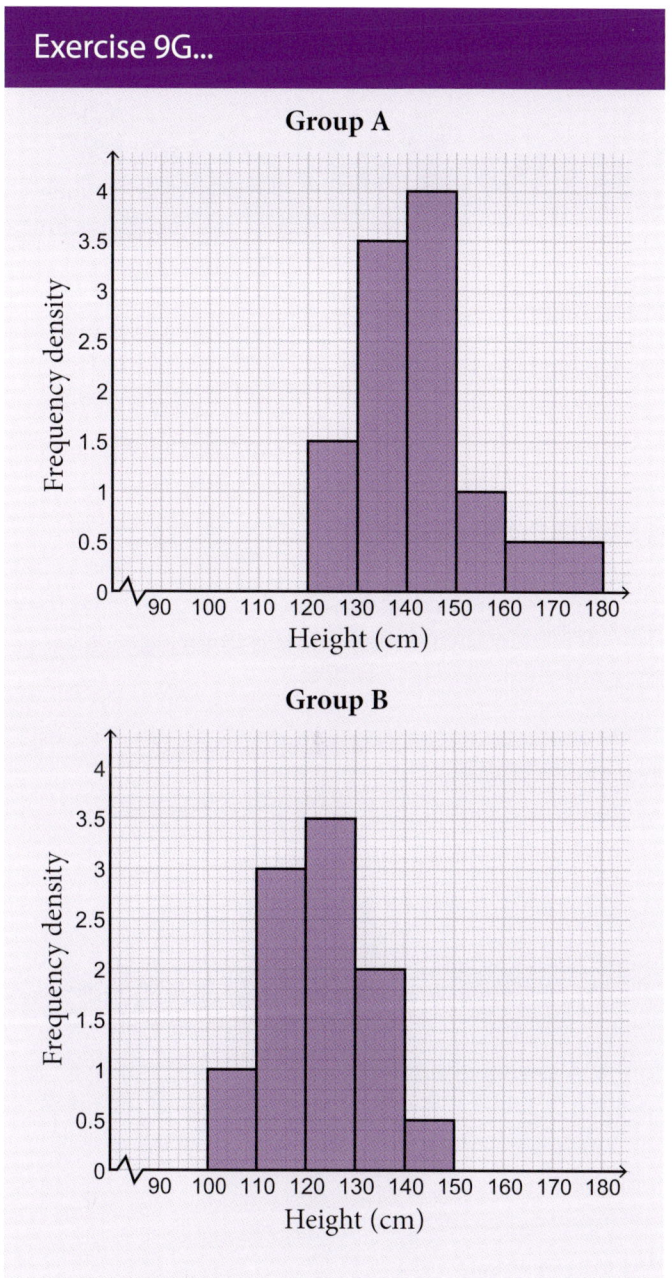

9.5 Summary

This chapter discussed methods of presenting single-variable data, including frequency tables for ungrouped and grouped data, box plots and stem-and-leaf diagrams.

You have learnt how to interpret these diagrams, including understanding that area in a histogram represents frequency. You have also seen some basic connections between histograms and probability distributions.

A key skill is the comparison of two datasets. The two datasets may be presented in different forms, for example one histogram and one box plot. Look at an average and a measure of spread for each set.

Chapter 10
Central Tendency and Variation

10.1 Introduction

Central tendency is the word statisticians use for an average, or a figure that is representative of an entire dataset. You have learnt about several measures of central tendency: **mean**, **median** and **mode**.

Variation is the word used for a measure of spread within a dataset. The word **dispersion** is also used. As measures of variation, you have learnt about the range and interquartile range.

In this chapter you learn about **standard deviation** and **variance**, more sophisticated measures of spread within a dataset.

Notation

Statisticians use the Greek letters μ and σ for the population mean and population standard deviation respectively.

For a sample mean $\bar{x}$ is used and s is used for the standard deviation based on a sample.

For a summation use Σ (upper case sigma). For example, the sum of all the x values is denoted Σx.

The letter n is used to denote the number of items of data in the dataset. For data in a frequency table, $n = \Sigma f$

Key words
- **Central tendency**: Central tendency is the phrase statisticians use for an average, or a figure that is representative of an entire dataset.
- **Mean**, **median** and **mode**: Three different averages, or measures of central tendency.
- **Variation**: The word used for a measure of spread within a dataset. The word 'dispersion' is also used.
- **Standard deviation**: One measure of spread, or variation, within a dataset. It is based on the average difference between each data point and the mean.
- **Variance**: Another measure of spread or variation. Variance is the square of the standard deviation.
- **Range** and **interquartile range**: Two measures of spread. Range is the difference between highest and lowest values in a dataset. Interquartile range is the difference between the upper and lower quartiles.

Before you start
You should:
- Understand the terms mean, median and mode.
- Understand the terms range and interquartile range.
- Know how to calculate the mean from a frequency table.
- Know how to estimate the mean from a grouped frequency table.

The mean of a set of data is defined as the sum of all the data divided by the number of items in the dataset, as shown in the following example.

Worked Example
1. Find the mean of the lengths listed below.

   6.4 cm, 3.8 cm, 4.7 cm, 5.8 cm, 4.8 cm

   $\Sigma x = 6.4 + 3.8 + 4.7 + 5.8 + 4.8 = 25.5$

   The mean $\mu = \dfrac{\Sigma x}{n} = \dfrac{25.5}{5} = 5.1$ cm

In the following example, the data are presented in a **frequency table**. It demonstrates how to find the mean, the median and the mode when the frequency table is ungrouped.

Worked Example
2. Charlie counted the number of text messages he received over 40 days. The results are shown in the following table.

Number of text messages	Frequency
12	8
13	11
14	6
15	2
16	4
17	9

   Find: **(a)** the mean, **(b)** the mode and **(c)** the median number of text messages.

Let the number of text messages be x and the frequency f.

Extend the table to include the column fx. We also require Σf and Σfx, so these are included in the final row of the table.

Number of text messages, x	Frequency, f	fx
12	8	96
13	11	143
14	6	84
15	2	30
16	4	64
17	9	153
	$\Sigma f = 40$	$\Sigma fx = 570$

(a) $\mu = \dfrac{\Sigma fx}{\Sigma f}$

> **Note:** This formula for calculating the mean from a frequency table is equivalent to the formula used for a list of values, $\mu = \dfrac{\Sigma x}{n}$

$\mu = \dfrac{570}{40} = 14.25$

(b) The mode is 13, since this is the value that has the highest frequency.

(c) The data are discrete, so the position of the median is given by:

$\dfrac{n+1}{2} = \dfrac{40+1}{2} = 20.5$

The median value is the mean of the 20th and 21st items. The first two frequencies, relating to x values of 12 and 13, are 8 and 11. The cumulative frequency for these two values is 19.

Therefore, the 20th and 21st items are both 14. The median, therefore, is 14.

> **Note:** Remember from Chapter 9 that, for grouped continuous data, or for data given in a cumulative frequency table, use $\dfrac{n}{2}$ for the position of the median item. For discrete data, grouped or ungrouped, use $\dfrac{n+1}{2}$.

10: CENTRAL TENDENCY AND VARIATION

What you will learn

In this chapter you will learn how to:
- Estimate the median using linear interpolation.
- Interpret measures of central tendency and variation, including standard deviation and variance.
- Calculate standard deviation and variance of a population or sample, including from summary statistics.

In the real world...

Biologists use statistics such as mean and standard deviation lifespan to compare populations of endangered fish, birds and land animals.

In 2012, a newspaper article suggested there were fewer than 100 adult cod left in the North Sea.

Cod populations have come under a lot of pressure in recent years, as have many other fish populations, due to overfishing and pollution.

But 100 proved to be a gross underestimate. Scientists analysed the data and found there were still several million adult cod in the North Sea at that time.

Exercise 10A (Revision)

1. Jake has a collection of pencils. He measures the lengths of each one. The data are recorded below in centimetres.

   10.2   8.0   7.5   9.4   5.6   6.8   7.6   7.7

   (a) Find the mean length.
   (b) Find the median length.

2. Barbara recorded the number of customers coming into her hairdressing salon for the month of April. The salon was open on 24 of the 30 days in April. The results are shown in the table below.

Number of customers	Frequency
12	1
13	5
14	11
15	7

   Calculate (a) the mean, (b) the mode and (c) the median number of customers.

10.2 Measures of Central Tendency

You should be familiar with mean, mode and median as measures of central tendency.

- The **mean** is used for quantitative data and uses all the data. It is therefore a value that represents the entire dataset, but it is affected by extreme values.
- The **median** is often used for quantitative data when there are extreme values in the dataset, since it is not affected by them.
- The **mode** is used for qualitative data, or quantitative data when there is a single mode, or sometimes when there are two (a bi-modal dataset). It can be useful since, unlike the mean and median, its value is always one of the values in the dataset.

In the next example, the data are given in a **grouped frequency table**.

Worked Example

3. The masses of various objects are measured to the nearest kilogram. The results are shown in the grouped frequency table below.
   (a) Estimate the mean mass.
   (b) What is the modal group?
   (c) Explain why your answer to part (a) is an estimate of the mean, rather than a true value?

Mass (kg)	Frequency
1 – 7	3
8 – 14	5
15 – 21	4
22 – 28	6
29 – 35	2

Note: the masses are rounded to the nearest kilogram, so these are discrete data.

Also note that the data are given in a grouped frequency table. In this case two extra columns should be added to the table, as shown below.

Mass (kg)	Frequency, f	Mid-point, x	fx
1 – 7	3	4	12
8 – 14	5	11	
15 – 21	4		
22 – 28	6		
29 – 35	2		
	$\Sigma f =$		$\Sigma fx =$

(a) The midpoint is the average of the lower and upper class limits. The first two midpoints have been calculated. The midpoint is then used as x in the calculation of fx.

This calculation is left as an exercise.

(b) The modal group is 22 – 28 kg, since this is the group with the highest frequency.

(c) The value calculated in part (a) is an estimate because the data are presented in a grouped frequency table. Therefore, the raw data are not known, only that 3 objects are in the range 1 to 7 kg, etc.

The following example demonstrates how to calculate a missing frequency in a frequency table if the mean is given.

Worked Example

4. The table below shows the weekly time spent travelling to work for a group of office workers.

Weekly time for journey to work (hours)	Frequency
0 – 1	17
1 – 2	13
2 – 5	
5 – 8	3

The mean travel time is 1.8 hours. Find the missing frequency in the frequency table.

Copy the table, letting the missing frequency be p.

> **Note:** Use of x, f or n for the missing frequency have been avoided as these variables are all used in other ways.

Also add columns for the midpoint x and fx. Complete these columns. The third entry in the fx column is an expression in terms of p.

Weekly time for journey to work (hours)	Frequency, f	Mid-point, x	fx
0 – 1	17	0.5	8.5
1 – 2	13	1.5	19.5
2 – 5	p	3.5	$3.5p$
5 – 8	3	6.5	19.5
	$\sum f = 33 + p$		$\sum fx = 47.5 + 3.5p$

$$\mu = \frac{\sum fx}{\sum f}$$

$$1.8 = \frac{47.5 + 3.5p}{33 + p}$$

$$1.8(33 + p) = 47.5 + 3.5p$$

$$59.4 + 1.8p = 47.5 + 3.5p$$

$$1.7p = 11.9$$

$$p = 7$$

Note: A missing frequency should always be a non-negative integer. If you get anything else, check your work!

Linear interpolation

The following example demonstrates the method of linear interpolation to estimate the median from a grouped frequency table.

Worked Example

5. The table shows how long each of a group of students takes in the shower one morning. The times are rounded to the nearest minute. Use linear interpolation to estimate the median time.

Length of time in shower (minutes)	Frequency
0 – 3	2
4 – 5	25
6 – 8	30
9 –	12

Step 1: Determine which group is the median group.

There are 69 students. Since the data are discrete, the position of the median student is given by:

$$\frac{n+1}{2} = \frac{69+1}{2} = 35$$

The 35th student falls in the 6 – 8 minute group.

Step 2:
Method 1. Draw a number line for the median class. Include the class boundaries and the median point above the line. Below the line include the cumulative frequencies.

Note: Q_2 is used for the median, since it is the second quartile.

```
 5.5       Q_2              8.5
  •─────────•────────────────•
 27        35                57
```

The median is the point that exactly half of the students lie below, so its cumulative frequency is 35. Then use proportion to estimate the median.

$$\frac{Q_2 - 5.5}{8.5 - 5.5} = \frac{35 - 27}{57 - 27}$$

$$Q_2 - 5.5 = 3\left(\frac{8}{30}\right)$$

$$Q_2 = 6.3 \text{ minutes}$$

Method 2. Use the formula:

$$Q_2 = L_1 + \frac{\left\{\frac{n+1}{2} - (\Sigma f)_1\right\}c}{f_{med}}$$

where:
- L_1 is the lower class boundary of the median class;
- n is the sum of all the frequencies;
- $(\Sigma f)_1$ is the sum of all the frequencies up to but not including the median group;
- c is the class width of the median group; and
- f_{med} is the frequency of the median group.

$$Q_2 = 5.5 + \frac{\left\{\frac{69+1}{2} - 27\right\}3}{30}$$

$$= 6.3 \text{ minutes}$$

Note: The equivalent formula for grouped continuous data is:

$$Q_2 = L_1 + \frac{\left\{\frac{n}{2} - (\Sigma f)_1\right\}c}{f_{med}}$$

Note: The two methods are exactly equivalent and should always give the same answer.

Exercise 10B

1. The members of a cycling club recorded how far they cycled in a week. The results are shown in the table below.

Distance cycled, d (km)	Frequency, f	Midpoint, x km	fx
$0 \leq d < 10$	5		
$10 \leq d < 20$	14		
$20 \leq d < 30$	24		
$30 \leq d < 40$	15		
$40 \leq d < 50$	3		

Exercise 10B...

(a) Copy the table and complete the midpoint and fx columns.
(b) Estimate the mean distance cycled, showing all your working.
(c) Use the method of linear interpolation to estimate the median.
(d) Explain why your answers to parts (b) and (c) are estimates rather than exact values.

2. This week Sean has recorded the number of steps he has taken each day using his smart phone. From Sunday to Friday the total is 61 560. On Saturday, his phone records 12 000 steps. What effect will this have on the mean for the week? Show your working.

3. Sixteen people are waiting in a doctor's waiting room. The times they have waited are shown in the frequency table below.

Waiting time, t (mins)	Frequency, f
$0 \leq t < 10$	7
$10 \leq t < 20$	5
$20 \leq t < 30$	3
$30 \leq t < 40$	1

Estimate the mean waiting time, giving your answer in minutes and seconds.

4. Which of the three averages mean, median and mode would be best to use in the following situations?
   (a) Jane's average score in her summer exams to be written on her report card.
   (b) Men's shirt sizes for all those shirts bought in one week in a department store.
   (c) Average annual salary for adults living in Belfast.

5. The mean age of the twenty members of a swimming club is exactly 28. The mean age increases by exactly 2 years when two new members join. What is the mean age of the two new members?

6. Saoirse runs a small gift shop. She records the number of customers who visit the shop every day over a period of time. The results are shown in the following table.

Exercise 10B...

Number of visitors	Frequency
14 – 16	2
17 – 19	
20 – 25	20
26 – 30	17
31 – 34	10

One of the numbers in Saoirse's table has been lost. She remembers that the mean number of visitors was 25. Find the missing frequency.

7. A shoe shop in Armagh records the sizes of each adult pair of shoes it sells during one week. The results are shown in the table below.

Shoe size	Frequency
5	2
5.5	5
6	8
6.5	6
7	5
7.5	6
8	7
8.5	7
9	10
9.5	8
10	7
10.5	5
11	3
11.5	1

Find for this set of data:
(a) The mode.
(b) The median.
(c) The mean.
(d) The shoe shop sells the data to a shoe manufacturer. Explain why the manufacturer might use the mode when planning production numbers.
(e) The following week, the manager of the shop suggests doing two separate surveys: one for men and one for women. By careful consideration of the data in the table above, do you think these two surveys would give different modes? Explain your answer.

10.3 Measures of Variation

Variance

The **variance** of a set of values is a measure of variation or dispersion, i.e. a measure of how much the data are spread out from the mean.

A large value for variance suggests that the data are widely spread out from the mean. However, a small value for the variance indicates that the data values are clustered around the mean.

If the data provided are from an entire population, the variance can be calculated using the following formula:

$$\text{Variance} = \frac{\sum x^2}{n} - \mu^2$$

Standard deviation

The **standard deviation** is another measure of spread and is defined as the square root of the variance. It is denoted by the Greek letter σ and is given by the formula:

$$\sigma = \sqrt{\frac{\sum x^2}{n} - \mu^2}$$

The units for the standard deviation are the same as the original quantity and for this reason it is often preferred to variance as a measure of spread.

When the data are provided in a frequency table, the following equivalent formula can be used:

$$\sigma = \sqrt{\frac{\sum fx^2}{\sum f} - \mu^2}$$

Note: If the data provided is from a **sample** of the population, a different formula is used. This formula is not provided here, since in CCEA AS Mathematics this **sample standard deviation** should be calculated on the calculator.

In some cases, summary statistics may be provided. The following example demonstrates finding the mean and standard deviation from summary statistics.

Worked Example

6. A population has the following summary statistics:

$$\sum f = 62, \sum fx = 1176, \text{ and } \sum fx^2 = 24694.5$$

Find the mean and standard deviation.

Calculate the mean first:

$$\mu = \frac{\sum fx}{\sum f}$$

$$= \frac{1176}{62}$$

$$= 18.96774\ldots = 19.0 \,(3\text{ s.f.})$$

The mean is used in the calculation of the standard deviation:

$$\sigma = \sqrt{\frac{\sum fx^2}{\sum f} - \mu^2}$$

$$= \sqrt{\frac{24694.5}{62} - (18.96774\ldots)^2}$$

$$= 6.2067\ldots = 6.21 \,(3\text{ s.f.})$$

The next two examples show calculation of the standard deviation from raw data and from a frequency table, respectively.

Worked Examples

7. The mean value was calculated for the following lengths: 6.4 cm, 3.8 cm, 4.7 cm, 5.8 cm, 4.8 cm
The mean μ was found to be 5.1 cm.
Find the standard deviation.

$$\sum x^2 = 6.4^2 + 3.8^2 + 4.7^2 + 5.8^2 + 4.8^2 = 134.17$$

Then using:

$$\sigma = \sqrt{\frac{\sum x^2}{n} - \mu^2}$$

$$= \sqrt{\frac{134.17}{5} - 5.1^2}$$

$$= 0.908 \,(3\text{ s.f.})$$

8. In Example 2, Charlie counted the number of text messages he received over 40 days, with the results presented in a frequency table reproduced below.

Number of text messages	Frequency
12	8
13	11
14	6
15	2
16	4
17	9

(a) Find the standard deviation.
(b) Is the value found in part (a) a true value or an estimate? Explain your answer.

(a) The mean μ was found to be 14.25 cm. Extend the table to include the columns fx and fx^2. We also require Σf, Σfx and Σfx^2. These are included in the final row of the table.

Number of text messages, x	Frequency, f	fx	fx^2
12	8	96	1152
13	11	143	1859
14	6	84	1176
15	2	30	450
16	4	64	1024
17	9	153	2601
	$\Sigma f = 40$	$\Sigma fx = 570$	$\Sigma fx^2 = 8262$

$$\sigma = \sqrt{\frac{\Sigma fx^2}{\Sigma f} - \mu^2}$$

$$= \sqrt{\frac{8262}{40} - (14.25)^2}$$

$$= 1.87 \text{ (3 s.f.)}$$

(b) The mean and standard deviation calculated are true values, not estimates, since the data are given in a frequency table, rather than a grouped frequency table. A frequency table contains raw data.

You may be asked to combine data sets.

Worked Example

9. The scores in an IQ test for 10 candidates give a mean of 117 and a standard deviation of 7.6. If another candidate has a score of 107, what is the mean and standard deviation of the 11 candidates?

Step 1: Find Σx^2 for the original 10 candidates.

$$\sigma = \sqrt{\frac{\Sigma x^2}{n} - \mu^2}$$

$$7.6 = \sqrt{\frac{\Sigma x^2}{10} - 117^2}$$

$$7.6^2 = \frac{\Sigma x^2}{10} - 117^2$$

$$\frac{\Sigma x^2}{10} = 7.6^2 + 117^2$$

$$\Sigma x^2 = 10(7.6^2 + 117^2) = 137467.6$$

Step 2: Find Σx^2 for all 11 candidates.

$$\Sigma x^2 = 137467.6 + 107^2 = 148916.6$$

Step 3: Find the mean for all 11 candidates.

For the original 10 candidates the mean is 117
So, the total for these 10 is $10 \times 117 = 1170$
The total for all 11 candidates is $1170 + 107 = 1277$

The mean for all 11 candidates $\mu = \dfrac{1277}{11}$

Step 4: Find σ for all 11 candidates.

$$\sigma = \sqrt{\frac{148916.6}{11} - \left(\frac{1277}{11}\right)^2}$$

$$\sigma = 7.80 \text{ (3 s.f.)}$$

Calculating mean and standard deviation on a calculator

As noted in the previous example, you can use your calculator to find the mean and standard deviation. In addition to recording the final result, you are expected to show the statistics from which they are calculated.

As we have seen, the population mean and population standard deviation are given by the following formulae:

	From a list of values	From a frequency table
Mean	$\dfrac{\Sigma x}{n}$	$\dfrac{\Sigma fx}{\Sigma f}$
Standard deviation	$\sqrt{\dfrac{\Sigma x^2}{n} - \bar{x}^2}$	$\sqrt{\dfrac{\Sigma fx^2}{\Sigma f} - \bar{x}^2}$

Therefore, when using a calculator, you are expected to show the following statistics as your working:

	From a list of values	From a frequency table
Mean	Σx and n	Σfx and Σf
Standard deviation	Σx^2, n and $\bar{x}$	Σfx^2, Σf and $\bar{x}$

Note: In addition, when working from a grouped frequency table you are expected to show the midpoints for each group.

10: CENTRAL TENDENCY AND VARIATION

Instructions for finding the mean and standard deviation on a calculator are given at the end of this chapter.

Comparing datasets

When comparing datasets, use two comparisons, one relating to a measure of central tendency, the other to a measure of variation. In the previous chapter comparisons were made using median and interquartile range. An acceptable alternative is to use the mean and standard deviation, as in the following example. The example also shows the summary statistics you should give when calculating mean and standard deviation on the calculator.

Worked Example

10. The data in the frequency table below relate to the times taken for the 100 metres race by two different groups of pupils in the school sports day.
    (a) Compare the times taken for the 100 metre race for Group A and Group B.
    (b) How can any differences in these two distributions be explained?

Time taken, t (seconds)	Group A frequency	Group B frequency
$10 \leq t < 15$	0	2
$15 \leq t < 20$	1	9
$20 \leq t < 25$	4	5
$25 \leq t < 30$	8	2
$30 \leq t < 35$	3	0
$35 \leq t < 40$	2	0

(a) The group midpoints are 12.5, 17.5, 22.5, 27.5, 32.5 and 37.5. Estimate the mean and standard deviation on your calculator, using these midpoints as x values.

> **Note:** Instructions for how to do this on a calculator are given at the end of this chapter.

The frequencies are entered firstly for the Group A pupils. The intermediate statistics from the calculator are: $\Sigma f = 18$, $\Sigma f x = 500$, $\Sigma f x^2 = 14362.5$. The population mean and standard deviation are $\mu = 27.8$ and $\sigma = 5.13$ (3 s.f.).

Next enter the frequencies for the Group B pupils. The intermediate statistics from the calculator are: $\Sigma f = 18$, $\Sigma f x = 350$, $\Sigma f x^2 = 7112.5$. The population mean and standard deviation are: $\mu = 19.4$ and $\sigma = 4.13$ (3 s.f.).

The mean time for Group A is higher. The standard deviation is also higher, indicating that the data are more widely spread from the mean.

(b) Since the mean time for Group B is lower, these may be older pupils, able to run the 100 metre race more quickly.

Sample standard deviation

As outlined above, the calculation of the standard deviation depends upon whether data from an entire population is being used, or whether the data are from a sample.

Standard deviation based on data from a population is denoted σ, while standard deviation based on data from a sample is denoted s.

Alternatively, sometimes σ_n and σ_{n-1} are used for the standard deviation based on a population and on a sample, respectively.

Worked Example

11. A delivery man delivers 290 parcels in a week. He takes a sample of 10% of these parcels and measures their masses. The results are shown in the table below.

Weight (g)	0 –	50 –	100 –	150 –
Frequency	12	8	5	4

Estimate
(a) the mean and
(b) the standard deviation of all 1000 parcels.

The data presented here are continuous data, since a parcel's weight can take any value in the weight range. Therefore, there is no gap between the upper limit of one group and the lower limit of the next.

The table has been copied and the upper limits have been added to each group. In addition, three rows have been added: midpoint x, fx and fx^2.

Weight (g)	0 – 50	50 – 100	100 – 150	150 – 200
Frequency	12	8	5	4
Midpoint x	25	75	125	175
fx	300	600	625	700
fx^2	7500	45 000	78 125	122 500

Find the summary statistics, as described at the end of this chapter:

$$\sum f = 29, \sum fx = 2225 \text{ and } \sum fx^2 = 253125$$

These are available on the calculator as n, $\sum x$ and $\sum x^2$ respectively.

Find the sample mean and sample standard deviation on the calculator. These are available on the calculator as $\bar{x}$ and s respectively.

The estimate for the mean of the population is the mean of the sample. Therefore, we estimate $\mu = 76.7$ grams (3 s.f.)

For the sample standard deviation $s = 54.3$ grams (3 s.f.)

Note: Remember to write down the group midpoints when working with a grouped frequency table.

The following example again demonstrates calculation of the sample standard deviation. In this example discrete data are being considered and special care is taken when calculating the group midpoints.

Worked Example

12. The 238 bus runs from Belfast to Newry. Sometimes it is very busy, other times less busy. The maximum capacity is 160 people. The frequency table below shows the number of people on a random sample of 30 bus journeys during one week in September.

Number of people on bus	0 –	40 –	80 –	120 –
Frequency	6	8	10	6

Estimate the mean number of people travelling and the standard deviation.

Note: The number of people on the bus is a discrete variable (i.e. it can only take certain values in the range 0 – 160, integer values in this case, and not anything in between, such as 9.75). This is important when calculating the group midpoints.

The table has been copied as follows. The lower and upper limits are shown for each group. In addition, the midpoint, fx and fx^2 rows are shown.

Number of people on bus	0 – 39	40 – 79	80 – 119	120 – 160
Frequency	6	8	10	6
Midpoint, x	19.5	59.5	99.5	140
fx	117	476	995	840
fx^2	2281.5	28322	99002.5	117600

Enter the data into the calculator, as described at the end of this chapter.

From the calculator, $\sum f = 30$, $\sum fx = 2428$ and $\sum fx^2 = 247206$.

These are available on the calculator as n, $\sum x$ and $\sum x^2$ respectively.

The mean and sample standard deviation are:
$\bar{x} = 80.9$ (3 s.f.)
$s = 41.8$ (3 s.f.)

Exercise 10C

1. The weekly incomes, rounded to the nearest pound, of households on a certain street in Lisburn are given below:

   299 334 344 363 347
   371 189 266 363 324

   Calculate the mean and standard deviation for these weekly incomes.

2. The wingspans of a random sample of butterflies were recorded. The results are summarised in the table below.

Wingspan (mm)	Frequency
$5 \leq w < 15$	6
$15 \leq w < 25$	11
$25 \leq w < 35$	18
$35 \leq w < 45$	15
$45 \leq w < 55$	4

   (a) Calculate the mean and standard deviation for these wingspans. Remember to show midpoints and summary statistics as your working.

Exercise 10C...

(b) If a similar investigation is carried out a week later, is it likely the results will be the same or different? Explain your answer.

3. The rector at a large church records the number of people attending the Sunday morning service every week for 3 years. His results are shown in the table below.

Number of people	0 –	50 –	100 –	150 –	200 – 300
Frequency	35	42	25	16	38

Find the mean and standard deviation of the number of worshippers.

4. A group of 26 adults are given a crossword to complete. The times they took are rounded to the nearest minute and are summarised in the table below.

Time (minutes)	Frequency
2 – 8	2
9 – 12	7
13 – 18	5
19 – 23	8
23 – 30	

(a) Find the missing frequency for the 23 – 30 minute group.
(b) Use linear interpolation to estimate the median time taken to complete the crossword.
(c) Estimate the mean time taken.
(d) Estimate the standard deviation of the times.
(e) Explain why your answers are estimates.

5. The lengths of a sample of salmon in a fish farm are measured and are shown in the grouped frequency table below. Estimate the mean length and the standard deviation.

Length (cm)	Frequency
1 – 7	3
8 – 14	5
15 – 21	4
22 – 28	6
29 – 35	2

Exercise 10C...

6. In a class of 5 girls and 8 boys, the mean scores in a maths test are 18.4 and 16.125 respectively and the standard deviations are 4.2 and 5.8 respectively. Calculate the mean and standard deviation for the whole class.

7. The labelling on bags of cement indicates that the bags weigh 20 kg. The masses of a random sample of 50 bags are summarised in the table below.

Mass, m (kg)	Frequency
17.6 – 17.8	1
17.8 – 18	2
18 – 18.5	4
18.5 – 20	18
20 – 20.2	17
20.2 – 20.4	6
20.4 – 21.0	2

Find the mean and standard deviation of the masses, showing all appropriate summary statistics.

8. Each of the 21 members of a piano class recorded the number of minutes, x, spent practising during a given day. The results are as follows:

$$\sum x = 1021 \qquad \sum x^2 = 50385$$

Find the mean and standard deviation for these data.

9. The 36 students on an A-Level course each recorded the number of whole minutes x they spent on social media during a given day. The results are summarised below.

$$\sum x = 1836 \qquad \sum x^2 = 94147$$

(a) Find the mean $\bar{x}$ and the standard deviation σ for these data.
(b) The times for two other students were 65 and 37 minutes. Without further calculation, explain the effect on the mean of including these two students.

Exercise 10C...

10. Johnny and Deborah have both done 20 park runs this year. Their times are summarised in the frequency table below.

Time, t (mins)	Frequency (Johnny)	Frequency (Deborah)
$23 \leq t < 24$	2	0
$24 \leq t < 25$	8	0
$25 \leq t < 26$	6	1
$26 \leq t < 27$	4	10
$27 \leq t < 28$	0	7
$28 \leq t < 29$	0	2

Compare the distribution of times for the two runners and comment on the statistics.

Calculator Methodologies

These instructions are for finding the mean, standard deviation and other statistics on the Casio fx-991EX. The method is similar for other Casio models.

Specify whether you are using a frequency table or a list of values:

1. Press **SHIFT-SETUP**.
2. Press the down arrow for more options.
3. Press **3** for Statistics.
4. Choose **1: On** to use a frequency table, or **2: Off** for a list of values.

Enter the data:

1. Press **MENU 6** for Statistics mode.
2. Press **1** for **1-Variable**.
3. Type an x value into the table and = to enter it. If the data come from a grouped frequency table, the x values are the class midpoints. Continue entering x values as required.
4. If using a frequency table, press the down arrow and the right arrow to navigate to the top of the frequency column.
5. Enter the frequencies.
6. Press **AC** to clear the screen.

Display summary statistics for your data:

1. Press **OPTN**, then **2** for **1-Variable Calc**.
2. The summary statistics are displayed, including:

   $\bar{x}$ the mean of the values

   $\sum x$ the sum of the values

   $\sum x^2$ the sum of the squares of the values

   σ the standard deviation, assuming the data are from a population

   s the standard deviation, assuming the data are from a sample

   n the number of items

Note: Whether you are using a frequency table or a list of values, the calculator uses the notation n, $\sum x$ and $\sum x^2$ (rather than $\sum f$, $\sum fx$ and $\sum fx^2$).

10.4 Summary

Measures of central tendency are averages, such as mean, median and mode.

Measures of variation or dispersion are measures of spread within a dataset, for example range, interquartile range, standard deviation or variance.

You may be asked to find the mean and standard deviation from a list of data.

Otherwise, you may have to find the mean from a frequency table. If the data are given in a grouped frequency table, this will be an estimate. The median can also be estimated using linear interpolation.

Mean and standard deviation can also be found from summary statistics.

When calculating standard deviation and variance, be aware that the calculation varies depending on whether the data represent a population or a sample.

All these calculations can be carried out on the calculator.

When asked to compare two datasets you should compare either the mean and standard deviation, or the median and interquartile range.

Chapter 11
Correlation and Regression

11.1 Introduction

Scatter diagrams are useful for visualising whether a relationship exists between two variables. However, we need a more accurate and reliable method to measure the strength of **correlation** between two variables.

If you have studied GCSE Further Maths or GCSE Statistics, you have learnt about Spearman's rank correlation coefficient.

In this section you will learn about another measure of correlation, the **product-moment correlation coefficient** (or **PMCC**).

Key words
- **Bivariate data**: A dataset in which two variables are involved.
- **Correlation**: A link or dependence between two variables.
- **Regression**: A method to determine the strength and character of the relationship between two variables.

Before you start
You should:
- Be able to interpret scatter graphs.
- Be familiar with the term **correlation**.
- Recognise the distribution of points on a scatter graph that shows (a) a positive correlation, (b) a negative correlation and (c) no correlation.
- Recognise the equation of a straight line and be able to write down the line's gradient and y-intercept.

What you will learn
In this chapter you will learn:
- How to interpret scatter diagrams and **regression lines** for **bivariate data**, including recognition of scatter diagrams that include distinct sections of the population.
- An informal interpretation of correlation.
- How to calculate and interpret the **product-moment correlation coefficient**.
- That correlation does not imply causation.

In the real world...

Starting a small business can be hard work, and sometimes risky, but it can also bring rewards, both in terms of job satisfaction and financially. To make the most from the decision to start a business, the business owner should make financial plans: what is the estimated income for the coming year, what are expected costs and what factors could affect or change these forecasts?

For example, the owner of an ice cream van may be interested in whether a change in the temperature affects her sales. If so, how much more, or less profit will she make? She may ask herself questions such as:

Is she prepared for a summer that is cooler than the average? How much will her income drop if summer temperatures are 5°C lower than the average? Will she make any money at all during the winter months?

The owner can get help answering these questions by looking at data on the correlation between ice cream sales and temperature.

Exercise 11A (Revision)

1. Look at the scatter graph below showing the age and value of 9 cars. Do you think the age of a car affects its value? If so, how?

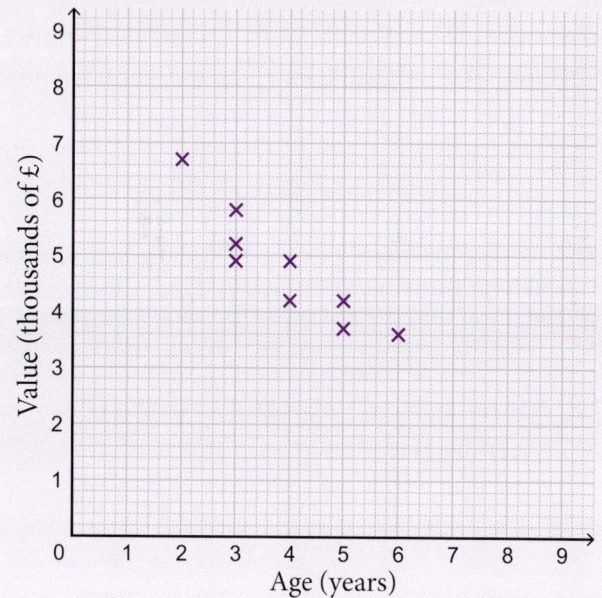

AS 2: APPLIED MATHEMATICS

Exercise 11A...

2. Look at the three graphs below. State which graph shows:
   (a) a positive correlation,
   (b) a negative correlation,
   (c) no correlation.

   Graph 1

   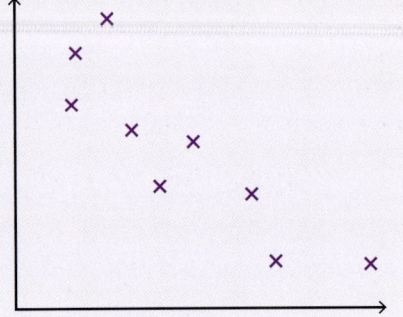

   Graph 2

   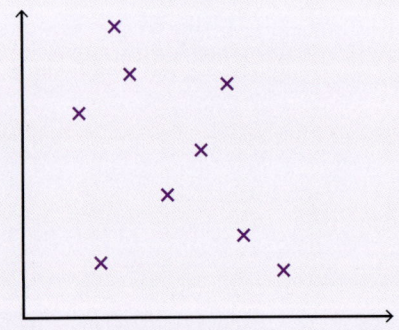

   Graph 3

   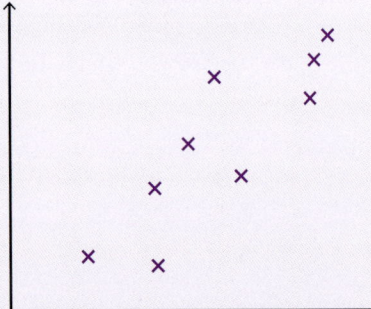

3. A straight line has the equation $y = 1.56 - 2.2x$
   Write down the line's gradient and y-intercept.

11.2 An Informal Interpretation of Correlation

Identifying correlation by eye and describing it

As in Exercise 11A above, you will need to identify the type of correlation between two variables plotted on a scatter graph: positive, negative or no correlation. Additionally, interpretation in the context of the question is expected.

Worked Example

1. Here is a simple set of bivariate data. x is the position of the fastest 10 runners in the Moira park run on Saturday 1st June. y is the position of the same runners in the same run on Saturday 8th June.

x	1	2	3	4	5	6	7	8	9	10
y	2	1	3	5	4	6	7	10	8	9

   A scatter graph is shown for these data.

   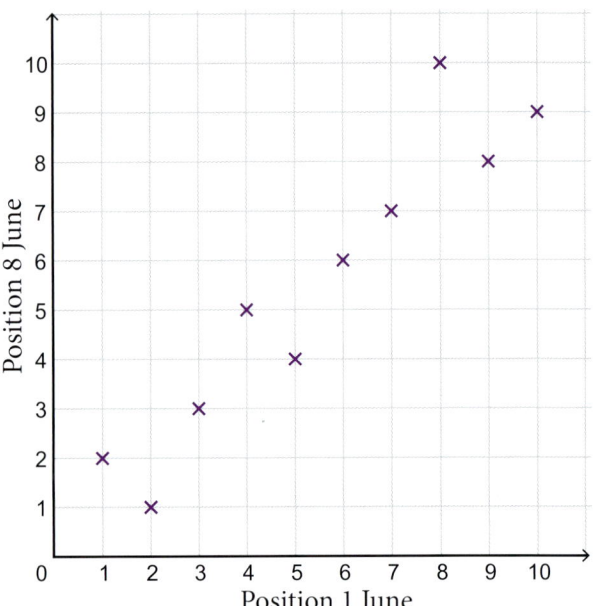

   (a) Suggest the type of correlation, if any, between these two variables. State which features of the graph lead you to this conclusion.
   (b) Suggest some reasons for this type of correlation between these two variables.

   (a) The scatter graph shows a strong positive correlation. This is evidenced by the facts that:
   - as x increases, y increases;
   - the points lie close to a straight line.

   (b) Each runner is likely to complete the race in a similar time from one week to the next; therefore, the runners' positions are not likely to change a lot.

Some scatter diagrams may include distinct sections of the population.

Worked Examples

2. The graph below shows the total sales of jam plotted against the sales of butter in Belfast during the years of World War II, 1939 to 1945. State the type of correlation between the sales of jam and butter and interpret your result.

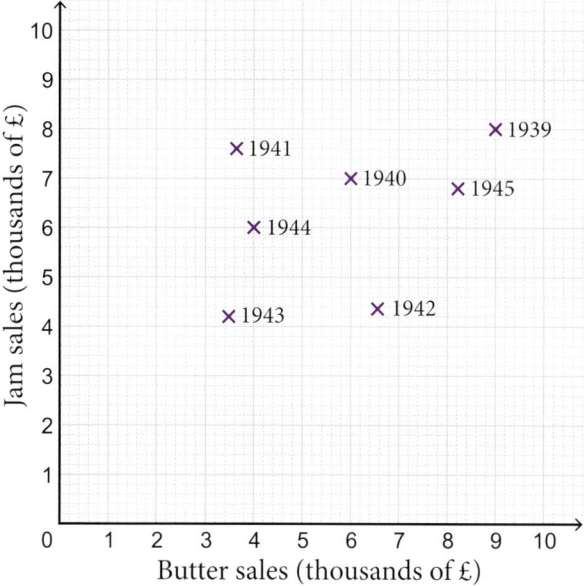

There appears to be no correlation. Higher sales of butter are not associated with higher sales of jam during these years.

3. The graph below again shows jam sales plotted against butter sales in Belfast, but for a wider range of years than in Example 2. What type of correlation does this graph show? Interpret your answer.

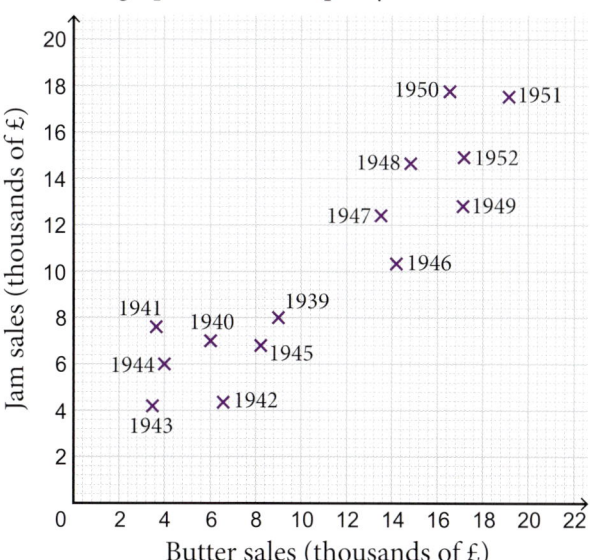

There now appears to be a weak positive correlation. During the war years, sales of butter and jam were both low. After the war, sales of butter and jam were both higher.

Note: This example shows that subsections of the population may follow a different trend to that of the whole population, or to a different sub-section.

Exercise 11B

1. The graph below shows the number of hours of recorded sunshine in two towns, Alfreton and Belper, for the first 9 days of June 2021. State the type of correlation between the hours of sunshine in the two towns and interpret your result.

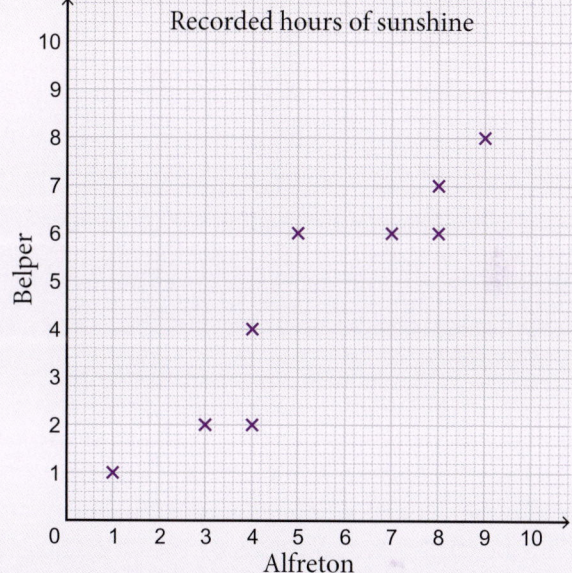

2. Máire and John both fly from Belfast to London regularly for meetings. The graph below shows the number of flights they each took for the years 2010 to 2019.

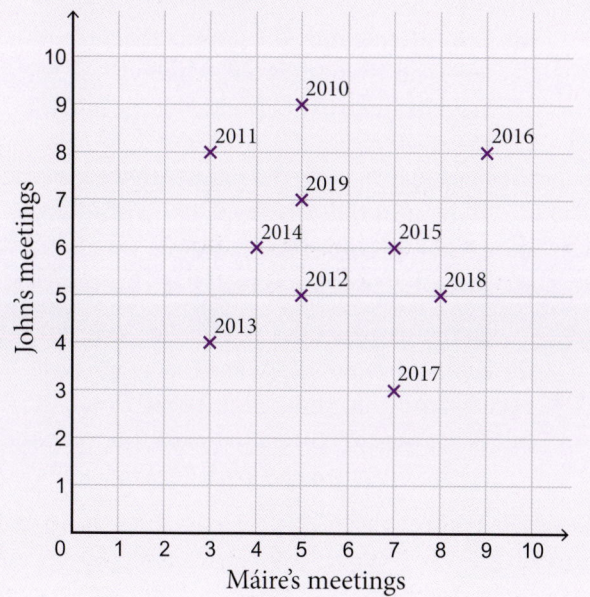

Exercise 11B...

  (a) State the type of correlation between the number of flights Máire takes and the number of flights John takes.
  (b) Interpret your answer to part (a).
  (c) Do you think Máire and John work together? Explain your answer.

3. A shop sells second hand mobile phones. The manager thinks there is a relationship between the battery life and the age of the phone. He collects the appropriate data for 10 phones and plots a graph.

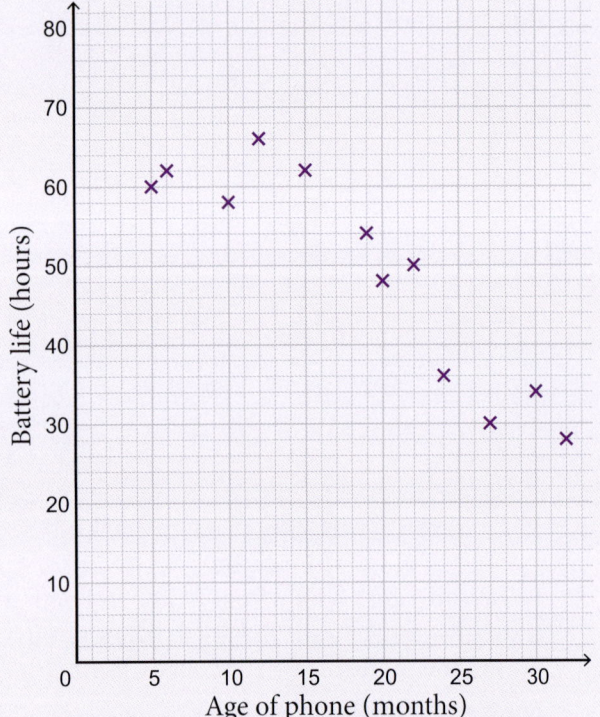

  (a) Considering only the phones that are less than 18 months old, what type of correlation exists between the age and the battery life?
  (b) Considering only the phones that are over 18 months old, what type of correlation exists?
  (c) Interpret your answers.

4. The graph in question 1 in Exercise 11A showed the ages and values of 9 cars. This graph has been extended to show a total of 30 cars.
  (a) What type of correlation exists between the age and value of cars over 20 years old?
  (b) Interpret your answer.

Exercise 11B...

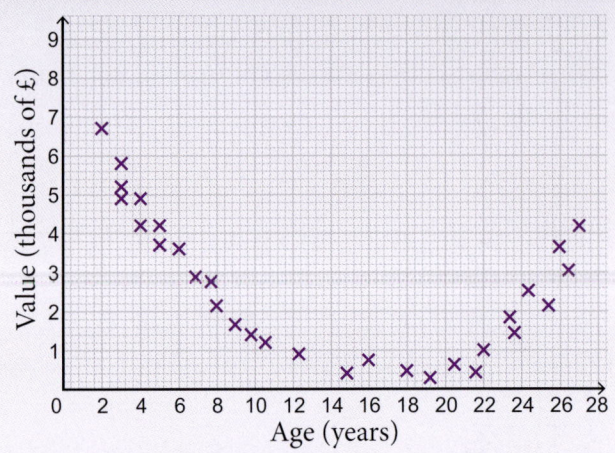

11.3 The Product-Moment Correlation Coefficient (PMCC)

If you have studied GCSE Further Mathematics or GCSE Statistics, you may have come across Spearman's rank correlation coefficient.

In this section you will learn about another correlation coefficient: the **product-moment correlation coefficient (PMCC)**.

The PMCC is denoted by the letter r and can take any value in the range between -1 and 1.

It gives an indication of how well two variables are correlated. A value close to -1 denotes a strong negative correlation; a value close to 1 indicates a strong positive correlation.

Calculating the PMCC

When plotting one variable (on the x-axis) against another (on the y-axis), the closer the points lie to a straight line, the stronger the correlation. If this straight line has a positive gradient, the value of r is positive; if the line has a negative gradient, the value of r is negative.

To calculate the PMCC, first calculate the values of the summary statistics S_{xx}, S_{yy} and S_{xy}. The following formulae are on the formula sheet:

$$S_{xx} = \Sigma x_i^2 - \frac{(\Sigma x_i)^2}{n}$$

$$S_{yy} = \Sigma y_i^2 - \frac{(\Sigma y_i)^2}{n}$$

$$S_{xy} = \Sigma x_i y_i - \frac{(\Sigma x_i)(\Sigma y_i)}{n}$$

and

$$r = \frac{S_{xy}}{\sqrt{S_{xx}S_{yy}}}$$

> **Note:** You may see the formulae for S_{xx}, S_{yy} and S_{xy} without the subscripts i, that is:
>
> $$S_{xx} = \Sigma x^2 - \frac{(\Sigma x)^2}{n}, \text{ etc}$$
>
> An alternative formula for calculating the PMCC is:
>
> $$r = \frac{n\Sigma xy - \Sigma x \Sigma y}{\sqrt{n\Sigma x^2 - (\Sigma x)^2}\sqrt{n\Sigma y^2 - (\Sigma y)^2}}$$
>
> This formula also appears on the formula sheet.

> **Note:** A high value of r alone does not imply a strong correlation. It is a necessary but not sufficient condition. You also need to inspect a scatter graph of the two variables in question to confirm the relationship is linear, since the PMCC only applies to variables that have a linear relationship.

The following example of PMCC being calculated uses the same data that was used in Example 1.

Worked Example

4. Given the bivariate data below, find the product-moment correlation coefficient for x and y. x is the position of the fastest 10 runners in the Moira park run on Saturday 1st June. y is the position of the same runners in the same run on Saturday 8th June.

x	1	2	3	4	5	6	7	8	9	10
y	2	1	3	5	4	6	7	10	8	9

The additional cells in the table below show how we calculate the various statistics required for the PMCC formula.

There are 10 pairs of data so $n = 10$.

x	1	2	3	4	5	6	7	8	9	10	$\Sigma x = 55$
y	2	1	3	5	4	6	7	10	8	9	$\Sigma y = 55$
xy	2	2	9	20	20	36	49	80	72	90	$\Sigma xy = 380$
x^2	1	4	9	16	25	36	49	64	81	100	$\Sigma x^2 = 385$
y^2	4	1	9	25	16	36	49	100	64	81	$\Sigma y^2 = 385$

Method 1

$$S_{xx} = \Sigma x_i^2 - \frac{(\Sigma x_i)^2}{n} = 385 - \frac{55^2}{10} = 82.5$$

$$S_{yy} = \Sigma y_i^2 - \frac{(\Sigma y_i)^2}{n} = 385 - \frac{55^2}{10} = 82.5$$

$$S_{xy} = \Sigma x_i y_i - \frac{(\Sigma x_i)(\Sigma y_i)}{n} = 380 - \frac{55 \times 55}{10} = 77.5$$

Then:

$$r = \frac{S_{xy}}{\sqrt{S_{xx}S_{yy}}} = \frac{77.5}{\sqrt{82.5 \times 82.5}} = 0.939 \text{ (3 s.f.)}$$

Method 2

$$r = \frac{n\Sigma xy - \Sigma x\Sigma y}{\sqrt{n\Sigma x^2 - (\Sigma x)^2}\sqrt{n\Sigma y^2 - (\Sigma y)^2}}$$

$$= \frac{10 \times 380 - 55 \times 55}{\sqrt{10 \times 385 - 55^2}\sqrt{10 \times 385 - 55^2}}$$

$$= 0.939 \text{ (3 s.f.)}$$

Calculating the PMCC from summary statistics

In some situations you may be given the summary statistics, but not the raw data.

Worked Example

5. The share prices of two companies are monitored over 30 days. Summary statistics for the 30 readings of each share price are shown below.

For company X: $\Sigma x = 3320$, $\Sigma x^2 = 368\,950$

For company Y: $\Sigma y = 9957$, $\Sigma y^2 = 3\,305\,009$

$$\Sigma xy = 1\,101\,975$$

$$n = 30$$

(a) Find the product-moment correlation coefficient, r.

(b) Comment on this value of r.

(a) Find the summary statistics S_{xx}, S_{yy} and S_{xy}.

$$S_{xx} = \Sigma x^2 - \frac{(\Sigma x)^2}{n} = 368\,950 - \frac{3320^2}{30} = 1536.66\ldots$$

$$S_{yy} = \Sigma y^2 - \frac{(\Sigma y)^2}{n} = 3\,305\,009 - \frac{9957^2}{30} = 280.7$$

$$S_{xy} = \Sigma xy - \frac{\Sigma x \Sigma y}{n} = 1\,101\,975 - \frac{3320 \times 9957}{30} = 67$$

$$r = \frac{S_{xy}}{\sqrt{S_{xx}S_{yy}}} = \frac{67}{\sqrt{1536.66\ldots \times 280.7}} = 0.102 \text{ (3 s.f.)}$$

> **Note:** Alternatively r could be calculated as in Method 2 of the previous example, using the formula $r = \dfrac{n\Sigma xy - \Sigma x\Sigma y}{\sqrt{n\Sigma x^2 - (\Sigma x)^2}\sqrt{n\Sigma y^2 - (\Sigma y)^2}}$

(b) The value of $r = 0.102$ indicates there is a very weak positive correlation between the two share prices. This could easily arise by chance.

> **Note:** Sometimes share prices show a strong positive correlation if the two companies operate in a similar sector. For example, the share prices of two airlines may rise at the same time and fall at the same time. This is because factors such as fuel price changes and variability in the demand for flying will affect all airlines.
>
> In this example the very weak correlation may indicate the two companies operate in completely different sectors.

Calculating the PMCC using a calculator

You may be asked to use a calculator to calculate the PMCC from a table of values. This section explains how to calculate the PMCC using a calculator. It is specifically written for the Casio fx-991EX 'Classwiz' calculator, but the process is similar for other Casio models.

Step 1: Set up an $x - y$ table without a frequency column:

1. Press: **SHIFT, MENU/SETUP**
2. Press the down arrow
3. Press **3** for Statistics
4. Press **2** (Frequency off)

Step 2: Enter linear regression mode:

1. Press **MENU**
2. Press **6** for Statistics mode (this may be a different number on other models)
3. Press 2 for $y = a + bx$ (linear regression mode)

Step 3: Now you should have an empty table with two columns for x and y values. Populate the table:

1. Enter these x values: 1, 2, 3, 4, 5, 6, 7, 8, 9, 10, pressing = after each one
2. Navigate to the top of the y column using the up, down, left and right arrow buttons.
3. Enter these y values: 2, 1, 3, 5, 4, 6, 7, 6, 8, 9, pressing = after each one.

Step 4: Display summary statistics for your data

1. Press **OPTN**, then **2** for 2-Variable Calc.
2. The summary statistics are displayed, including:

   $\sum x$ the sum of the x values
   $\sum x^2$ the sum of the squares of the x values
   $\sum y$ the sum of the y values
   $\sum y^2$ the sum of the squares of the y values
   $\sum xy$ the sum of the xy values
   n the number of pairs of values

3. Write down these 6 values as working.

Step 5: Display regression statistics for your data

1. Press **OPTN**, then **3** for Regression Calc.
2. The regression statistics are displayed:

   a the y-intercept of the regression line
   b the gradient of the regression line
   r the product-moment correlation coefficient

Interpreting the product-moment correlation coefficient

The following table summarises how to interpret the r value generated by the PMCC formula:

r range		Strength
$r \leq -0.9$	Negative correlation	Very strong
$-0.9 < r \leq -0.7$		Strong
$-0.7 < r \leq -0.3$		Moderate
$-0.3 < r < 0.3$	Zero or weak correlation	
$0.3 \leq r < 0.7$	Positive correlation	Moderate
$0.7 \leq r < 0.9$		Strong
$r \geq 0.9$		Very strong

If you are asked to interpret a value of r, you are expected to do two things:

1. State the strength of the correlation between the two variables, including terms such as positive, negative, strong and weak.
2. Also give an explanation in the context of the question.

Worked Examples

6. In Example 1 above, the finishing positions of 10 runners in two consecutive park runs were analysed. It was found that the product-moment correlation coefficient $r = 0.939$. Comment on this value of r.

The value of r indicates a **very strong positive correlation** between x and y. This suggests that each runner is likely to finish in a similar position in both races.

7. In a study of the age and value of ten cars, it is found that there is a correlation coefficient $r = -0.935$ between age in years and value in pounds (£). Comment on this value or r.

State that:

1. There is a **strong negative correlation**; and
2. For the cars in this survey, an older car is more likely to have a lower value.

Exercise 11C

1. The percentage unemployment rate u and number of house repossessions h in a town were recorded over 9 years. These data are shown in the table.

Unemployment u (per cent)	3.2	4	5	4.5	3.9	3.5	3.1	4.7	6
Number of house repossessions h	550	610	700	650	390	410	370	580	800

   (a) Calculate the product-moment correlation coefficient between u and h.
   (b) Comment on the value obtained in part (a).

2. Five competitors perform in a gymnastics event. Each competitor is given a mark out of 15 by two judges, P and Q, as shown in the table.

Competitor	A	B	C	D	E
Judge P	12	10	8	11	13
Judge Q	13	8	10	14	12

   (a) Find the product-moment correlation coefficient between the scores of the two judges.
   (b) Comment on the value obtained in part (a).

3. In school examinations, the marks of six pupils who each study chemistry and geography are given in the table.

Pupil	A	B	C	D	E	F
Chemistry	11	14	13	6	10	17
Geography	16	9	18	15	11	7

Exercise 11C...

   (a) Find the product-moment correlation coefficient between the chemistry and geography scores of the pupils.
   (b) Comment on the value obtained in part (a).

4. In an outfitter's shop, the collar size c and the shoe size s are measured for each of eight customers.

Customer	A	B	C	D	E	F	G	H
Shoe size s	10	9	9	10.5	8	8.5	11	9.5
Collar size c	17.5	15.5	17	15	15	16.5	16	17

   (a) Find the product-moment correlation coefficient between the customers' shoe sizes and collar sizes.
   (b) Comment on the value obtained in part (a).

5. Ten entrants for a piano exam recorded the number of hours they spent practising. The table gives the number of practice hours together with the mark each of them scored in the exam.

Entrant	A	B	C	D	E	F	G	H	I	J
Number of hours practice	14	23	17	29	20	17	21	28	17	22
Exam Score	77	96	75	92	85	84	83	89	78	84

   (a) Find the product-moment correlation coefficient between the number of hours of practice and the exam score for the entrants.
   (b) Comment on the value obtained in part (a).

6. There are 16 students in an IT class. Each student completes 2 tests, one on programming, the other in web design. The marks in the programming test p and the web design test w are calculated. The summary statistics are as follows:

   $\Sigma p = 1177$
   $\Sigma w = 1069$
   $\Sigma p^2 = 90775$
   $\Sigma w^2 = 75481$
   $\Sigma pw = 81056$

   (a) Calculate the product-moment correlation coefficient.
   (b) Interpret your result.

AS 2: APPLIED MATHEMATICS

Exercise 11C...

7. A rise in interest rates often results in a rise in the value of the pound. This happens because more investors buy the British currency when the interest available is increased. An increase in demand for the pound can then increase its value. The tables below show the level of interest rates and the value of the pound in January of each year from 2010 to 2021.

	Jan 2010	Jan 2011	Jan 2012	Jan 2013	Jan 2014	Jan 2015
UK interest rates	0.5	0.5	0.5	0.5	0.5	0.5
Pound vs Dollar	1.6	1.55	1.53	1.59	1.65	1.53

	Jan 2016	Jan 2017	Jan 2018	Jan 2019	Jan 2020	Jan 2021
UK interest rates	0.5	0.25	0.5	0.75	0.75	0.1
Pound vs Dollar	1.45	1.22	1.39	1.29	1.31	1.36

   (a) Find the product-moment correlation coefficient r between the level of UK interest rates and the value of the pound.
   (b) Interpret this value of r.

11.4 Linear Regression for Bivariate Data

The regression line

The regression line for bivariate data is a line of best fit through the points on a scatter graph.

AS Mathematics only requires knowledge of **linear regression**, in which the line of best fit is a straight line.

Therefore, in AS Mathematics a regression line has the form $y = a + bx$. We say this is the **line of regression of y on x**.

You **do not** need to know how to form the regression equation; any regression lines used will have the equation given.

However, you will have to use and interpret the regression equation. This may include recognition of scatter diagrams that include distinct sections of the population.

Interpreting and using the regression equation

The line of regression of y on x is given by $y = a + bx$ where a and b are constant values representing the y-intercept and gradient of the line respectively.

A linear regression model can be used when the points on a scatter graph lie close to a straight line and there appears to be a moderate or strong positive or negative correlation between the two variables.

...

Worked Example

8. Geoff is a postman. He walks to a lot of front doors delivering letters and parcels. Geoff thinks there may be a relationship between the value of a house and the distance from the road to the front door. Geoff does some research and plots the scatter graph below.

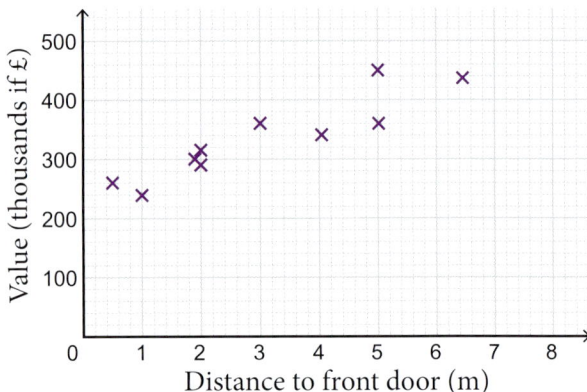

Is a linear regression model appropriate? Justify your answer.

The points on the graph appear to follow a straight line fairly closely. There appears to be a positive correlation between the two variables. Therefore, a linear regression model is appropriate.

...

The following example helps you to understand the significance of the a and b values in the linear regression equation.

It also demonstrates using the equation of the regression line to estimate the value of one of the variables.

...

Worked Example

9. As you move further from Belfast city centre, the population density falls. Therefore, there is a negative correlation between the population density D (measured in people per hectare) and the distance from the city centre d (measured in kilometres). Several parts of Belfast are surveyed and the population density is plotted against distance on a scatter graph. The equation of the linear regression line is: $D = -15d + 60$
   (a) What does the value −15 represent?
   (b) What does the value 60 represent?
   (c) What type of correlation exists between the two variables?
   (d) Use the equation to estimate the population density of an area that lies 2.5 km from Belfast city centre.

Comparing the regression equation with $y = mx + c$, observe that D is plotted on the vertical axis and d on the horizontal axis.
   (a) The value −15 is the gradient of the line. The value reflects the fact that as d increases by 1 km, D decreases by 15 people per hectare.
   (b) The value 60 is the y-intercept. This is where the regression line passes through the vertical axis. In other words, this is population density where $d = 0$ (i.e. in the city centre).
   (c) There is a negative correlation since the gradient of the regression line is negative, indicating that the line is from top left to bottom right.
   (d) Use the regression equation:
   $D = -15d + 60$
   When $d = 2.5$:
   $D = -15(2.5) + 60 = 22.5$
   The area in question may have a population density of about 22.5 people per hectare.

Interpolation and extrapolation

In the context of a regression line, **interpolation** refers to the use of the graph to estimate a y-value in a region of the graph where we know the straight-line relationship holds.

Extrapolation is the assumption that the straight line relationship continues to hold for a wider range of values.

The following example demonstrates use of interpolation. It also shows the limitations associated with extrapolation. Extrapolation where it is not appropriate can lead to incorrect conclusions.

Worked Example

10. In Exercise 11A, question 1, a scatter graph demonstrated the relationship between the age of a car and its value. The graph is shown again here, with the linear regression line shown.

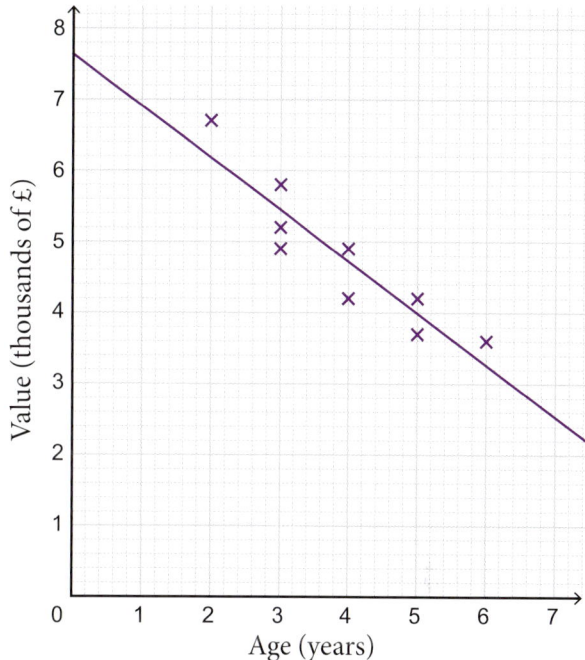

   (a) Use the regression line to estimate the value of a car that is 4 years and 6 months old.
   (b) Do you think it would be possible to use the regression line to estimate the value of a car that is 20 years old? Explain your answer.

   (a) Using an age of 4.5, use the regression line to estimate the value of the car, roughly £4400.
   (b) No, extrapolation would be inappropriate in this situation. If the regression line were continued and used to estimate the value of a 20 year old car, the value would be negative, clearly unrealistic. Older cars, sometimes known as vintage cars, can increase in value with age, as shown in Exercise 11B, question 4. The regression line shown appears to fit for cars up to 6 years old, but it would not be a suitable model up to 20 years.

Exercise 11D

1. A regression line has the equation
   $y = 2.5x - 1.5$
   (a) Estimate the value of y when $x = 2$
   (b) Estimate the value of x when $y = 8.5$

2. For each of these scatter graphs, state whether a linear regression model is appropriate and give a reason.

   (a)
   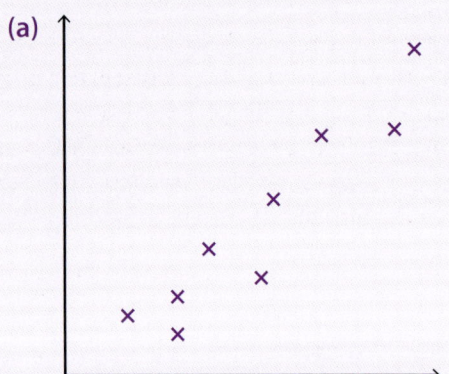

   (b)
   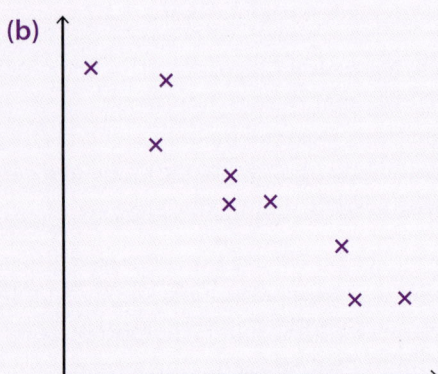

   (c)
   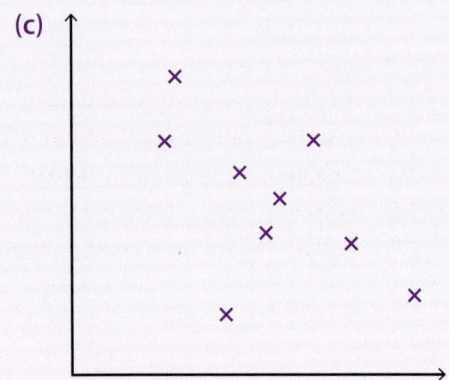

   (d)
   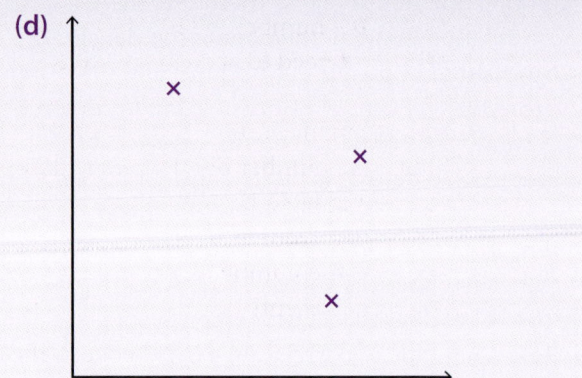

3. Different masses are attached to a spring and the length of the spring is measured for each one. The results from the experiment are shown below.

Mass m (g)	20	40	60	80	100
Length of spring y (cm)	48	55.1	56.3	61.2	68

   The equation of the regression line is:
   $y = 43.89 + 0.2305m$
   (a) Use the equation of the regression line to estimate the length of the spring when a mass of 70 g is attached.
   (b) Estimate the mass attached to the spring when the length of the spring is 50 cm.
   (c) In the regression equation, what do the figures of 0.2305 and 43.89 represent? In both your answers refer to the length of the spring.
   (d) Do you think it would be wise to use the regression equation to find the length of the spring when a mass of 1.5 kg is attached?

4. Last year a hotel had 50 staff while 287 guests were staying. This year they have a staff shortage. With 40 staff there are 231 guests. The hotel manager looks at the staffing and guest numbers from several previous years and plots them on a scatter graph. She then finds that the equation of the regression line of g on s is:
   $g = 6s - 10$
   where g is the number of guests and s is the number of staff.
   (a) Estimate the number of guests the hotel can accommodate if staff numbers fall to 25.

Exercise 11D...

(b) Estimate the number of staff members the hotel would need to accommodate 260 guests.

(c) Could the manager use her formula to estimate the number of guests that could be accommodated if she hired in total 100 staff? Explain your answer.

(d) What does the number 6 in the regression equation represent?

11.5 Correlation and Causation

You need to understand the concept of independent and dependent variables. In bivariate data, the **independent variable** is plotted on the x-axis and the **dependent variable** on the y-axis.

Worked Example

11. Consider the value of a car and its age.
    (a) Which of these is the dependent variable and which the independent variable?
    (b) Which one is plotted on the x-axis?

(a) The car's age is the independent variable. The car's value is the dependent variable, since it is dependent on the age.
(b) The independent variable, the car's age, is plotted on the x-axis.

In the previous example, there is clear dependence, or **causality**: the aging of a car causes its value to fall. This may not always be the case, even if there is a strong correlation between two variables.

It is important to note that correlation does not necessarily imply causation.

Worked Examples

12. Example 3 showed that, between 1939 and 1952 there was a weak positive correlation between sales of jam and sales of butter in Belfast. Is there any causality associated with this correlation?

No. It is not possible to infer that increasing butter sales **causes** increased jam sales, or the other way round.

The fact that butter and jam sales increased together after the war may have been brought about by other factors: increased supplies, greater affluence and the end of rationing, among other things.

13. "The higher the wind speed, the faster the blades of a windmill rotate. Therefore, windmills whose blades rotate rapidly cause higher wind speeds." Discuss this statement and suggest an alternative causation.

The causation is the wrong way around. In reality, higher wind speeds cause the blades to rotate more quickly.

14. "There is a positive correlation between ice cream sales and the number of people drowning in the sea. (That is, as ice cream sales increase, the number of people drowning also increases.) Therefore, ice cream consumption causes drowning." Discuss this statement and suggest an alternative cause for the increase in the number of people drowning.

The statement that ice cream consumption causes drowning is false.

There is a third factor that causes both the increase in the number of ice creams sold and the number of people drowning: temperature. As temperature rises, the number of ice cream sales increases and the number of people swimming in the sea increases, leading to more accidents.

Exercise 11E

1. In each of the following examples, state which is the dependent variable and which the independent variable.
    (a) The size of a house and its value.
    (b) The length of Mr Walker's daily walk and the time he takes for it.
    (c) A child's height and their age in years.

2. The principal of a school took a survey on the number of hours the pupils watch TV per week. She noticed a negative correlation between the number of hours spent watching TV and a pupil's performance in the end of year exams. The principal concluded that watching TV causes children to do less well at school. Discuss the principal's conclusion.

3. "Since the 1950s, both the level of CO_2 in the atmosphere and obesity levels have increased sharply. Hence, atmospheric CO_2 causes obesity." Discuss this conclusion.

Exercise 11E...

4. Twenty people were surveyed at random in a doctor's surgery. They were asked two questions:
   - How many times have you had a cold or flu in the last year?
   - How many times have you taken medication for cold or flu symptoms in the last year?

   The chart overleaf is a scatter graph displaying these data, with the number of colds plotted on the y-axis and the number of cold remedies on the x-axis.

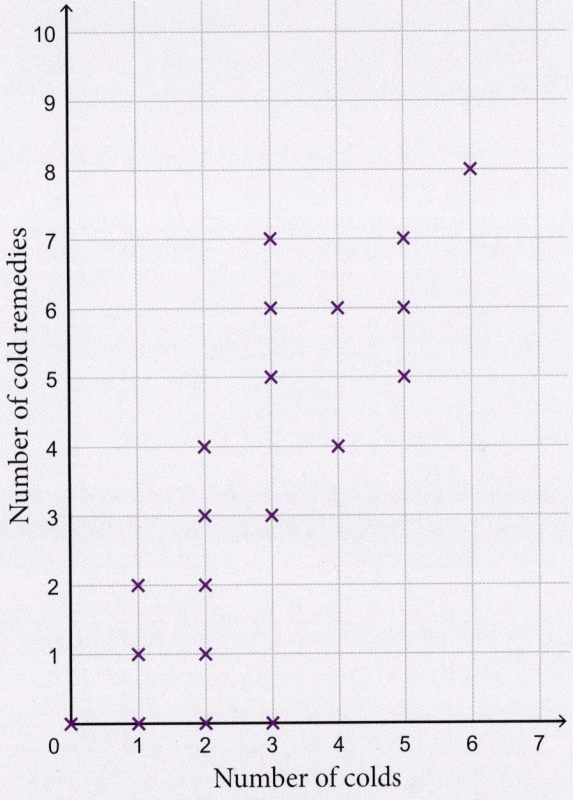

   There is a positive correlation between the two variables. Is any causation taking place here and, if so, which way does it work?

5. Another doctor in the same practice is carrying out a survey on sleep patterns. She gives each of her patients a 'fitness score' out of 20, based on their overall health. She also asks these patients to estimate on average how many hours of sleep they get each night. For all those patients getting less than 7 hours' sleep, she plots a scatter graph and notices there is a negative correlation between a patient's fitness score and their average number of hours of sleep. She concludes that sleeping less than 7 hours each night can cause ill health. Comment on her conclusion.

6. Recently I visited two hospitals with my six-year-old son. The first one was a small hospital in a small town. The second hospital was a much busier city hospital. There were many more patients and staff. After the second visit, my son asked me why there were more sick people in the second hospital: are the doctors and nurses making more people sick? Discuss my son's question.

11.6 Summary

The product-moment correlation coefficient (PMCC) is a measure of the correlation between two variables. You should know how to calculate the PMCC, both using the formulae (which are provided on the formula sheet) and on the calculator.

This measure of correlation can take values between -1 and 1, with -1 being a perfect negative correlation and 1 being a perfect positive.

However, a high level of correlation does not necessarily mean that an increase in one variable **causes** the increase in the other.

The linear regression equation will be given in the form $y = a + bx$. Calculation of the coefficients a and b is not required but interpretation of their values in the context of the question is expected.

You will also be asked to use the regression equation to estimate one of the variables in question.

Chapter 12
Data Cleaning

12.1 Introduction

In this chapter you will learn how to handle errors, **outliers** and missing data in a dataset.

Key words
- **Outlier**: An extreme value that lies outside the overall pattern of the data.

Before you start
You should know how to:
- Calculate quartiles and interquartile range.
- Find the mean and standard deviation of a set of numbers, or from a set of summary statistics.
- Interpret various statistical diagrams.

Notation
We use Q_1, Q_2 and Q_3 for the lower quartile, median and upper quartile respectively.

Worked Example

1. Windspeed measurements are taken on Rathlin Island. The daily maximum gust speed in knots is shown in the stem and leaf diagram below for the first 20 days in April.

    | 1 | 3 6 8 8 9 9 |
    | 2 | 0 0 3 3 4 4 4 5 6 7 8 9 |
    | 3 | 7 |
    | 4 | 0 |

    Key: 1|3 means 13 knots

    (a) Calculate the median value.
    (b) Calculate the interquartile range.

    (a) There are 20 items in the dataset, so $n = 20$
    The position of the median is:
    $$\frac{n+1}{2} = \frac{20+1}{2} = 10.5$$
    Q_2 is the average of the 10th and 11th items:
    $$Q_2 = \frac{23 + 24}{2} = 23.5$$

    (b) The position of the lower quartile is:
    $$\frac{n+1}{4} = \frac{20+1}{4} = 5.25$$
    Q_1 is the 5th item, i.e. $Q_1 = 19$
    The position of the upper quartile is:
    $$3\left(\frac{n+1}{4}\right) = 3\left(\frac{20+1}{4}\right) = 15.75$$
    Q_3 is the 16th item, i.e. $Q_3 = 27$
    The interquartile range is $Q_3 - Q_1 = 27 - 19 = 8$

    Note: In this example the data are ordered from smallest to largest in the stem and leaf diagram. Remember to order the data, if it has not already been done, before calculating the quartiles and the median.

What you will learn
In this chapter you will learn how to:
- Recognise and interpret possible outliers in data sets and statistical diagrams.
- Handle missing data and errors in datasets.

In the real world...

In 1981, an article, written by Professor Terence Hamblin, appeared in the British Medical Journal, containing this paragraph:

"*A statue of Popeye in Crystal City, Texas, commemorates the fact that singlehandedly he raised the consumption of Spinach by 33%. ... German chemists reinvestigating the iron content of spinach had shown in the 1930s that the original workers had put the decimal point in the wrong place and made a tenfold overestimate of its value. Spinach is no better for you than cabbage, Brussels sprouts, or broccoli.*"

It seems Hamblin was right about one thing, but wrong on two other counts.

Firstly, in the cartoon, Popeye once claimed that he ate spinach for its Vitamin A content, but never for its iron content.

Secondly, the decimal point story is a myth. The true iron content of spinach was measured by a scientist called Bunge in 1892. Earlier erroneously high measures, made by von Wolff in 1871, were explained in 1907 as resulting from iron contamination from heating dishes and other bad science; nothing to do with decimal points. Some 20th century textbooks continued to cite the poor 19th century science of von Wolff as though it were true.

The one thing that Professor Hamblin got right was that spinach, while very good for you, is probably no better than many other veg!

Exercise 12A (Revision)

1. Find the median, quartiles and interquartile range of these salaries:

   £10 000   £10 250   £11 000   £18 000
   £22 500   £25 000   £27 000   £56 000

2. A student studies the number of people travelling in cars on the Westlink in Belfast during the rush hour by taking a sample of 100 cars. The results are shown in the table below. Find the mean number of people in the cars in this sample and the standard deviation.

Number of occupants	Frequency
1	56
2	27
3	10
4	6
5	1

3. The distances jumped by ten long jump competitors are recorded, to the nearest 0.1 m, in the stem and leaf diagram below. What is the mean?

   ```
   4 | 6
   5 | 1 3 7
   6 | 2 4 7 9
   7 | 0 2
   ```
   Key: 4|6 means 4.6 metres

12.2 Outliers and Errors in Data

Identifying outliers
An outlier is an extreme value that lies outside the overall pattern of the data.

You may have to identify outliers in data sets and statistical diagrams. There are a number of different methods to identify outliers. The method may depend on the nature of the data and the calculations that are to be carried out.

Any rule needed to identify outliers will be specified in the question.

Method 1

This involves calculation of these **critical values**:

$$Q_3 + 1.5(Q_3 - Q_1)$$
$$Q_1 - 1.5(Q_3 - Q_1)$$

where Q_1 and Q_3 are the lower and upper quartiles respectively.

You may see these written as:

$$Q_3 + 1.5 \text{ IQR}$$
$$Q_1 - 1.5 \text{ IQR}$$

where the IQR is the interquartile range.

Any data that does not lie between these values is considered an outlier.

Method 2

This involves calculation of these **critical values**:

The mean ±2 standard deviations

Any data that does not lie between these values is considered an outlier.

Worked Example

2. A group of students is given an IQ test. The students' scores are shown in the stem and leaf diagram below. An outlier is defined as a score that is less than $Q_1 - 1.5(Q_3 - Q_1)$ or greater than $Q_3 + 1.5(Q_3 - Q_1)$. Identify any outliers.

   ```
    7 | 0
    8 | 1 9
    9 | 5 6 9
   10 | 1 3 5 5 7 9
   11 | 0 2 3 7
   12 | 1 6
   13 | 9
   ```
   Key: 8|1 means 81

There are 19 values in the dataset, i.e. $n = 19$

The position of the lower quartile is $\dfrac{n+1}{4} = 5$

The position of the upper quartile is $3\left(\dfrac{n+1}{4}\right) = 15$

The lower quartile is the 5th item, i.e. $Q_1 = 96$
The upper quartile is the 15th item, i.e. $Q_3 = 113$
The interquartile range is $(Q_3 - Q_1) = 113 - 96 = 17$
The critical values are:
$Q_1 - 1.5(Q_3 - Q_1) = 96 - 1.5 \times 17 = 70.5$
$Q_3 + 1.5(Q_3 - Q_1) = 113 + 1.5 \times 17 = 138.5$

The values 70 and 139 are outliers, since they lie outside the range 70.5 – 138.5.

Handling outliers

Sometimes outliers are legitimate values that could still be correct. For example, in Example 2 above, there could be students scoring 70 and 139 in the IQ test. These might not be errors.

However, there are occasions when an outlier should be removed from the data since it is clearly an error and it would be misleading to keep it in. Such erroneous data values are known as **anomalies**.

Having highlighted anomalies within a dataset, it is important that these values are removed before any further statistical analysis is performed.

Worked Example

3. Scientists conduct an experiment to determine the mass of a proton, a sub-atomic particle. They conduct the experiment 13 times to gain a good understanding. The 13 results are shown below.

Measured mass of a proton ($\times 10^{-27}$ kg)
1.47, 1.83, 1.97, 1.75, 1.01, 1.51, 1.55, 1.63, 1.87, 1.67, 1.67, 1.59, 1.63

(a) Highlight any outliers in this dataset by assuming outliers lie outside of the range of values given by: the mean ±2 standard deviations
(b) The scientists know that any data values lying outside this range are extremely likely to be errors caused by instrumentation problems. Clean the dataset by removing outliers and find the mean mass based on the cleaned dataset.

(a) Calculate the standard deviation:
$n = 13$
$\sum x = 21.15$
$\sum x^2 = 35.0725$
$\mu = \dfrac{\sum x}{n} = 1.6269 ...$
$\sigma = 0.2258 ...$

The critical values are:
$1.6269 \pm 2 \times 0.2258$
$= 1.1753, 2.0785$

The value 1.01×10^{-27} is an outlier, since it lies outside the range $1.1753 - 2.0785 \times 10^{-27}$.

(b) The value 1.01 is removed from the dataset and the mean is calculated for the remaining 12 values:
$n = 12$
$\sum x = 20.14$
$\mu = \dfrac{\sum x}{n} = 1.68$ (3 s.f.)

The scientists conclude that the mass of a proton is roughly 1.68×10^{-27} kg.

> **Note:** There should be some attempt to justify the removal of outliers. In this case the scientists justify the removal of the value 1.01 by noting that it was extremely likely to be caused by instrumentation problems.

In certain situations, it may not be necessary to use one of the formal methods described above to highlight errors in a dataset. A data value may be clearly erroneous without the need for any analysis.

Worked Example

4. Miss Brown compiles a list of those pupils in her P6 class who may sit the Northern Ireland transfer test. The dataset includes 16 names and dates of birth, as shown below.

Name	Date of Birth
Elizabeth Armstrong	01/09/2012
Caoimhe de Brun	18/01/2013
Reuben Buchanan	01/08/2012
Jay Ward	20/02/2012
Aoife Clarke	17/11/2013
Aine O'Sullivan	16/01/2013
Seth Gibson	18/11/2012
Anna Graham	14/10/2012
Jessica Kelly	09/12/2012
Padraig Doyle	25/05/2013
Mia Reid	18/06/1983
Owen Murphy	08/02/2013
Oisin Mac Carthaigh	01/06/2013
Seán McCollum	14/05/2013
Elsie McCracken	24/09/2012
Bella Parkes	01/02/2013

AS 2: APPLIED MATHEMATICS

The dataset is sent to a clerical officer at the examination board. His job is to calculate the average age of the pupils in each class.

(a) Indicate one clear error in Miss Brown's dataset.
   Note: Ordinarily, in this year group, pupils have a date of birth between 1 July 2012 and 30 June 2013, although it is possible for a pupil to have a slightly earlier date of birth if, for example, they repeated a school year.

(b) The clerical officer's computer does not allow a blank date of birth. Suggest how he could clean the data.

(c) Calculate the maximum error in the average age of the pupils after the data are cleaned using the method you have chosen in part (b).

(a) Mia Reid's date of birth is wrong, since it is roughly 30 years earlier than the others.

(b) The clerical officer could replace Mia Reid's date of birth, using an average of the other 15 ages. Alternatively, he could replace the erroneous date with the date that lies halfway between 1 July 2012 and 30 June 2013. Note that, in this example, both methods give the same date, 1 January 2013.

(c) The maximum error in Mia Reid's age is 6 months, or 0.5 years. The maximum error in the average age is $\frac{0.5}{16}$, or 0.03125 years, which is 11.4 days (3 s.f.)

The following example shows an error in a table of bivariate data. Two strategies are discussed for cleaning the data.

Worked Example

5. The table below shows the ages and heights of 12 trees recently planted in a park.

Age (months)	Height (m)	Age (months)	Height (m)
2	2.1	14	13.6
4	2.35	16	4.05
6	2.75	18	3.95
8	3.0	20	4.2
10	3.25	22	4.3
12	3.75	24	4.1

(a) One of the recorded height values appears to be incorrect. Show that the height in question is an outlier. You may assume that outliers are defined as lying above $\mu + 2\sigma$ or below $\mu - 2\sigma$, where μ and σ are the mean and standard deviation respectively.

(b) (i) When calculating the mean height of the trees, explain why it would be appropriate to remove the tree whose height is an outlier, i.e. to calculate the mean of the remaining 11 trees.
   (ii) Estimate the mean height in this way.

(c) Further investigation reveals that the equation of the regression line is: $h = 0.1a + 2.2$ where h is the height and a is the age.
   (i) Use this equation to estimate the missing height.
   (ii) Hence estimate the mean height of all 12 trees.

(a) $\sum x = 51.4$, $n = 12$, Mean $= \frac{51.4}{12} = 4.283\ldots$

$\sum x^2 = 321.025$, $\sigma = 2.899\ldots$

The mean + 2 standard deviations
$= 4.283 + 2 \times 2.899 = 10.081$

$13.6 > 10.081$, therefore the tree with a recorded height of 13.6 metres is an outlier.

(b) (i) Either of these explanations is sufficient:
   - The tree is roughly in the middle of the age range. There appears to be a positive correlation between age and height, so the height should also lie roughly in the middle of the range of heights. Therefore, leaving this value out in the calculation of the mean should not introduce much bias.
   - Outliers should be removed before performing any statistical analysis.

   (ii) If h is the height, then
   $\sum h = 37.8$ and $n = 11$
   $\bar{h} = \frac{37.8}{11} = 3.44$ m (3 s.f.)

(c) (i) Calculate the missing height using the linear regression equation:
   $h = 0.1a + 2.2$
   $h = 0.1(14) + 2.2 = 3.6$

   (ii) $\sum h = 41.4$ and $n = 12$
   $\bar{h} = \frac{41.4}{12} = 3.45$ m

Note: It is reassuring that the two different estimates of the mean height are very similar.

Exercise 12B

1. A medical student records the blood sodium levels of a group of adult men. The results, in milliequivalents per litre (mEq/L), are shown in the box plot diagram below.

   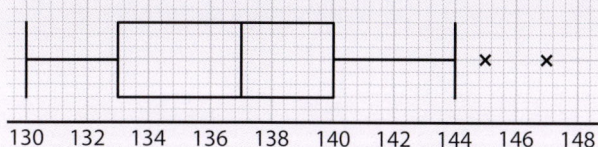

   The student decides to clean the data by removing any outliers. Each outlier is shown with an X on the diagram. If the outliers are removed from the original set of results, describe the effect this will have on:
   (a) the median,
   (b) the standard deviation.

2. The weekly incomes, rounded to the nearest pound, of households on a certain street in Lisburn are given below:

   247   299   334   344   363   347
   371   189   266   363   324

   (a) Calculate the mean and standard deviation for these weekly incomes.
   (b) Outliers are defined to be more than $1.5(Q_3 - Q_1)$ above Q_3 or more than $1.5(Q_3 - Q_1)$ below Q_1. Determine whether there are any outliers in this set of values. Show all your working.

3. Jamie is a supporter of Liverpool football team. For the last 12 years he has recorded the number of goals the team scores and the team's final position in the league. His results are shown in the scatter diagram opposite.
   (a) Jamie's friend Conor says one of the points looks wrong because Liverpool have never finished below 10th in the league. Which point has been recorded wrongly?
   (b) Verify that, for the point in question, the league position is an outlier. Assume that outliers are defined as being outside of the range:

   The mean ±2 standard deviations

   Show all your working.

Exercise 12B...

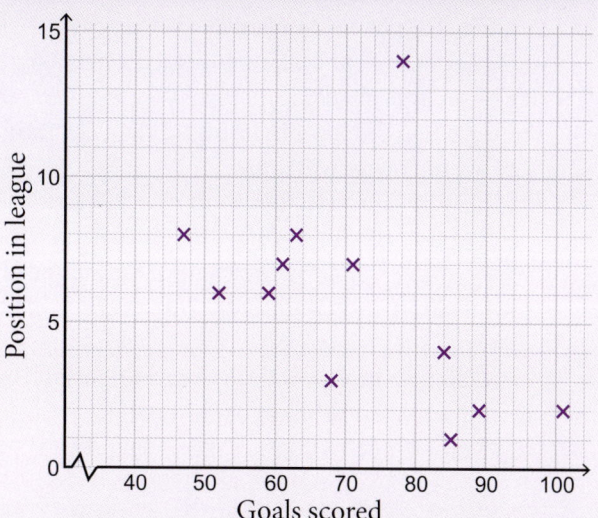

   (c) The linear regression equation for the data is: $p = -0.125g + 13.7$
   where p is the league position and g is the number of goals scored. Use this equation to estimate the league position for the point in question, assuming that the number of goals is correct.
   (d) State also how much confidence you would have in this estimate.
   (e) What type of correlation exists between the goals scored and the league position, stating whether it would be a strong or weak correlation.

4. A party organiser writes down the ages of the people attending a birthday party. They are:

   10, 10, 11, 110, 9, 11, 12, 10, 11, 10, 9

   (a) Explain why data cleaning is necessary.
   (b) Suggest two possible methods for cleaning the data and calculate a mean age after using one of them.

5. Here are the ages of people playing in an orchestra, as reported in the concert programme:

   12, 12, 13, 14, 16, 20, 25, 37, 38, 42, 165

   (a) Show that the age of 165 is an outlier, assuming that outliers are defined as any data that does not lie between the two values:
   $Q_1 - 1.5$ IQR
   $Q_3 + 1.5$ IQR
   where the IQR is the interquartile range.

Exercise 12B...

A journalist wishes to publish a newspaper story about the concert, including the average age of the musicians.
(b) Give two different methods of data cleaning the journalist could use.
(c) The journalist decides to remove the age of 165 to clean the data. Then he calculates the mean age of the musicians, the standard deviation and the interquartile range using the cleaned data. What age did the journalist report as the average age of the musicians in his newspaper article?
(d) Further investigations suggest that the outlying age should in fact have been 65. The journalist adjusts the mean age in his article before it goes to print. Without carrying out any further calculations, state what effect, if any, the inclusion of this extra piece of data has on the mean.
(e) State also how it would affect
  (i) the standard deviation,
  (ii) the interquartile range.

6. The percentage exam scores obtained by a small class of mathematics students are given below.

  65   58   72   61

There was one additional score, but this has been removed for data cleaning purposes. The mathematics teacher sends the cleaned dataset, comprising these four scores, to the school principal. The mean score fell by 8 marks as a result of the data cleaning. What was the fifth score and was the teacher justified in removing it from the dataset? Justify your answer.

7. Here are the ages of the candidates who sat their driving test at a particular test centre during one week:

```
0 | 9
1 | 8 9 9
2 | 1 2 5 6 6 8
3 | 2 3 3 5 9
4 | 1 2 6 6
```
Key: 1|8 means 18

(a) One of the ages has been recorded incorrectly. Which one?

Exercise 12B...

(b) The test centre records show that the mean age of all the candidates was 30. Clean the data and find the median age and interquartile range for the cleaned dataset.

8. A university is conducting a survey into traffic on the A1 road. A researcher questions 100 motorists leaving Belfast, asking each one their final destination. The results are shown in the frequency table below.

Destination	Frequency
Dunmurry	13
Lisburn	24
Hillsborough	11
Dromore	8
Banbridge	17
Newry	15
Dublin	11
Vladivostok	1

(a) Look up distances in miles from Belfast to each of these locations.
(b) Do you think that the data should be cleaned? Justify the data cleaning that you think should take place.
(c) Find the mean distance travelled by this sample of motorists, after any data cleaning has taken place.

9. The stem and leaf diagram shows the ages of people living in Homely Street.

```
1 | 2 6 9
2 | 2 5 7 9
3 | 1 3 3 8
4 | 2 2 2 6 7
5 | 0 4 5 6 9
6 | 1 1 3 7
7 | 0 6 7
8 | 1
```
Key: 1|2 means 12

(a) Thirty people live in Homely Street. The oldest person has been missed out from the stem and leaf diagram. The range of the ages is 70. How old is the oldest person?
(b) Find the mean and standard deviation of the residents of Homely Street, based on the entire dataset, including the oldest resident.

Exercise 12B...

10. Each member of a class of 23 students is asked to weigh his or her school bag before they come to school the next day. In the next day's maths class, the teacher collects the data, shares the dataset with the class and asks the students to draw a box plot of the masses. The two box plots shown below are the most common responses.

You may assume that the students who include an outlier are using the following definition:
An outlier is any value that is either:
- Greater than $Q_3 + 1.5$ IQR; or
- Less than $Q_1 - 1.5$ IQR

Which box plot is correct? Explain your answer.

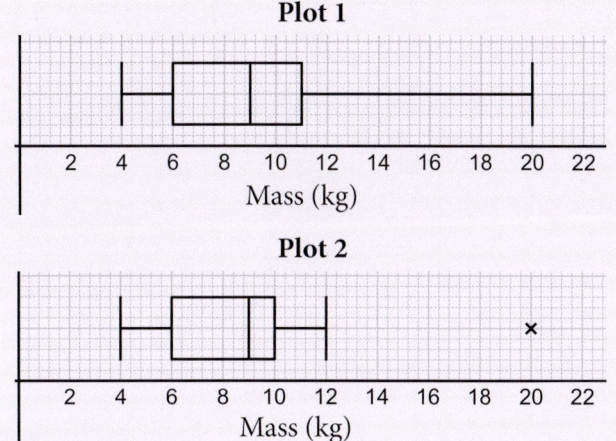

Plot 1

Plot 2

12.3 Missing Data

In this section the options for cleaning a dataset with a missing value or values are presented.

In the case of missing data, one approach is to remove the missing record and not include this record in statistical calculations.

Worked Example

6. The number of hours of sunshine (**insolation**) is measured daily on Rathlin Island. The data for June are shown in the stem and leaf diagram.
(a) How many of the recorded days were completely cloudy?
(b) There are five missing data values in this table. Without doing any calculations, suggest a strategy for cleaning the data if:
  (i) the monthly mean is required;
  (ii) an annual mean is to be calculated.

0	0	0	1	4	7	
1	2	4	4	6	8	
2	0	1	5	6	8	9
3	0	7	9			
4	5	6				
5	5	7				
6	4	9				

Key: 1|2 means 1.2 hours of sunshine

(a) On two days 0.0 hours of sunshine were recorded.
(b) (i) A monthly mean for June can be estimated by taking the mean of the 25 values in the table. That is, the missing data points can be ignored or removed.
However, if similar insolation data were available from another nearby location, it may be possible to use these values to estimate the insolation on Rathlin Island.
(ii) When calculating the annual mean, removal of 5 days in June is likely to create bias, since June is probably one of the sunniest months of the year. In this case, it is important to estimate the missing values, either by taking an average of the existing 25 values, or by using similar data from a nearby location.

The next example shows a slightly more sophisticated approach when handling a missing item of data.

Worked Example

7. A small company employs 12 people. The table below lists the employees' names, ages and salaries.

Name	Age	Salary
Zoe Salmon	22	£22 000
Padraig O'Leary	33	£31 000
Shena Bradley	40	£35 500
Jenny Williams	50	£41 000
Matt Taylor	23	
Steve McIntyre	35	£33 250
Tadhg Byrne	42	£31 000
Thomas Dampier (Deputy Director)	51	£45 000
Callum Shea	24	£21 500
Duncan Ross	36	£32 000
Saoirse Daly	46	£37 000
Chris Dampier (Director)	52	£45 000

Thomas, the Deputy Director, wishes to calculate the mean salary for the company's employees, but there is one salary missing from the table. He considers removing the missing record and finding the mean of the remaining 11 figures.

(a) Suggest a more sophisticated approach, using an appropriate subset of the workers to estimate the missing salary.
(b) Later, Thomas discovers he had under-estimated Matt's salary by £7920. Find the percentage error in the **total** wage bill.
(c) Would this percentage error be larger or smaller if the company had 120 workers, rather than 12? Show all your working.

(a) Matt Taylor's salary is missing. We know he is 23 years old. It is reasonable to estimate his salary as the mean of the other salaries in his age group. Two other employees are in their early twenties: Zoe Salmon and Callum Shea. The mean of their salaries is:
$$\frac{22000 + 21500}{2} = £21750$$
Using this as an estimate of Matt Taylor's salary, estimate the mean for the 12 employees:
Estimated total = 22000 + 21750 + 21500 + 31000 + 33250 + 32000 + 35500 + 31000 + 37000 + 41000 + 45000 + 45000 = £396 000
$$\text{Estimate of mean salary} = \frac{396000}{12} = £33000$$

(b) Percentage error: $\frac{7920}{396000} \times 100\% = 2\%$

(c) If the company had 120 workers rather than 12, the total wage bill would be roughly 10 times higher, so roughly £3 960 000.
The percentage error is then:
$$\frac{7920}{3960000} \times 100\% = 0.2\%$$
The percentage error is smaller.

Exercise 12C

1. A scientist measures the lengths of a sample of 50 baby turtles on a beach in Queensland, Australia. One value is missing from the dataset, so the box plot below shows the results for 49 measurements. All values are in millimetres.

   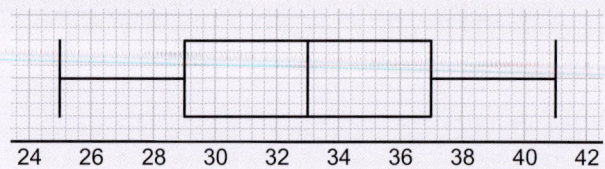

   It is known that the range for the entire sample is 18 mm.
   (a) One possible value for the missing length is 43 mm. Find the other possible value.
   (b) Assume the missing length is 43 mm and this is added to the dataset. Explain what effect this has on:
      (i) the median value,
      (ii) the standard deviation.

2. The stem and leaf diagram below shows the ages of people entering a polling station to vote in an election between 7pm and 8pm.

1	8 9
2	0 1 3 3 6 9
3	0 0 2 2 3 5 7 7 9
4	1 3 4 5 5 7
5	2 7
6	1 8
7	2

   Key: 1|8 means a voter age of 18 years

   (a) Thirty people were included in the survey, but two of the ages have been lost. The range of the 30 ages is known to be 60. The median age of the voters is known to be 36.5. Clean the dataset by finding the two missing ages.
   (b) Find the mean age using the cleaned dataset.

3. Seth's dad measures his height regularly. The table shows his height between the ages of 320 weeks and 410 weeks (roughly the ages of 6 and 8 years old).

Exercise 12C...

Date	Weeks	Height (cm)
15/07/2019	321	120
09/08/2019	325	121
06/09/2019	329	122
24/10/2019	336	123
07/12/2019	342	124
10/01/2020	347	125
09/03/2020	355	126
06/05/2020	364	127
23/07/2020	375	129
17/09/2020	383	130.5
12/12/2020	395	
12/03/2021	408	134
06/05/2021	416	135
02/07/2021	424	136

Seth's dad knows he measured Seth's height on 12 December 2020, when he was 395 weeks old, but he has lost the measurement.

(a) For the purposes of plotting a graph of Seth's height against his age, suggest a possible method for handling the missing data point.

(b) After plotting the graph, Seth's dad then finds that the equation of the linear regression line is:
$y = 0.153x + 71.5$
where y is Seth's height and x is his age in weeks. Use this equation to estimate the missing height value.

4. A survey of 10 basketball players is taken. The number of points scored by each player during one season is recorded. However, one of the player's scores has been lost. The results of the remaining 9 players are summarised as follows:

$\sum x = 405; \sum x^2 = 29312; n = 9$

The standard deviation for the scores of all 10 players is required. Two different strategies that could be used to clean the data and estimate the standard deviation are given below:

(1) The missing data point could be ignored and the standard deviation calculated for the remaining 9. This would then be used as an estimate of the standard deviation for the 10 players.

Exercise 12C...

(2) The tenth player's score could be estimated as the mean of the other 9. Then the standard deviation could be calculated using 10 data points.

(a) Which of the strategies (1) or (2) above would result in a lower standard deviation? Explain your answer.

(b) Estimate the standard deviation using strategy (1).

5. An estate agent is asked to calculate the average house price in a small town and how this has changed from April 2020 to April 2021. She uses the sale price of all the houses sold in the town during these two months. In April 2020, only 10 houses were sold in the town, but one sale price is missing. The nine available prices are shown below.

April 2020 – prices of houses sold (thousands of pounds)
160 182 160 189 230 140 209 412 280

In April 2021, 100 houses were sold in the town. Again, the sale price of one of the houses is missing. The sale prices for the remaining 99 houses are summarised in the frequency table below.

April 2021 – prices of houses sold

House sale price (thousands of pounds)	Frequency
100 –	6
150 –	38
200 –	27
250 –	13
300 –	9
350 –	5
400 – 450	1

(a) Find the mean of the April 2020 sale prices using the 9 values available.

(b) Estimate, using the frequency table, the mean of the April 2021 sale prices using the 99 values available.

The estate agent knows that all sale prices were between £100 000 and £450 000 during both these months.

Exercise 12C...

(c) What is the largest possible error in the mean value calculated for April 2020?
(d) What is the largest possible error in the mean value estimated for April 2021?
(e) Does the estate agent conclude that house prices rose between April 2020 and April 2021? Discuss briefly how much confidence she should have in her figures.

6. Hourly air temperature readings are collected from a buoy in the Irish Sea. The readings for 24 hours in July are summarised in the frequency table below.

Recorded temperature, T (°C)	Frequency, f
$0 \leq T < 5$	4
$5 \leq T < 10$	4
$10 \leq T < 15$	5
$15 \leq T < 20$	3
$20 \leq T < 25$	1
$95 \leq T < 100$	7

(a) Explain how it is clear from the frequency table that there is a problem with the readings from the buoy.

Having spotted that there is a problem, Pedro looks at the raw data. These are the readings:

Time	0000	0100	0200	0300	0400	0500	0600	0700
T (°C)	3.1	2.6	2.5	3.2	5.0	5.8	7.6	8.8

Time	0800	0900	1000	1100	1200	1300	1400	1500
T (°C)	10.3	11.5	12.0	13.1	14.2	16.4	19.6	20.3

Time	1600	1700	1800	1900	2000	2100	2200	2300
T (°C)	17.5	99.99	99.99	99.99	99.99	99.99	99.99	99.99

(b) Describe the problem.
(c) Suggest a way in which the data could be cleaned.

7. Nathan is studying for his GCSE exams and takes 10 subjects. His form teacher is preparing his school report. It includes a score out of 20 for each of his 10 subjects. However, the form teacher has lost Nathan's maths score. She must give a mean score for his GCSE subjects, but she only has 9 scores out of 10. Since there is one missing value, the dataset must be cleaned.

For her first attempt, she decides to omit the maths score and calculate the mean based on the other 9 subjects. She finds that, when cleaning the data in this way, the mean score is 16 and the variance is 4.

(a) The teacher then remembers that she could estimate Nathan's maths score based on his class tests. Using this method for cleaning the data, she estimates his maths score to be 11. Find the mean value for all 10 subjects that she writes on the report card. Find also the variance for all 10 subjects.
(b) After the report cards have been sent home, the teacher discovers that Nathan's true score in maths was 18. What were the true mean and variance of Nathan's ten scores?

8. A company has 50 employees. Michael White in the human resources department tracks the mean salary for the company's employees. According to his figures, the mean salary for these 50 employees is £25 000.
(a) Calculate the total annual wage bill for the company.
(b) A new employee, Faye Perks, starts work in the IT department. Michael must file a report stating the company's total wage bill. He has not yet been told Faye's salary, so this is missing data. He knows that the employees in the IT department usually earn between £22 000 and £35 000. Explain why removing the missing data would not be the best strategy for cleaning the data.
(c) Suggest a suitable alternative strategy and use it to estimate Faye's salary.
(d) Later Michael learns that he over-estimated Faye's salary by £2545. This caused the figure in his report to be 0.2% higher than the true figure for the total wage bill. Find Faye's salary.
(e) If the company had 500 employees rather than 50, would the percentage error in the total wage bill be higher or lower than 0.2%. Explain your answer.

12.4 Summary

An outlier is an extreme value that lies outside the overall pattern of the data.

You may have to identify outliers in data sets and statistical diagrams.

There are a number of different methods to identify outliers and a suitable method will be given in an exam question.

Data cleaning may be necessary when a dataset includes erroneous values or missing values.

There are various ways to clean data. The preferred method depends upon the type of data in question and on the type of analysis that is planned.

It may be possible to safely remove the offending data from the dataset.

Alternatively, it may be more appropriate to replace the data with average figures, such as an appropriate mean. It may be possible to use a mean value obtained from an appropriate subset of the data, or using data from a similar survey.

Chapter 13
Probability

13.1 Introduction

Key words
- **Random variable**: A variable whose value is dependent on a random phenomenon.
- **Mutually exclusive**: Two events are mutually exclusive if they cannot both occur at the same time.
- **Exhaustive events**: Two or more events are exhaustive if at least one of the events must occur.
- **Statistical dependence** and **independence**: Two events are independent if the occurrence of one does not affect the probability of of the other event occurring. Otherwise they are dependent.
- **Venn diagram**: A diagram in which circles are used to represent the occurrence of events and overlapping circles represent two or more events occurring.
- **Tree diagram**: A diagram representing a sequence of events, with each branch representing the occurrence of an event.

Before you start
You should have come across probability at GCSE and know about:
- Calculating probabilities involving selection of items, both with and without replacement.
- Probability tree diagrams.
- Listing possible outcomes.

What you will learn
In this chapter you will:
- Learn how to use the addition and multiplication laws.
- Gain understanding of the following concepts:
  – mutually exclusive events;
  – exhaustive events;
  – statistical dependence and independence.
- Learn how to calculate combined probabilities of up to three events, using tree diagrams, Venn diagrams and two-way tables.

In the real world...
The National Lottery is a game of chance. If you buy one ticket the probability of winning the jackpot is infinitesimally small. But when millions of people play, the chances of somebody winning are much greater. It could be anyone!

Exercise 13A (Revision)

1. A toolbox contains three washers, four nuts and two screws. An item is chosen at random from the toolbox. Write down the probability that the item is:
   (a) a screw;   (b) a nut;
   (c) not a washer;   (d) a nail.

2. Three members of the same class are chosen at random to read in assembly. The class comprises 15 boys (B) and 15 girls (G). List all the possible outcomes. The first one has been done for you:
   - GGG
   - ...

3. Caz rolls a die. She keeps rolling until she gets a 6. Using a probability tree diagram if required, find the probability that Caz rolls the die:
   (a) exactly 3 times;
   (b) fewer than 3 times;
   (c) more than 3 times.

13.2 Definitions and Notation

An **experiment** is a repeatable process that could give rise to a number of different **outcomes**. Rolling a regular die is an example of an experiment, with the possible outcomes being 1, 2, 3, 4, 5 and 6.

An **event** is a collection of one or more outcomes. For example, if a die and a coin are thrown together, getting a 6 and a head is one possible event.

A **sample space** is the set of all possible outcomes.

Where outcomes are equally likely, the probability of an event is the number of outcomes in the event divided by the total number of possible outcomes. For example, when rolling a die, the probability of getting a number that is a factor of 30 is $5/6$, because there are 5 outcomes in the event (1, 2, 3, 5 and 6 are all factors of 30) and 6 possible outcomes altogether.

Probability notation

In this chapter and the next one, you will come across the following notation:

$P(...)$

This means the probability of ... happening.

For example, when a fair 6-sided die is thrown, $P(6) = \dfrac{1}{6}$

Random variables

A random variable is a variable whose values depend on the outcomes of a random phenomenon.

In this chapter and the next one, you will come across the following notation:

$P(X = x)$

This means the probability that the random variable X takes a particular value.

In the example above, relating to the throw of a single 6-sided die, if X is the random variable "the outcome of the throw", then $P(X = 6) = \dfrac{1}{6}$

Worked Example

1. Give some examples of random variables in the following situations.
   (a) Padraig throws a die 10 times.
   (b) Lucy stands beneath a tree in Autumn while the leaves are coming down, with her hands spread out.

   (a) X could be the random variable "The number of sixes Padraig gets".
   (b) Y could be the random variable "The number of leaves Lucy catches in one minute".
   There are many other possible answers.

13.3 Calculating Probabilities

When an experiment has two parts (for example, if it comprises throwing two dice, or a die and a coin), a sample space diagram can help you to visualise all the possible outcomes.

Providing the outcomes are all equally likely, you can count the outcomes in an event to find the probability of that event.

Worked Examples

2. Two coins are tossed.
   (a) Draw a sample space diagram to show the possible outcomes.
   (b) Find the probability of getting at least one tail.

(a)
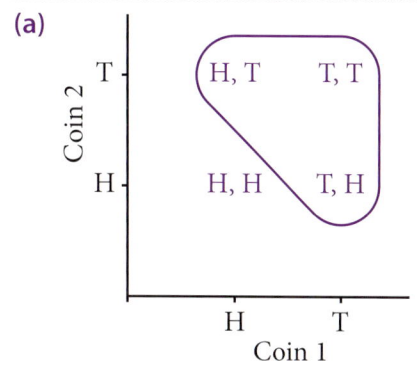

(b) There are three outcomes in the event "at least one tail", as shown in the diagram:

∴ $P(\text{at least one tail}) = \dfrac{3}{4}$

3. A fair spinner has four sections numbered 1 to 4. Two of these spinners are spun together and the sum of the two numbers is recorded. Find the probability that the sum is:
   (a) exactly 6;
   (b) more than 6.

The sample space diagram below shows all possible outcomes. There are 16 possible outcomes, each of them equally likely, since both spinners are fair.

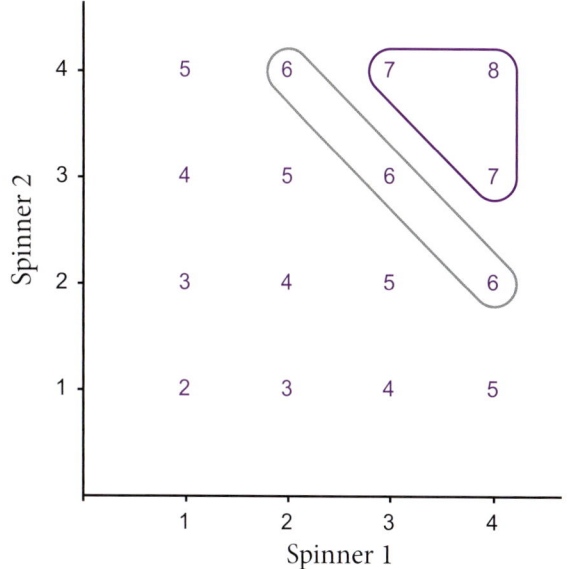

(a) There are 3 outcomes resulting in a sum of 6, shown inside the grey loop on the diagram:

$\therefore P(6) = \dfrac{3}{16}$

(b) There are also 3 outcomes resulting in a sum greater than 6, shown inside the purple loop:

$\therefore P(\text{more than } 6) = \dfrac{3}{16}$

---

4. Dan does a crossword in his newspaper every day for 30 days. The frequency table below shows his times to complete the crossword.

Time t (minutes)	Frequency
$5 \leq t < 7$	4
$7 \leq t < 9$	11
$9 \leq t < 11$	10
$11 \leq t < 13$	3
$13 \leq t < 15$	2

Find the probability that, on a day chosen at random, Dan completed the crossword in:
(a) under 7 minutes;
(b) over 10 minutes.

(a) There are four outcomes in this event, and a total of 30 possible outcomes:

$P(< 7 \text{ minutes}) = \dfrac{4}{30} = \dfrac{2}{15}$

(b) There are 10 times in the third group. The time 10 minutes lies exactly halfway through the group. Therefore, assume that 5 times were less than 10 minutes and 5 were greater than 10 minutes. Then the total number of times above 10 minutes is approximately $5 + 3 + 2 = 10$. So:

$P(> 10 \text{ minutes}) \approx \dfrac{10}{30} = \dfrac{1}{3}$

> **Note:** An approximately equals sign has been used here because of the assumption made.

Exercise 13B

1. Two coins are tossed.
   (a) Draw a sample space diagram to show the possible outcomes.
   (b) Using your sample space diagram, calculate the probability of getting one head and one tail.

Exercise 13B...

2. Two fair six-sided dice are thrown. The **product** X of the two numbers is recorded.
   (a) Draw a sample space diagram to show all the possible outcomes.
   (b) Using your sample space diagram, find
      (i) $P(X = 18)$
      (ii) $P(X > 18)$
      (iii) $P(8 \leq X \leq 9)$

3. The heights of 100 sunflowers in a field are measured and the results are shown in the frequency table.

Height h (cm)	Frequency
$70 \leq h < 80$	10
$80 \leq h < 90$	19
$90 \leq h < 100$	44
$100 \leq h < 110$	21
$110 \leq h < 120$	6

One sunflower is chosen at random.
(a) Find the probability that this sunflower is over 1 metre in height.
(b) Find the probability that the sunflower is between 80 cm and 110 cm in height.
(c) The probability that a sunflower is smaller than 85 cm is 0.2. Find the probability that the sunflower chosen at random is between 85 cm and 100 cm in height.

4. Youssef has two sets of cards: Set A and Set B. He picks one card from set A and one card from set B. He calculates the **product** Y of the two numbers.

Use the following clues to find the value of x.
• $P(Y \leq 40) = 1$
• $P(Y < 15) = P(Y \text{ is odd})$

Exercise 13B...

5. The histogram shows the distribution of masses, in tonnes, of 60 family cars.

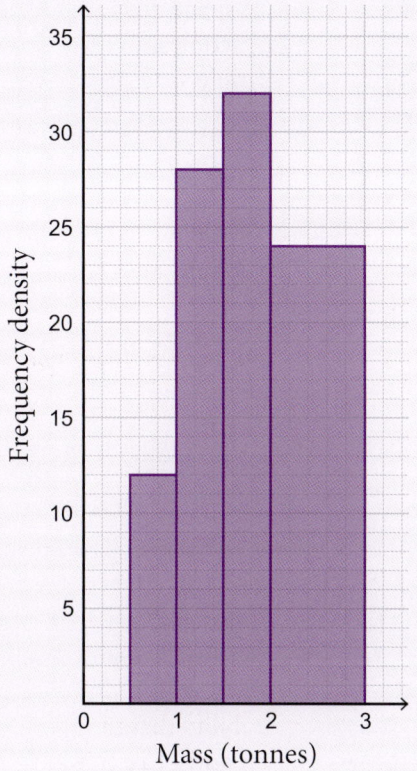

One car is chosen at random.
(a) Find the probability the car has a mass of more than 2.5 tonnes.
(b) Estimate the probability the car has a mass less than 1.25 tonnes.

13.4 Venn Diagrams

A Venn diagram can be used to represent events graphically. The regions of a Venn diagram can be filled with frequencies or with probabilities.

A rectangle represents the sample space. Inside it are circles (or ovals, etc) that represent events.

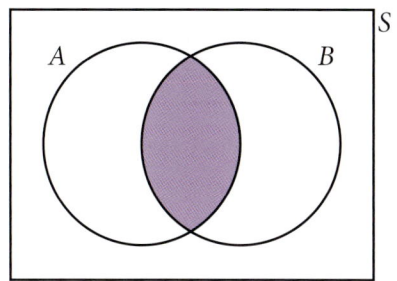

The shaded region above shows the event A **and** B.

This event is also called the **intersection** of A and B.
It represents the event that both A and B occur.
It is denoted $A \cap B$.

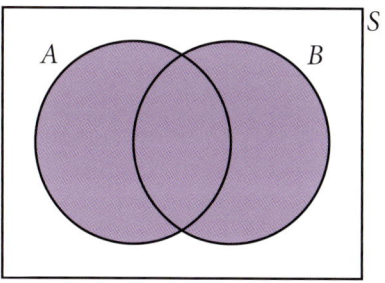

The shaded region above shows the event A **or** B.
This event is also called the **union** of A and B.
It represents the event that either A or B (or both) occur.
It is denoted $A \cup B$.

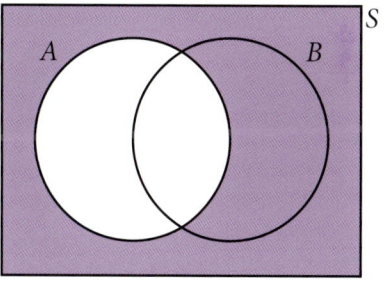

The shaded region above shows the event **not** A.
It is also called the **complement** of A.
It represents the event that A does not occur.
It is denoted A' or $\overline{A}$.

Worked Examples

5. Twenty school children were asked about their packed lunches.

   Twelve of them said they had brought a piece of fruit.
   Eleven said they had a packet of crisps.
   Five said both.

   (a) Draw a Venn diagram to represent this information.

   A child is chosen at random.
   (b) Use the Venn diagram to find the probability that:
   (i) The child's packed lunch contains a piece of fruit.
   (ii) The child's packed lunch contains neither a piece of fruit, nor a packet of crisps.

AS 2: APPLIED MATHEMATICS

(a) Begin by putting the number bringing both fruit and crisps (5) in the intersection.
- If 12 brought fruit, there must be 7 in the 'fruit only' section of the diagram.
- Likewise there must be 6 in the 'crisps only' section.
- 5 + 7 + 6 = 18, so there must be 2 bringing neither. This number lies inside the sample space, but outside of the circles.

We can draw the Venn diagram:

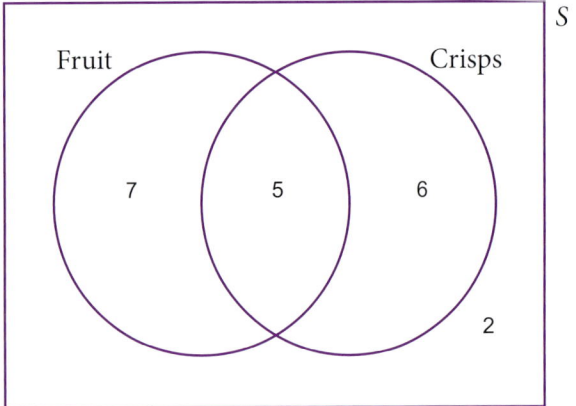

(b) (i) $P(\text{Fruit}) = \dfrac{12}{20} = \dfrac{3}{5}$

(iii) $P(\text{No fruit or crisps}) = \dfrac{2}{20} = \dfrac{1}{10}$

6. In a survey of A-Level students, 64 are taking Maths, 94 Chemistry and 58 Physics. In addition, 28 take both Maths and Physics, 26 take Maths and Chemistry and 22 take Chemistry and Physics. 14 students take all three subjects. Find how many take only one of these subjects.

This is an example of a Venn diagram with three circles.
- Begin by placing 14 at the centre.
- Since the intersection between Maths and Chemistry is 26 in total, there are 26 − 14 = 12 taking only Maths and Chemistry.
- Likewise calculate that 14 take Maths and Physics only; 8 take Chemistry and Physics only.

We can mark these on the Venn diagram:

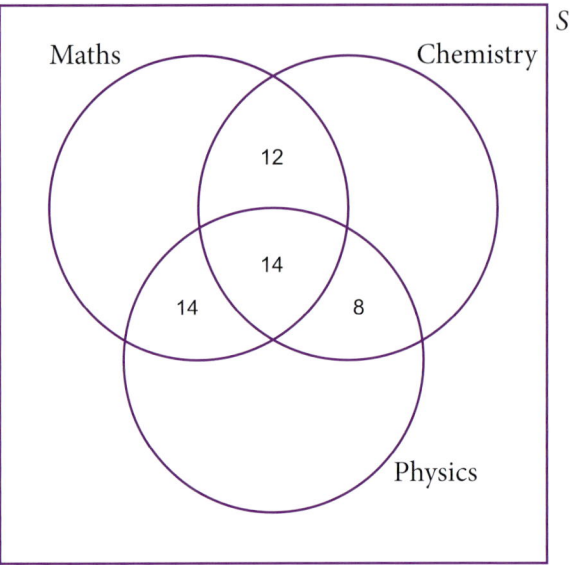

- Finally, those taking only Maths can be calculated: 64 − (12 + 14 + 14) = 24
- Those taking only Chemistry: 94 − (12 + 14 + 8) = 60
- Those taking only Physics: 58 − (14 + 14 + 8) = 22

We can complete the Venn diagram:

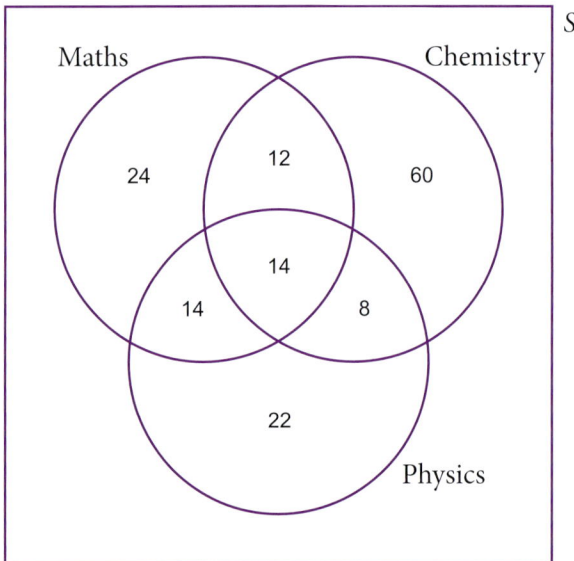

The number of students only taking one subject is: 24 + 60 + 22 = 106

....

The previous two examples used Venn diagrams with frequencies. The next example uses probabilities.

....

Worked Examples

7. In a school canteen, the probability of a teacher choosing sausages is 0.6. In the same school canteen,

the probability of a teacher choosing chips is 0.8. There is a 0.15 probability of a teacher choosing neither sausages or chips. A teacher is chosen at random. With the help of a Venn diagram, find:
(a) The probability that the teacher chooses both sausages and chips.
(b) The probability that the teacher chooses sausages or chips (or both).
(c) The probability that the teacher does not choose sausages.
(d) Given that there are 60 teachers in the canteen, how many choose neither sausages or chips?

Adding the three probabilities given:
$0.6 + 0.8 + 0.15 = 1.55$

The probabilities in the sample space must always add up to 1. The excess of 0.55 must appear in the intersection:

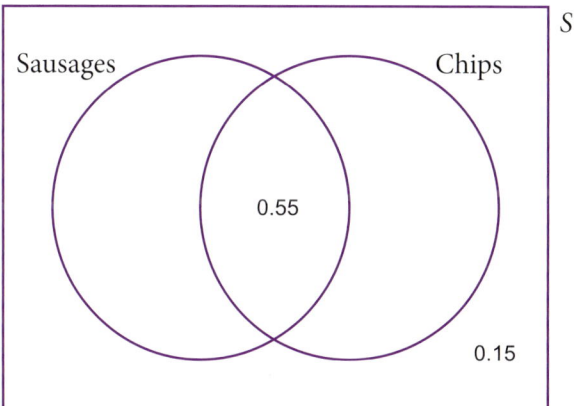

Next find the probability in the 'sausages only' region of the diagram. The two probabilities in the sausages circle must add up to 0.6, so the missing probability is 0.05.

Likewise, the two probabilities in the chips circle add up to 0.8, so the probability in the chips only section is 0.25.

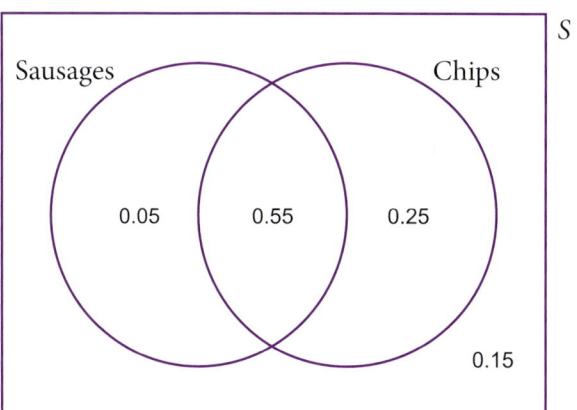

(a) P(Sausages and chips) = 0.55
(b) P(Sausages or chips) = 0.05 + 0.55 + 0.25 = 0.85
(c) P(Not sausages) = 0.25 + 0.15 = 0.4
(d) P(Neither) = 0.15
Number of teachers choosing neither
= 0.15 × 60 = 9

8. A and B are two events. $P(A) = 0.5$ and $P(B) = 0.2$ and $P(A \cap B) = 0.1$
Find:
(a) $P(A \cup B)$ (b) $P(B')$
(c) $P(A \cap B')$ (d) $P(A \cup B')$

Although there are no instructions to draw a Venn diagram, it is useful here.

There are two events A and B. We know the circles intersect because we are told $P(A \cap B) = 0.1$.

We can also work out:
- the number in the 'A only' section: $0.5 - 0.1 = 0.4$;
- the number in the 'B only' section: $0.2 - 0.1 = 0.1$;
- the number outside the circles, since all the probabilities must add up to 1: $1 - (0.4 + 0.1 + 0.1) = 0.4$

We can draw the Venn diagram:

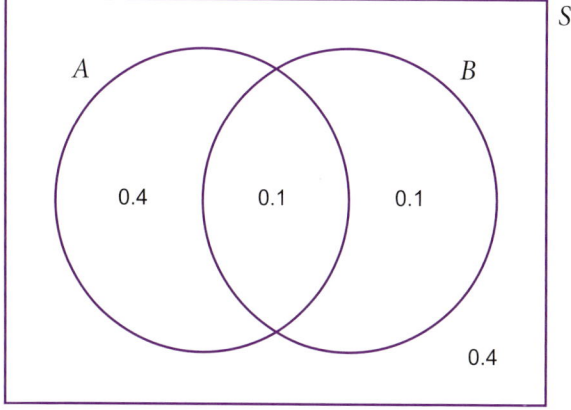

(a) $P(A \cup B)$ is the probability of events A or B (or both) taking place. It can be found by adding the probabilities inside both circles: $P(A \cup B) = 0.4 + 0.1 + 0.1 = 0.6$
(b) $P(B')$ is the probability of B not taking place. It can be found by adding all the probabilities outside the B circle: $P(B') = 0.4 + 0.4 = 0.8$
(c) $P(A \cap B')$ is the probability of A **and** not B. It is the region that is outside the B circle, but also inside the A circle: $P(A \cap B') = 0.4$

(d) $P(A \cup B')$ is the probability of A **or** not B. On the diagram, combine the A circle with everything outside the B circle. The only region not included is 'B only'.
$P(A \cup B') = 0.4 + 0.1 + 0.4 = 0.9$

Exercise 13C

1. A man places two bets. He estimates that the probability of winning the first bet is 0.4, the probability of winning the second is 0.5 and the probability of winning both is 0.1.
   (a) Draw a Venn diagram, showing the events 'Wins bet 1' and 'Wins bet 2'.
   (b) Find the probability that he does not win either bet.
   (c) Find the probability that he wins exactly one bet.

2. There are 100 passengers on a flight arriving at Belfast International Airport.
   65 of the passengers have been spending time in England.
   20 were on holiday in France.
   15 were on holiday in Germany.
   10 of the passengers spent some time in England and some in France.
   5 of the passengers spent some time in England and some in Germany.
   2 of the passengers spent time in France and Germany.
   1 of the passengers has been visiting all three countries.
   (a) Draw a Venn diagram to represent this information.
   A passenger is chosen at random.
   (b) Find the probability this passenger has been in Germany.
   (c) Find the probability the passenger did **not** spend any time in England, France or Germany.

3. Forty teenagers are at a youth club. One teenager is chosen at random. The Venn diagram shows the probability he or she plays table tennis, pool, table football or some combination of these.

Exercise 13C...

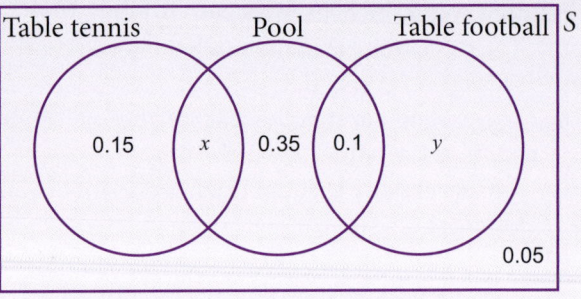

The probability the teenager plays table football is ¼.
(a) Find y.
(b) Find x.
(c) How many teenagers play table tennis?
(d) How many teenagers play table tennis and table football?

4. On a recipe website there are 100 recipes. The Venn diagram shows the number of recipes that have cheese, eggs and chocolate as ingredients.

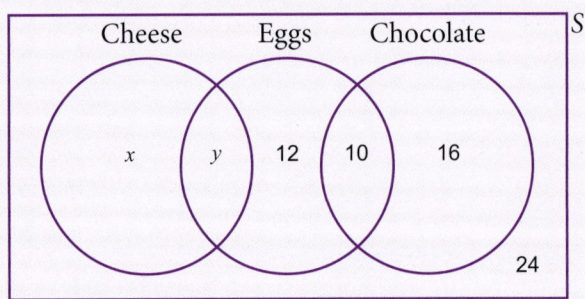

One recipe is picked at random.
(a) Given that $P(\text{cheese}) = P(\text{eggs})$, find the values of x and y.
(b) How many recipes involve both cheese and chocolate?

5. C and D are two events. Given that $P(D) = 0.4$, $P(C \cap D) = 0.2$ and $P(C \cup D) = 0.9$, find:
   (a) $P(C' \cap D)$ (b) $P(C \cap D')$
   (c) $P(C)$ (d) $P(C' \cap D')$

6. Fifty countries of the world are chosen at random. Their flags are printed out. One of the 50 flags is chosen at random. The Venn diagram shows the three events: the flag includes the colours red, white and green.

Exercise 13C...

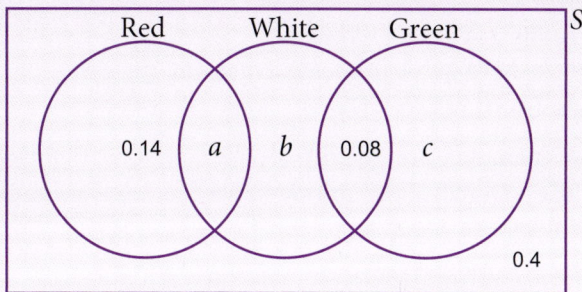

(a) Given that $P(\text{White}) = 2P(\text{Red})$ and that $P(\text{Not green}) = 0.82$, find the three probabilities a, b and c.
(b) How many of these 50 flags feature green, but not red or white.
(c) The flag of Malawi is red, black and green. Was Malawi one of the 50 countries chosen? Explain your answer.

Two-way tables

More complex information is sometimes displayed in a two-way table.

Worked Example

9. There are 90 students on a summer camp. On the last night, they have a choice of activity: attending a concert, going to a disco or going on a late night walk.
42 students attend the concert, two-thirds of them being girls.
11 girls go to the disco.
In total 25 people go on the walk, two fifths of them being boys.
(a) What percentage of the students on the camp are boys?
(b) A boy is chosen at random. What is the probability he chooses to go to the disco?

This type of question is best answered using a two-way table, as shown below.

	Boys	Girls	Total
Concert	14	28	42
Disco	12	11	23
Walk	10	15	25
Total	36	54	90

The numbers in bold have been calculated.

To calculate the total number at the disco, for example, note that the three numbers in the Total column must add up to 90, so the disco total is 23.

Using this, and the fact that 11 girls attend the disco, it is possible to work out the number of boys attending the disco, 12.

The remaining numbers in bold are calculated in a similar way.

(a) There are 36 boys out of 90 students.
This is $\frac{36}{90} = \frac{2}{5} = 40\%$ of the students.

(b) 12 out of 36 boys go to the disco. So:
$P(\text{disco}) = \frac{12}{36} = \frac{1}{3}$

Exercise 13D

1. There are 200 passengers on a ferry. Some of them came in cars; others are foot passengers. Some passengers choose to buy a meal in the ferry's restaurant. The two-way table below summarises the information.

	Meal	No meal	Total
Car passengers			114
Foot passengers		56	
Total			200

(a) Of the 114 car passengers, two-thirds choose to have a meal in the restaurant. Copy and complete the table.
(b) A passenger is chosen at random. What is the probability he or she takes a meal in the restaurant?

2. Here is a table of the planets in the Solar System and the number of moons orbiting each one.

Planet	Number of Moons
Mercury	0
Venus	0
Earth	1
Mars	2
Jupiter	79
Saturn	82
Uranus	27
Neptune	14

AS 2: APPLIED MATHEMATICS

Exercise 13D...

The four planets closest to the Sun (Mercury, Venus, Earth and Mars) are small, rocky planets. The four planets furthest from the Sun (Jupiter, Saturn, Uranus and Neptune) are gas giants.

(a) Copy and complete the following two-way table, giving the **number** of planets in each cell.

	Rocky planets	Gas giants	Total
Has at least one moon	2		
Has no moons			
Total			8

(b) One of the planets in the solar system is chosen at random. Using your two-way table, find the probability it is:
  (i) A rocky planet with at least one moon.
  (ii) A gas giant with no moons.
(c) One of the planets with moons is chosen at random. Find the probability it is a gas giant.

3. The entire year group of 120 Year 13 students in a school in Armagh is surveyed about their lunch on school days.

In total 50 students said they have a school lunch, of which 24 were girls.
Two fifths of the girls have a school lunch.
One boy and no girls said they go home.

(a) Copy and complete the table below.

	Girls	Boys	Total
School lunch			
Packed lunch			
Home for lunch			
Total			

(b) One of the students taking a school lunch is chosen at random. What is the probability this student is a girl?
(c) One of the boys is chosen at random. What is the probability this student has a packed lunch?
(d) A Year 13 student is chosen at random. Find the probability this student is a girl who takes a packed lunch.

13.5 Mutually Exclusive, Exhaustive and Independent Events

Mutually exclusive events

If events have no outcomes in common, they are **mutually exclusive**.

On a Venn diagram, the two circles do not overlap. You can use a simple addition rule to work out the probability of both events occurring:

For mutually exclusive events, $P(A \cup B) = P(A) + P(B)$

Independent events

When one event has no effect on another, they are **independent** events.

Therefore, if A and B are independent, the probability of A happening is the same, whether or not B happens.

For independent events, $P(A \cap B) = P(A) \times P(B)$

This rule is often used to determine whether two events are independent.

Exhaustive events

For a set of exhaustive events, at least one of the events must occur. For example, when rolling a single die, the event 'getting an even number' and the event 'getting an odd number' are exhaustive, since one of them has to occur. On a Venn diagram, exhaustive events would fill the entire sample space.

The addition law

For two mutually exclusive events, A and B, there is no intersection of the two circles on a Venn diagram. In this case $P(A \cup B) = P(A) + P(B)$.

In cases where A and B are **not** mutually exclusive, there is an intersection between the circles, as shown in the following Venn diagram.

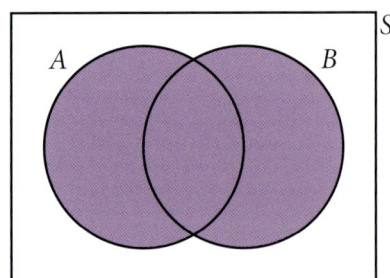

For events A and B that are not mutually exclusive:

$$P(A \cup B) = P(A) + P(B) - P(A \cap B)$$

This is sometimes known as the **addition law**.

The multiplication law

For independent events, it has been established that

$P(A \cap B) = P(A) \times P(B)$

Alternatively:

$P(A \text{ and } B) = P(A) \times P(B)$

This is a simplified version of the **multiplication law**. It is used extensively in the next section on tree diagrams.

If you later work towards A2 Mathematics, you will extend your understanding of the multiplication law and use it in more advanced ways.

Worked Examples

10. Events A and B are mutually exclusive.
$P(A) = 0.3$ and $P(B) = 0.5$
Find:
(a) $P(A \text{ or } B)$
(b) $P(A \text{ but not } B)$
(c) $P(\text{neither } A \text{ nor } B)$

Note: These could be written as:
(a) $P(A \cup B)$ (b) $P(A \cap B')$ (c) $P(A' \cap B')$

A Venn diagram can help you to answer this question. Since A and B are mutually exclusive, the two circles do not intersect:

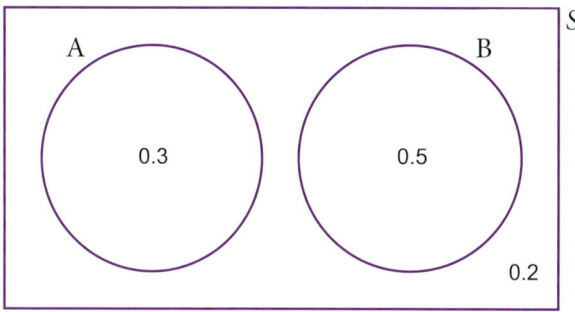

(a) Using the addition rule for mutually exclusive events:
$P(A \text{ or } B) = P(A) + P(B)$
$= 0.3 + 0.5 = 0.8$
(b) Everything in circle A but not in circle B:
$P(A \text{ but not } B) = 0.3$
(c) Everything outside both circles:
$P(\text{neither } A \text{ nor } B) = 0.2$

11. Events X and Y are independent.
$P(X) = \dfrac{1}{2}$ and $P(Y) = 0.2$
Find $P(X \cap Y)$.

$P(X \cap Y)$ means $P(X \text{ and } Y)$

Since events X and Y are independent,
$P(X \cap Y) = P(X) \times P(Y)$
$\therefore P(X \cap Y) = \dfrac{1}{2} \times 0.2 = 0.1$

12. A group of 21 students was asked whether they had played computer games P, Q and R. The results are shown in the Venn diagram.

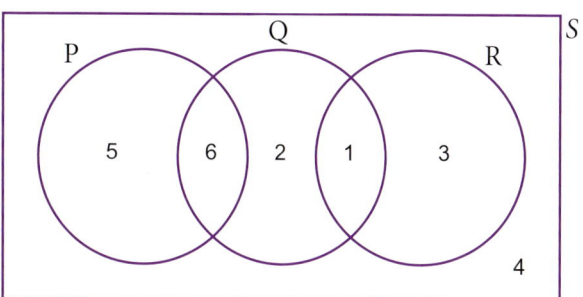

A student is chosen at random.
(a) Find the probability the student has played Q or R, or both.
(b) Determine whether the events 'playing game P' and 'playing game Q' are statistically independent.
(c) State, giving a reason, whether the events 'playing game P', 'playing game Q' and 'playing game R' are exhaustive.

(a) The number of students who have played Q, R or both is $6 + 2 + 1 + 3 = 12$

$P(Q \cup R) = \dfrac{12}{21} = \dfrac{4}{7}$

(b) The number of students who have played P is:
$5 + 6 = 11$
The number of students who have played Q is $6 + 2 + 1 = 9$

$\therefore P(P) = \dfrac{11}{21}$ and $P(Q) = \dfrac{9}{21} = \dfrac{3}{7}$

$P(P) \times P(Q) = \dfrac{11}{21} \times \dfrac{3}{7} = \dfrac{11}{49}$

The number of students who have played both P and Q is 6.

$\therefore P(P \cap Q) = \dfrac{6}{21} = \dfrac{2}{7}$

$P(P) \times P(Q) \neq P(P \cap Q)$

Therefore the two events playing game P and playing game Q are not independent.

(c) The three events are not exhaustive. The Venn diagram shows that four people have never played any of the games.

Exercise 13E

1. $P(F) = 0.6$ and $P(G) = 0.8$. Given that F and G are independent events, find $P(F \cap G)$.

2. $P(Q) = 0.13$ and $P(Q \cap R) = 0.052$. Given that Q and R are independent events, find $P(R)$.

3. The events A and B are independent such that: $P(A) = 0.13$ and $P(B) = 0.26$. Find:
   (a) $P(A \cap B)$
       Hint: Use the multiplication law for independent events for this.
   (b) $P(A \cup B)$
       Hint: Use the addition law for this.

4. Events A and B are mutually exclusive. $P(A) = 0.1$ and $P(B) = 0.8$.
   (a) Draw a Venn diagram to show this information.
   (b) Find $P(A \cup B)$
       Hint: This means $P(A$ or $B)$.
   (c) Find $P(A' \cap B')$
       Hint: This means $P($neither A nor $B)$.

5. Look at the two-way table below. Of 100 pairs of sunglasses on sale in a shop, it shows the number of pairs offering UV protection and polarisation.

	UV protection	No UV protection	Total
Polarisation	16	14	
No polarisation		28	
Total			

   (a) Copy and complete the table.
   (b) A pair of sunglasses is chosen at random. If P is the event that this pair of sunglasses is polarising and U is the event that the sunglasses offer UV protection, determine whether:
       (i) P and U are mutually exclusive events.
       (ii) P and U are independent events.

6. Declan tries his luck at two stalls on a fairground. Look at the Venn diagram that follows, which shows the probability of him winning each game.
   By calculation, determine whether winning at the coconut shy and winning at the whack-a-mole are independent events.

Exercise 13E...

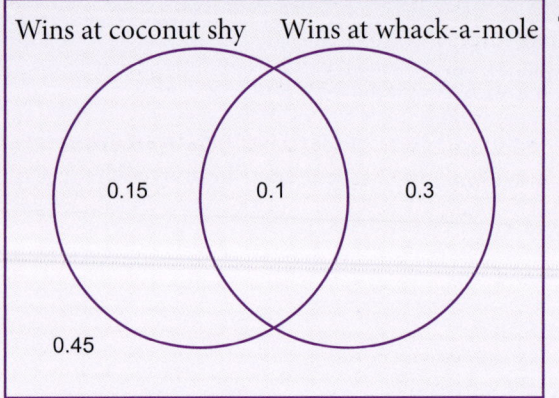

7. In Sandy's burger restaurant, the probability that the waiter smiles at you is 0.05. The probability he or she speaks to you is 0.08. Only one in a hundred customers experience both these things. Event A is 'the waiter smiles'. Event B is 'the waiter speaks'.
   (a) With the help of a Venn diagram, find the probability the waiter smiles or speaks, or does both.
   (b) Giving a reason or appropriate calculations in each case, determine whether:
       (i) Events A and B are mutually exclusive.
       (ii) Events A and B are independent.

8. Is it possible for two events to be both:
   (a) Mutually exclusive and independent?
   (b) Exhaustive and mutually exclusive?
   Explain your answers.

9. Two fair, six-sided dice are rolled. Event A is 'the two dice land on the same number'. Event B is 'the sum of the numbers is 6'. Explain why events A and B are not mutually exclusive.

10. Mr and Mrs Johnston have three children: Freddy, Grace and Harriet. The events F, G and H represent Freddy, Grace and Harriet waking up in the night, respectively. The Venn diagram shows, over 100 nights, the number of times each of these three events occurred.

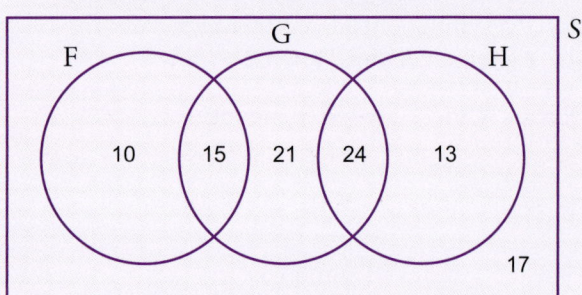

Exercise 13E...

(a) Are the events F and H mutually exclusive? Explain your answer.
(b) Are the events F and G independent? Explain your answer.
(c) Are the events G and H independent? Explain your answer.
(d) Two of the children share a bedroom, while the third child has their own room. Using your answers to parts (b) and (c), which child probably has their own room? Explain your answer.

11. Events A and B, shown in the Venn diagram, are independent.

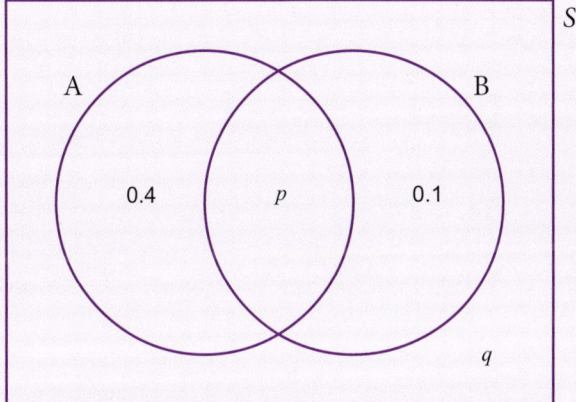

Find the two possible values for p and the two corresponding values of q.

12. A and B are two events such that $P(A) = 0.4$, $P(B) = 0.5$ and $P(A \cup B) = 0.8$.
(a) Find $P(A \cap B)$.
(b) Using your answer to part (a), state whether:
  (i) A and B are mutually exclusive events.
  (ii) A and B are independent events.
Find also:
(c) $P(A')$
(d) $P(A \cup B')$
(e) $P(A' \cup B)$

13.6 Probability Tree Diagrams

A **tree diagram** can be used to show the probability of two or more events happening in succession.

They are often used for selection of two or three items or people from a larger group.

Tree diagrams fall into two categories: **with replacement** and **without replacement**.

Selection with replacement

Worked Example

13. There are 10 marbles in a pot, four of them green and six red. A marble is selected at random, its colour is noted and the marble is replaced in the pot. Two more marbles are selected in this way.
(a) Draw a tree diagram to display this information.
Find the probability that, out of the three marbles chosen,
(b) 2 of the marbles are red;
(c) at least one marble is green;
(d) no marbles are red.

(a) The tree diagram is shown below. Note that the probability of choosing a green ball remains constant at $\frac{4}{10}$ for each selection.

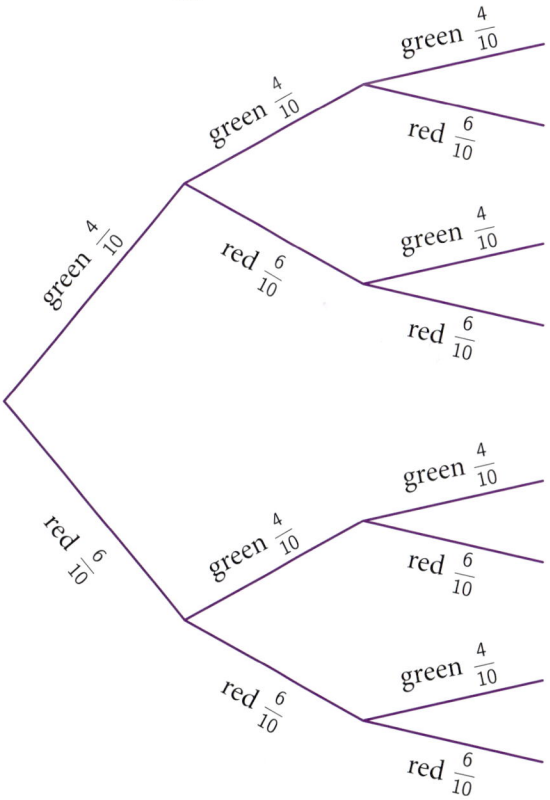

Note: You don't need to simplify the fractions in the tree diagram. The calculations often work out easier with unsimplified fractions. However, always remember to give your final answer in its simplest form.

(b) There are three ways to get two red balls: GRR, RGR and RRG.

Follow the branches of the tree diagram and multiply the probabilities to find the overall probability for this combination:

$$P(\text{GRR}) = \frac{4}{10} \times \frac{6}{10} \times \frac{6}{10} = \frac{144}{1000}$$

$$P(\text{RGR}) = \frac{6}{10} \times \frac{4}{10} \times \frac{6}{10} = \frac{144}{1000}$$

$$P(\text{RRG}) = \frac{6}{10} \times \frac{6}{10} \times \frac{4}{10} = \frac{144}{1000}$$

All three ways to get two red marbles have the same probability.

$$\therefore P(2 \text{ red}) = 3 \times \frac{144}{1000} = \frac{432}{1000} = \frac{54}{125}$$

(c) To find the probability that at least one marble is green it is easiest to calculate the probability of this **not** happening, which involves getting three red marbles:

$$P(\text{RRR}) = \frac{6}{10} \times \frac{6}{10} \times \frac{6}{10} = \frac{216}{1000}$$

So:

$$P(\text{at least one green}) = 1 - \frac{216}{1000} = \frac{784}{1000} = \frac{98}{125}$$

(d) There is only one way to get no red marbles, which is getting three greens:

$$P(\text{GGG}) = \frac{4}{10} \times \frac{4}{10} \times \frac{4}{10} = \frac{64}{1000} = \frac{8}{125}$$

Note: In Chapter 14 on the binomial distribution, you will learn a better way to calculate probabilities in 'with replacement' situations, where the probability of 'success' is constant.

Selection without replacement

The following example is a 'without replacement' problem, in which the probabilities change at each stage.

Note: The binomial distribution cannot be used in this type of problem.

Worked Example

14. A tennis club consists of 6 girls and 9 boys. Three members of the club are chosen at random. The number of boys chosen is denoted by the random variable X. Show that:

(a) $P(X = 3) = \dfrac{4}{91}$

(b) $P(X = 2) = \dfrac{216}{455}$

This can be done using a tree diagram with three stages, since three children are selected. Remember, this situation is **without replacement**, so when a child is chosen from the original 15, they cannot be chosen again.

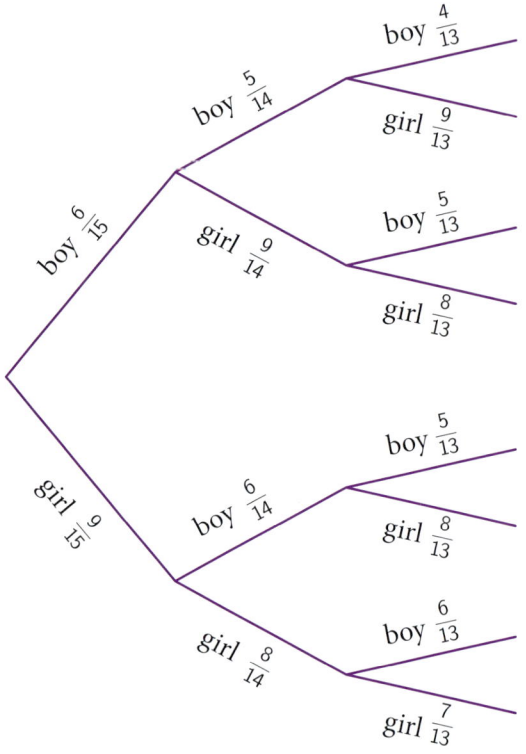

(a) $P(X = 3) = \dfrac{6}{15} \times \dfrac{5}{14} \times \dfrac{4}{13} = \dfrac{4}{91}$

(b) There are three ways to choose 2 boys: BBG, BGB and GBB.

$$P(\text{BBG}) = \frac{6}{15} \times \frac{5}{14} \times \frac{9}{13} = \frac{9}{91}$$

$$P(\text{BGB}) = \frac{6}{15} \times \frac{9}{14} \times \frac{5}{13} = \frac{9}{91}$$

$$P(\text{GBB}) = \frac{9}{15} \times \frac{6}{14} \times \frac{5}{13} = \frac{9}{91}$$

Each way has the same probability. Therefore:

$$P(X = 2) = 3 \times \frac{9}{91} = \frac{27}{91}$$

Note: As you gain confidence answering this type of question, you will no longer need to use a tree diagram.

13: PROBABILITY

Dependent events

The previous example demonstrates dependent events, since the probability of choosing a boy in the second selection is dependent on the outcome of the first selection.

Dependent events also occur in many other settings, as shown in the next example.

Worked Example

14. The probability of it raining when Sean leaves for school is 0.23. If it is raining, the probability of him taking the bus is 0.9. If it is dry, the probability he takes the bus is 0.25. If Sean doesn't take the bus, he cycles. Draw a tree diagram, showing all four possibilities and use it to calculate the probability that, on a day chosen at random, Sean takes the bus.

The tree diagram should be constructed with the weather first and the method of transport second, since the decision about transport is the dependent event:

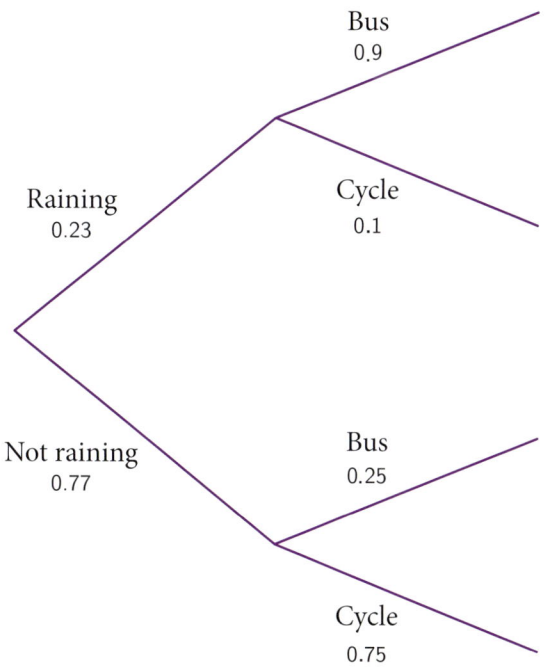

There are two possibilities: raining and bus; not raining and bus.

Work out the probability for each way, multiplying probabilities as you move through the tree diagram. Add the two answers:

$P(\text{bus}) = (0.23 \times 0.9) + (0.77 \times 0.25)$
$= 0.3995$

Exercise 13F

1. From a group of 7 boys and 3 girls, 2 children are chosen at random to speak to the mayor, who is visiting the school. Find the probability:
   (a) both children are boys;
   (b) both children are girls;
   (c) one boy and one girl are chosen.

2. Charlie has two six-sided dice, each with faces numbered from 1 to 6. One of the dice is fair, but the other is not; it will land on numbers 2 to 6 with equal probability, but lands on 1 with a different probability.
   When Charlie rolls the dice the probability that he gets a total of 2 is ¹⁄₂₄. What is the probability that he gets a total of 12 when he rolls the dice?

3. A set of 5 cards is numbered from 1 to 5. A card is chosen at random. The number is recorded and the card is not replaced. A second card is chosen and the number recorded. Use a tree diagram to find the probability that:
   (a) the cards are both even;
   (b) the sum of the two numbers is odd.

4. There are 104 pupils in a sixth-form college, of whom 55 are studying only arts subjects, 24 only science subjects and the rest a mixture of both. Three students are selected at random, without replacement. Find the probability that:
   (a) all three students are studying only arts subjects;
   (b) exactly one of the three students is studying only science subjects.

5. An unbiased die has faces numbered 1 to 6 inclusive. The die is rolled and the number that appears on the uppermost face is recorded.
   (a) State the probability of not recording a 4 in one roll of the die.
   The die is thrown until a 4 is recorded.
   (b) Find the probability that a 4 occurs for the first time on the third roll of the die.

6. A biased coin is tossed three times and the outcome is recorded each time. It is known that the coin lands on heads 25% of the time.
   (a) Draw a tree diagram to show this information.

Exercise 13F...

(b) Find the probability that the coin lands on heads all three times.
(c) Find the probability that the coin lands on heads only once.

The whole experiment is repeated a second time.

(d) Find the probability of obtaining either 3 heads or 3 tails in **both** trials.

7. Teagan rolls a fair, six-sided die repeatedly until she gets a 6. Her brother Malachy says that if she gets a 6 in her first three throws, she wins the game; if she doesn't he wins. Find the probability that Teagan wins.

8. A bus driver knows that if it is raining, he will be late 60% of the time. If it's dry he's only late 40% of the time. It rains on 40% of days. The driver's supervisor says "You are late more than half the time!" Is the supervisor correct? Justify your answer with a tree diagram and appropriate calculations.

13.7 Mixed Section

In many problems there are no instructions about which type of diagram to draw. You will need to decide upon the most suitable strategy, including any diagram required, and how to use it to solve the problem.

Exercise 13G

1. Fiachra changes one of his socks, at random, every day. What is the probability that he wears the same sock on his right foot for an entire week?

2. A pet shop has 4 female hamsters and x male hamsters for sale. A customer buys 2 of the hamsters, chosen at random. Each hamster is equally likely to be chosen. The probability that both the chosen hamsters are male is $\frac{1}{3}$. What is the value of x?

3. During summer activities week, 120 students each choose one activity from swimming, football and tennis.

   46 of the students were girls.
   36 of the students choose tennis, and $\frac{2}{3}$ of these were boys.

Exercise 13G...

25 girls choose swimming.
27 students choose football.

A boy is picked at random. What is the probability that he chooses swimming?

4. In a population, $\frac{3}{5}$ of the adults are overweight. The probability of an overweight adult having Type 2 diabetes is $\frac{9}{50}$. This probability is 6 times the probability of an adult who is not overweight having the disease. An adult is chosen at random from the population. What is the probability the chosen adult has Type 2 diabetes?

5. A survey of all the households in Carrickfergus shows that 70% have a laptop computer, 20% have a tablet device and 85% have a laptop or a tablet or both.
   (a) Find the probability that a household chosen at random has both types of device.
   (b) Determine whether having a laptop and having a tablet are independent events.

6. In a school sixth form there are 119 pupils.
   47 of them study Art.
   48 of them study Biology.
   50 of them study Chemistry.
   14 students take both Art and Biology.
   13 students take both Art and Chemistry.
   19 students take both Biology and Chemistry.
   6 students take all three subjects.
   (a) Draw a Venn diagram to show this information.

   A sixth form student is chosen at random.
   (b) What is the probability the student studies Chemistry, but not Art or Biology?
   (c) What is the probability the student studies Biology or Chemistry or both?
   (d) Determine whether studying Biology and Chemistry are independent events.

7. Two events P and Q are such that $P(T) - P(Q) = 0.3$ and $P(T) + P(Q) = 0.7$. Given also that $P(T \cup Q) = 0.5$, find:
   (a) $P(T \cap Q)$
   (b) $P(T)$
   (c) $P(Q)$
   (d) $P(T' \cap Q')$

Exercise 13G...

8. A biased die in the shape of a cube has the numbers 1 to 6 on its faces. When the die is thrown:
   - the probability of getting 1 is equal to the probability of getting 2;
   - the probability of getting 4 is twice the probability of getting 2;
   - the probability of getting 3 is twice the probability of getting 4; and
   - the probability of getting 4, 5 and 6 are equal.

   When this die is thrown find the probability of getting a 2.

13.8 Summary

In this chapter you have learnt probability notation and how to calculate probabilities.

You have learnt the meanings of the terms **mutually exclusive**, **exhaustive** and **independent** events.

You have learnt the addition law:
$P(A \cup B) = P(A) + P(B) - P(A \cap B)$

In the case of mutually exclusive events, this simplifies to:
$P(A \cup B) = P(A) + P(B)$

For independent events,
$P(A \cap B) = P(A) \times P(B)$

This is a simplified form of the multiplication law. It is often used to test whether two events are independent. The multiplication law is covered in more detail in A2 Mathematics.

Exhaustive events are a set of events in which at least one of the events must happen.

You have learnt how to calculate combined probabilities of up to three events, using tree diagrams, Venn diagrams and two-way tables.

Chapter 14
Binomial Distribution

14.1 Introduction

Imagine an experiment that results in success or failure, with a fixed, known probability of success.

Now imagine repeating the experiment, say 20 times.

The binomial probability distribution allows calculation of the probability of, say, 15 successes out of 20.

The binomial probability distribution is an example of a **discrete** probability distribution, since the number of successes is an integer.

Key words
- **Binomial**: Relating to two numbers. In the binomial probability distribution these two numbers are the probabilities of success and failure.

Before you start
You should:
- Have studied the binomial expansion in AS Pure Mathematics.
- Understand factorials and combinatorial notation.
- Have an understanding of probability and the notation relating to it.

Factorials and combinatorials
The factorial $n!$ is defined as:
$$n! = n \times (n-1) \times (n-2) \times \ldots \times 1$$

You will see a combinatorial written as either ^nC_r or $\binom{n}{r}$.

The definition is as follows:
$$^nC_r = \frac{n!}{(n-r)!\,r!}$$

Worked Examples
1. Find:
   (a) $5!$ (b) $3!$ (c) $2!$ (d) $\binom{5}{3}$

   (a) $5! = 5 \times 4 \times 3 \times 2 \times 1 = 120$
   (b) $3! = 3 \times 2 \times 1 = 6$
   (c) $2! = 2 \times 1 = 2$
   (d) $\binom{5}{3} = \dfrac{5!}{(5-3)!\,3!} = \dfrac{120}{2 \times 6} = 10$

2. Out of a class of 15 pupils, a PE teacher must choose 11 of them to play for the school team. How many different ways could she choose the team?

   Method 1 – using factorials on the calculator
   $$^{15}C_{11} = \frac{15!}{(15-11)!\,11!} = 1365$$

   Method 2 – using the combinatorial button on the calculator

   There is a button labelled nCr on all calculators. On most Casio models you press SHIFT ÷
   - Type 15 followed by nCr followed by 11.
   - This comes up on the screen as 15C11.
   $^{15}C_{11} = 1365$

What you will learn
In this chapter you will learn how to:
- Use the terminology and notation relating to the binomial distribution.
- Use the binomial distribution, an example of a discrete probability distribution.
- Calculate probabilities using the binomial distribution.
- Link binomial probabilities to the binomial expansion and tree diagrams.
- Use the Binomial Cumulative Distribution Function table given in the CCEA formula booklet and use the calculator to determine cumulative probabilities.

In the real world...
A data storage company may keep multiple, say ten, backups of important data, such as customer emails or website configuration data.

Experts on data security will estimate the probability that any one copy of the data will become corrupt or accidentally deleted, or in the worst case, be hacked.

14: BINOMIAL DISTRIBUTION

What is the probability that two of the copies are compromised in one of these ways? What is the probability that all ten of them are?

These questions can be answered using the binomial distribution.

Exercise 14A (Revision)

1. Calculate:
   (a) $4!$ (b) $\dfrac{8!}{4!}$

2. Evaluate:
   (a) $\binom{7}{3}$ (b) $^6C_3$

3. Find the binomial expansion for $(3 + 2x)^3$.

4. A bag contains two red beads and three black beads. A bead is drawn from the bag and replaced. A second bead is drawn. Find the probability of drawing a red bead both times.

Requirements

There are four requirements that must be met when using the binomial distribution:

- There is a fixed number of trials.
- The trials are all independent of each other.
- Each trial has only two possible outcomes, sometimes labelled 'success' and 'failure'.
- The probability of success is constant.

Notation

When answering a written problem, decide first whether a situation can be modelled using the binomial distribution.

If so, it is good practice to begin your solution by stating that this is the case, using the notation $\text{Bin}(n, p)$. You may also see $B(n, p)$.

For example, in a question involving 10 adults and a probability of 0.6 that an adult is carrying an organ donor card, begin by stating:

Let X be the number of adults carrying an organ donor card.

$X \sim \text{Bin}(10, 0.6)$ or $X \sim B(10, 0.6)$

This means X can be modelled using a binomial distribution, with 10 trials and a fixed probability of success of 0.6 for each trial.

14.2 Calculating Probabilities Using the Binomial Distribution Formula

If n is the number of trials of an experiment, p is the fixed probability of success for each trial and x is the number of successes, then:

$$P(X = x) = {}^nC_x\, p^x\, (1 - p)^{n-x}$$

Note: This formula is on the formula sheet.

Worked Examples

3. There are 20 cows in a field. The farmer knows that there is a probability of 0.05 that a cow has a particular disease.
   (a) Find the probability that exactly 4 cows have the disease.
   (b) The farmer is told that if any of his cows have the disease, the whole farm must go into quarantine for 2 weeks. Find the probability of this happening.

Let X be the random variable the number of cows that have the disease.

X follows a binomial distribution, because all the conditions are met:

- For each cow there are two possible outcomes (disease or no disease)
- There is a fixed probability of success (having the disease), which is 0.05.
- There is a fixed number of independent trials, 20.

$\therefore X \sim \text{Bin}(20, 0.05)$

$P(X = x) = {}^nC_x\, p^x\, (1 - p)^{n-x}$

(a) $P(X = 4) = {}^{20}C_4 (0.05)^4 (0.95)^{16}$
$\qquad\qquad\quad = 0.013327\ldots$

The probability of 4 cows having the disease is 0.0133 (3 s.f.).

Note: You can check this on the calculator. See the method for finding $P(X = x)$ at the end of this chapter.

(b) The farm will go into quarantine if at least one cow has the disease. The most straightforward approach is to find the probability that no cows have the disease and subtract this from 1.

$P(X \geq 1) = 1 - P(X = 0)$
$\therefore P(X \geq 1) = 1 - {}^{20}C_0 (0.05)^0 (0.95)^{20}$
$\qquad\qquad\quad = 1 - 0.3584\ldots$
$\qquad\qquad\quad = 0.642\ (3\ \text{s.f.})$

Note: One of the requirements for use of the binomial distribution is independent trials. In this case, there is a possibility this requirement is not met, since one cow having the disease may increase the likelihood of others having it. However, if the disease is not transmissible from one cow to another (for example if they can only contract it through their food), then the requirement is satisfied.

4. The probability that a woman wins at bingo is constant at 0.058.
   (a) State two requirements associated with the use of the binomial distribution.
   (b) The woman has 10 games of bingo. Find the probability that:
      (i) she wins none of the games;
      (ii) she wins fewer than three of the games.
   (c) Let n be the smallest number of games she must play before there is at least a 90% chance that she will have won a game.
      (i) Show that n satisfies $0.942^n \leq 0.1$
      (ii) Find the value of n.

(a) Requirements:
   - There is a fixed number of trials.
   - The trials are all independent of each other.
   - Success and failure are the only two possible outcomes of each trial.
   - The probability of success is constant.

(b) (i) Let X be the random variable representing the number of wins. Then:
$X \sim \text{Bin}(10, 0.058)$
$P(X = 0) = {}^{10}C_0 (0.058)^0 (0.942)^{10}$
$= 0.550185$
$= 0.550$ (3 s.f.)

Note: You can check your answer on the calculator. See the method for finding $P(X = x)$ at the end of this chapter.

(ii) $P(X < 3) = P(X = 0) + P(X = 1) + P(X = 2)$
$= 0.550185 + {}^{10}C_1 (0.058)^1 (0.942)^9$
$+ {}^{10}C_2 (0.058)^2 (0.942)^8$
$= 0.550185 + 0.338755 + 0.093859$
$= 0.983$ (3 s.f.)

(c) (i) $X \sim \text{Bin}(n, 0.058)$
$P(X > 0) \geq 0.9$
$\Rightarrow P(X = 0) \leq 0.1$
$\Rightarrow {}^n C_0 (0.058)^0 (0.942)^n \leq 0.1$
$0.942^n \leq 0.1$

(ii) Since the unknown is one of the indices, this equation requires logarithms.
$0.942^n \leq 0.1$
Take logs of both sides:
$\log(0.942^n) \leq \log(0.1)$
Use the power law of logs on the LHS:
$n \log(0.942) \leq \log(0.1)$
$n \geq \dfrac{\log(0.1)}{\log(0.942)}$

Note: The inequality sign changes when dividing or multiplying by a negative number. $\log(0.942)$ is negative, so the sign changes from $\leq$ to $\geq$.

$\therefore n \geq 38.5 \ldots$
$n = 39$

She must play 39 games to be 90% sure of a win.

Vertical line graphs can be used to visualise the probabilities in a binomial distribution.

Worked Example

5. Fifteen conkers fall from a tree one day in September. Each conker has a 0.1 chance of germinating and becoming a horse chestnut sapling.
   (a) Find the probability that at least 2 of the conkers germinate.
   (b) Draw a vertical line chart to illustrate the shape of this distribution.
   (c) For what value of x is $P(X = x)$ largest?

Let X be the random variable 'the number of conkers germinating'. X follows a binomial distribution, since there are a fixed number of trials, 15, with a fixed probability of success in each trial, 0.1. The trials are independent of each other.
$\therefore X \sim \text{Bin}(15, 0.1)$

(a) To find the probability that at least 2 of the conkers germinate, find the probability of 0 and 1 conkers germinating, and subtract these from 1.
$P(X = 0) = {}^{15}C_0 (0.1)^0 (0.9)^{15} = 0.20589 \ldots$
$P(X = 1) = {}^{15}C_1 (0.1)^1 (0.9)^{14} = 0.34315 \ldots$
$P(X \geq 2) = 1 - P(X = 0) - P(X = 1)$
$= 1 - 0.20589 - 0.34315$
$= 0.45096$
$= 0.451$ (3 s.f.)

(b) $P(X = 2) = {}^{15}C_2(0.1)^2(0.9)^{13} = 0.267$
$P(X = 3) = {}^{15}C_3(0.1)^3(0.9)^{12} = 0.129$
...
$P(X = 15) = {}^{15}C_{15}(0.1)^{15}(0.9)^0 = 1 \times 10^{-15}$

Note: You can calculate all these probabilities at once on your calculator. See the calculator tips at the end of this chapter.

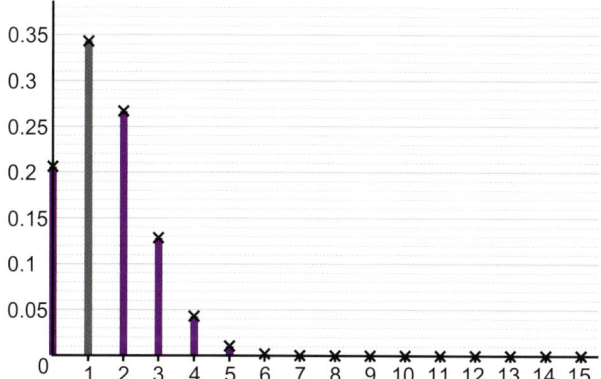

(c) From the vertical line chart it can be seen that $P(X = x)$ is largest when $x = 1$, shown in grey in this chart.

Exercise 14B

1. State two of the conditions required for a random variable to be modelled by a binomial distribution.

2. Four coins are tossed together.
   (a) Daniel says 'The probability of getting exactly 2 heads is 50%.' Is he correct? Justify your answer by working out the relevant binomial probability.
   (b) Draw a vertical line chart to show $P(X = 0)$, $P(X = 1)$, ... $P(X = 4)$.

3. There are 4 lifts in a large office building. The probability of each lift breaking down in a day is 0.05. Find the probability that only one lift is working today.

4. A cat has a litter of 4 kittens. There is a 50% chance a kitten will be male. Find:
   (a) The probability the cat has 4 female kittens.
   (b) The cat has at least one female kitten.

5. A survey reveals that 10% of the population is left-handed. In a family of 4 children, what is the probability that 3 of the children are left-handed?

Exercise 14B...

6. The Thames flood barrier was built in 1982 and has been closed over 200 times since then to protect London from flooding. It is mainly used in the winter months of December, January and February, when the risk of flooding is higher. When it was built, the probability of the barrier being closed at least once during any one winter month was 0.5. During the winter of 2021 – 2022, the probability was 0.7.
   (a) Find the probability of the barrier being closed during each of the three winter months when it was first built.
   (b) Find the probability of the barrier being closed during each of the three months of winter 2021 – 2022.

7. A roulette wheel has 18 black holes, 18 red holes and one green hole. The wheel is spun and the ball is launched.
   (a) What is the probability of the ball landing in a red hole on one spin?

   The wheel is spun three times.
   (b) What is the probability that the ball lands in a red hole on **exactly one** of those three spins?
   (c) What is the probability that the ball lands in a red hole on **at least one** of those three spins?

8. According to data from an airline, the probability of a flight arriving on time is 0.8. During the year Rearden takes 4 flights.
   (a) What is the probability Rearden arrives on time on all 4 flights?
   (b) What is the probability Rearden arrives on time on exactly 2 occasions?
   (c) What is the probability Rearden arrives on time on at least 1 occasion?

9. A factory produces components for the electronics industry. In a recent quality control inspection, it was discovered that 1% of the components coming off the production line are faulty. Ten components are packaged and shipped to Alpha Electronics. Find the probability that Alpha Electronics receive at least one faulty component in the batch.

Exercise 14B...

10. Nine cars are taking part in a race. The probability of a car crashing during the race is p. If X is the random variable 'the number of cars that crash', find an expression, in terms of p, for:
(a) $P(X = 1)$
(b) $P(X = 2)$

It is known that $P(X = 1) = P(X = 2)$.
(c) Find p.
(d) Hence find the probability that exactly 3 cars crash.

11. When a biased coin is tossed, the probability of getting a head is ¾. When a biased die is thrown, the probability of getting a 6 is 7/12. All other outcomes are equally likely. A trial consists of tossing the coin and throwing the die. Find the probability:
(a) of getting a head and a 4,
(b) of getting a tail and a prime number,
(c) of getting, in 4 trials, a head together with a 6 only once.

14.3 Links Between Binomial Probabilities, the Binomial Expansion and Tree Diagrams

Pascal's Triangle
Consider the following:
$(1 + x)^0 = 1$
$(1 + x)^1 = 1 + x$
$(1 + x)^2 = 1 + 2x + x^2$
$(1 + x)^3 = 1 + 3x + 3x^2 + x^3$
$(1 + x)^4 = 1 + 4x + 6x^2 + 4x^3 + x^4$

All the expansions above are examples of the **binomial expansion**, so-called because there are two terms inside the brackets. You learnt how to expand brackets using the binomial expansion in AS Mathematics.

Looking at the coefficients only, an interesting pattern emerges:

```
          1
        1   1
      1   2   1
    1   3   3   1
  1   4   6   4   1
        ...
```

Each number is the sum of the two numbers above it. This pattern is known as **Pascal's Triangle**.

The numbers in Pascal's Triangle are also the numbers obtained from $\binom{n}{x}$, if n is the row number and x the column (both starting from 0):

$$\binom{0}{0}$$
$$\binom{1}{0} \quad \binom{1}{1}$$
$$\binom{2}{0} \quad \binom{2}{1} \quad \binom{2}{2}$$
$$\binom{3}{0} \quad \binom{3}{1} \quad \binom{3}{2} \quad \binom{3}{3}$$
$$\binom{4}{0} \quad \binom{4}{1} \quad \binom{4}{2} \quad \binom{4}{3} \quad \binom{4}{4}$$
...

Using Pascal's triangle, we could work out all the coefficients in any expansion of $(a + b)^n$.

The formula for the binomial expansion of $(a + b)^n$ is:
$$(a + b)^n = \binom{n}{0}a^n + \binom{n}{1}a^{n-1}b + \binom{n}{2}a^{n-2}b^2 + \cdots + \binom{n}{n}b^n$$

> **Note:** This formula appears under Binomial Series on the Pure Mathematics section of the formula sheet.

Worked Example

6. Use Pascal's triangle to expand $(a + b)^4$.

$$(a + b)^4 = \binom{4}{0}a^4 + \binom{4}{1}a^3b + \binom{4}{2}a^2b^2 + \binom{4}{3}ab^3 + \binom{4}{4}b^4$$

The coefficients can be found on row 4 of Pascal's triangle (remembering the top row is row 0).

$(a + b)^4 = 1a^4 + 4a^3b + 6a^2b^2 + 4ab^3 + 1b^4$
$ = a^4 + 4a^3b + 6a^2b^2 + 4ab^3 + b^4$

Consider a situation that can be modelled using a binomial distribution, with n trials and a probability a of success in each trial. Let b be the probability of failure in each trial, so $b = 1 - a$.

Each term in the expansion above gives the probability of

some number of successes. Look at the power of a in each term.

$\binom{n}{0} a^n$ gives the probability of n successes and 0 failures

$\binom{n}{1} a^{n-1} b$ gives the probability of $n - 1$ successes and 1 failure

$\binom{n}{2} a^{n-2} b^2$ gives the probability of $n - 2$ successes and 2 failures

...

$\binom{n}{n} b^n$ gives the probability of 0 successes and n failures

Worked Example

7. A boy plants ten sunflower seeds. There is a fixed probability of ⅛ that a seed will grow into a sunflower. Using the binomial expansion for $(a + b)^n$, where a and b are the probabilities of success and failure for one trial, find:
(a) the probability that exactly two seeds will become sunflowers;
(b) the probability that at least 80% of the seeds will become sunflowers;
(c) the probability that less than 3 seeds will become sunflowers.

Consider a seed becoming a sunflower as success.

Then $a = \dfrac{1}{8}$ and $b = \dfrac{7}{8}$

The general expansion formula is:

$(a + b)^n = \binom{n}{0} a^n + \binom{n}{1} a^{n-1} b + \binom{n}{2} a^{n-2} b^2 + \cdots + \binom{n}{n} b^n$

When $n = 10$ (ten seeds are planted in this case):

$(a + b)^{10} = \binom{10}{0} a^{10} + \binom{10}{1} a^9 b + \binom{10}{2} a^8 b^2 + \cdots + \binom{10}{10} b^{10}$

(a) The term relating to 2 successes is the term involving a^2.

$P(X = 2) = \binom{10}{8} a^2 b^8$

$= 45 \times \left(\dfrac{1}{8}\right)^2 \times \left(\dfrac{7}{8}\right)^8$

$= 0.242$ (3 s.f.)

(b) We are interested in 8, 9 or 10 successes. Look for the terms involving a^8, a^9 and a^{10}:

$P(X \geq 8) = \binom{10}{0} a^{10} + \binom{10}{1} a^9 b + \binom{10}{2} a^8 b^2$

$= 1 \left(\dfrac{1}{8}\right)^{10} + 10 \left(\dfrac{1}{8}\right)^9 \left(\dfrac{7}{8}\right)^1 + 45 \left(\dfrac{1}{8}\right)^8 \left(\dfrac{7}{8}\right)^2$

$= 2.12 \times 10^{-6}$ (3 s.f.)

So that's very unlikely.

(c) Less than 3 means 0, 1 or 2 successes. Look for the terms involving a^0, a^1 and a^2:

$P(X < 3) = \binom{10}{8} a^2 b^8 + \binom{10}{9} a^1 b^9 + \binom{10}{10} b^{10}$

$= 45 \left(\dfrac{1}{8}\right)^2 \left(\dfrac{7}{8}\right)^8 + 10 \left(\dfrac{1}{8}\right)^1 \left(\dfrac{7}{8}\right)^9 + 1 \left(\dfrac{7}{8}\right)^{10}$

$= 0.880$ (3 s.f.)

Exercise 14C

1. Using Pascal's triangle to find the coefficients, find the binomial expansion for $(p + q)^3$.

2. (a) Using Pascal's triangle, write out the expansion of $(a + b)^6$.
   (b) A scientist has a container holding a large number of fruit flies. The probability that a fruit fly, chosen at random from the container, is female is ⅘. The scientist picks 6 fruit flies, chosen at random, from the container. Using your answer to part (a), find the probability that:
      (i) none of the flies is male;
      (ii) at least 2 of the flies are male.

3. (a) Write down the binomial expansion of $(a + b)^5$ using Pascal's triangle to help.
   There are 50 tickets in a school raffle and 10 prizes.
   (b) Find the probability that a ticket bought at random will be a winning ticket.
   (c) Ella buys five tickets. Using your answer to part (a), find the probability that:
      (i) Ella doesn't win any prizes,
      (ii) Ella wins five prizes,
      (iii) Ella wins at least one prize.

4. (a) Write down the binomial expansion of $(a + b)^{10}$ using Pascal's triangle to help.
   (b) A single die is thrown 10 times. What is the probability of scoring a prime number on every throw?

14.4 Binomial Cumulative Distribution Function

A table of the Binomial Cumulative Distribution Function is given in the formula booklet. You may use this table or your calculator to determine cumulative probabilities, for example $P(X \leq 4)$.

This is quicker and easier than finding and summing $P(X = 0), P(X = 1), \ldots, P(X = 4)$.

Worked Example

8. An icosahedron is a solid with 20 identical triangular faces, like the one shown.

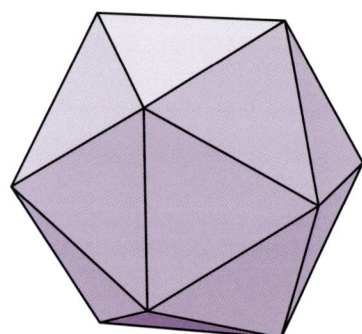

A die is the shape of an icosahedron with the numbers 1 to 20 on the faces. The die is rolled 25 times. Find the probability of getting:
(a) less than 2 scores of twenty;
(b) 2 or more scores of 20.

(a) Method 1: Using the Binomial Cumulative Distribution tables.

For each roll of the die, there is a fixed probability of success of $\frac{1}{20}$ or 0.05.

Less than 2 means 0 or 1.
$P(X < 2) = P(X \leq 1)$
$\qquad = 0.6424$

This value was obtained from the Binomial Cumulative Distribution table where we find that $P(X \leq 1) = 0.6424$ when $p = 0.05$ and $n = 25$:

$p =$	0.05	0.10	0.15
$n = 25, x = 0$	0.2774	0.0718	0.0172
1	0.6424	0.2712	0.0931
2	0.8729	0.5371	0.2537
3	0.9659	0.7636	0.4711
4	0.9928	0.9020	0.6821
5	0.9988	0.9666	0.8385

Method 2: Alternatively, calculation of $P(X \leq 1)$ can be done on the calculator. See the calculator tips for finding $P(X \leq x)$ at the end of this chapter.

(b) $P(X \geq 2) = 1 - P(X \leq 1)$
$\qquad = 1 - 0.6424$
$\qquad = 0.3576$
$\qquad = 0.358 \text{ (3 s.f.)}$

In the next example we cannot use the binomial cumulative distribution table, because the probability of success is not one of the values listed.

Worked Examples

9. Susan rolls a standard six-sided die 25 times. What is the probability of Susan getting 5 or more sixes?

If X is the random variable representing the number of sixes thrown on a standard die, then:

$$X \sim \text{Bin}\left(25, \frac{1}{6}\right)$$

The probability of getting 5 or more sixes can be worked out:
$P(X \geq 5) = 1 - P(X \leq 4)$

Method 1: We cannot use the binomial cumulative tables, since the probability of $\frac{1}{6}$ is not listed.

$P(X = 0) = {}^{25}C_0 \left(\frac{1}{6}\right)^0 \left(\frac{5}{6}\right)^{25} = 0.0105$

$P(X = 1) = {}^{25}C_1 \left(\frac{1}{6}\right)^1 \left(\frac{5}{6}\right)^{24} = 0.0524$

$P(X = 2) = {}^{25}C_2 \left(\frac{1}{6}\right)^2 \left(\frac{5}{6}\right)^{23} = 0.1258$

$P(X = 3) = {}^{25}C_3 \left(\frac{1}{6}\right)^3 \left(\frac{5}{6}\right)^{22} = 0.1929$

$P(X = 4) = {}^{25}C_4 \left(\frac{1}{6}\right)^4 \left(\frac{5}{6}\right)^{21} = 0.2122$

$\therefore P(X \leq 4) = 0.0105 + 0.0524 + 0.1258 + 0.1929 + 0.2122 = 0.5938$

$P(X \geq 5) = 1 - 0.5938$
$\qquad = 0.4062$
$\qquad = 0.406 \text{ (3 s.f.)}$

Method 2: See the calculator tips at the end of this chapter for finding $P(X \leq x)$.

From the calculator, $P(X \leq 4) = 0.5937\ldots$
$P(X \geq 5) = 1 - 0.5937\ldots$
$= 0.4062\ldots$
$= 0.406$ (3 s.f.)

10. There are a large number of coins in a box, some of which are fake. A coin expert takes 12 coins from the box at random. The probability she has chosen more than 5 fakes is $\dfrac{1}{2000}$. Find the probability that any one coin is a fake.

There are 12 trials, so the section of the table under consideration relates to $n = 12$:

$p =$	0.05	0.10	0.15
$n = 12, x = 0$	0.5404	0.2824	0.1422
1	0.8816	0.6590	0.4435
2	0.9804	0.8891	0.7358
3	0.9978	0.9744	0.9078
4	0.9998	0.9957	0.9761
5	1.0000	0.9995	0.9954
6	1.0000	0.9999	0.9993
7	1.0000	1.0000	0.9999

Let X be the number of fakes chosen.

The probability that more than 5 fakes are chosen is $\dfrac{1}{2000}$, i.e.

$P(X > 5) = \dfrac{1}{2000} = 0.0005$

$\therefore P(X \leq 5) = 1 - 0.0005 = 0.9995$

Looking along the $x = 5$ row, this probability appears in the column for $p = 0.1$ (highlighted in the table).

Therefore, the probability that any one coin is a fake is 0.1.

11. Some coins are tipped out of a pot. Niamh notices that eighteen of the coins land on heads. 'That's incredible', she says to herself, 'The chances of getting more than 17 heads was only 1 in 5000.' How many coins were in the pot?

Let X be the number of heads.

$P(X > 17) = \dfrac{1}{5000} = 0.0002$

$\Rightarrow P(X \leq 17) = 0.9998$

Let n be the number of trials (the number of coins). The probability of success for each trial (getting a head) is $p = 0.5$.

Looking down the $p = 0.5$ column of the tables, it is in the $n = 20$ section that we find $P(X \leq 17) = 0.9998$:

$p =$	0.40	0.45	0.50
$n = 20, x = 0$	0.0000	0.0000	0.0000
1	0.0005	0.0001	0.0000
2	0.0036	0.0009	0.0002
3	0.0160	0.0049	0.0013
4	0.0510	0.0189	0.0059
5	0.1256	0.0553	0.0207
6	0.2500	0.1299	0.0577
7	0.4159	0.2520	0.1316
8	0.5956	0.4143	0.2517
9	0.7553	0.5914	0.4119
10	0.8725	0.7507	0.5881
11	0.9435	0.8692	0.7483
12	0.9790	0.9420	0.8684
13	0.9935	0.9786	0.9423
14	0.9984	0.9926	0.9793
15	0.9997	0.9985	0.9941
16	1.0000	0.9997	0.9987
17	1.0000	1.0000	0.9998
18	1.0000	1.0000	1.0000

So there were 20 coins in the pot.

Exercise 14D

1. A fair tetrahedron-shaped die has its four sides coloured red, green, blue and yellow. It is rolled 8 times. Find, using the tables for the Binomial Cumulative Distribution Function, the probability of
   (a) the die landing on red less than 3 times,
   (b) the die landing on blue more than 3 times.

2. A spinner is designed so that the probability it lands on green is 0.3. Jasper has 12 spins. Find the probability that Jasper obtains
   (a) no more than 2 greens,
   (b) at least 5 greens.

   Jasper wants to use the spinner in a class competition. He wants the probability of winning a prize to be less than 0.05. Each member of the class has 12 spins and the number of greens is recorded.
   (c) Find the number of greens needed to win a prize.

Exercise 14D...

3. Ten adults are waiting for a train. Find the probability that at least half of them have high blood pressure. You may use the fact that one in four of the adult population has high blood pressure.

4. (a) Customers at a bookshop can pay by cash or by credit card. The probability that a customer pays by cash is 0.2. Eight customers are selected at random. Calculate the probability that exactly three customers pay by cash.
   (b) n customers are selected at random. If the probability that at least one of the customers pays by cash is greater than 0.8, find the least possible value of n.

5. A small school has 20 members of staff. The probability of any member of staff, chosen at random, being off sick throughout the school year is p. Let X be the random variable 'the number of members of staff off sick throughout the school year'.
   (a) Find an expression, in terms of p, for $P(X = 2)$.
   (b) It is known that $P(X = 3) = 4P(X = 2)$. Prove that $p = 0.4$
   (c) Hence find the probability that at least three members of staff are off sick throughout the school year.

6. A standard 6-sided die is rolled 8 times. Find the probability of getting
   (a) exactly one six,
   (b) no more than 3 sixes,
   (c) at least 4 sixes.
   Hint: It is not possible to use the binomial cumulative distribution tables in this question, since the probability of success is not listed.

7. Grainne walks to her friend's house every Monday for 10 weeks. The probability that she gets wet more than 3 times out of 10 is 0.3504.
   (a) Find the probability that it rains on a Monday.
   (b) What assumptions have been made?

Exercise 14D...

8. Several divers are searching a stretch of the River Lagan for unexploded WW2 bombs. The probability of any one diver finding an unexploded bomb is 0.05. Four of the divers find unexploded bombs. When they all return to the surface, their supervisor says 'We weren't expecting that! The probability of more than three of you finding bombs was 1 in 10 000!' How many divers are there?

9. There are 5 balls in a bag, 2 of them red.
   (a) Find the probability of drawing a red.
   A ball is withdrawn, and its colour is recorded. It is then replaced in the bag. This action is performed 9 times in total. Let X be the number of times a red ball is drawn.
   (b) Find the largest value of x such that $P(X > x)$ is greater than 0.5.

10. The vertical line chart shown gives binomial probabilities for $P(X = x)$ for $n = 11$. The value of p is not given.

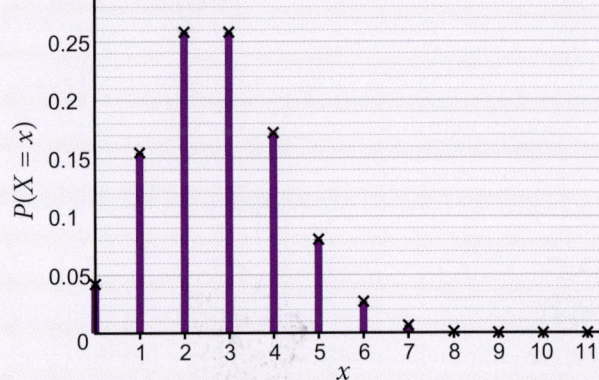

Which of these statements are true and which false? Use the chart to answer these questions; a calculator should not be needed.
(a) $P(X = x)$ is greatest when $x = 1$.
(b) $P(X = 2) = P(X = 3)$
(c) $P(X \leq 5) > P(X \leq 6)$
(d) $P(X \geq 10) > P(X = 0)$

14: BINOMIAL DISTRIBUTION

Calculator Methodologies

Some calculators have functions allowing calculation of the Binomial Probability Distribution $P(X = x)$ and the Binomial Cumulative Probability $P(X \leq x)$. The tips here are written for the Casio FX-991EX.

Binomial probability distribution, $P(X = x)$

1. Press **MENU** then **7** for Distribution.
2. Press **4** for Binomial PD (Binomial Probability Distribution).
3. Press **1** for **list** or **2** for **variable**.
4. If using a list, enter all values of x for which you require a probability. Press **=** to enter each one, then **=** again to finish. Then enter values for N and p and press **=**.
5. If using a single x value, enter values for x, N and p and press **=**.

Worked Example

12. Using the calculator, find the probability of 5 successes from 10 trials, with a probability of 0.2 of success in each trial.

    Enter Menu 7, Binomial PD, then 2 for variable.

    Entering $x = 5$, $N = 10$ and $p = 0.2$ gives:
    $P = 0.0264...$
    $P = 0.0264$ (3 s.f.)

Calculator Methodologies

Binomial cumulative distribution, $P(X \leq x)$

1. Press **MENU** then **7** for Distribution.
2. Press the down arrow to get more options, then **1** for Binomial CD (Binomial Cumulative Distribution).
3. Press **1** for **list** or **2** for **variable**.
4. If using a list, enter all values of x for which you require a probability. Press **=** to enter each one, then **=** again to finish. Then enter values for N and p and press **=**.
5. If using a single x value, enter values for x, N and p and press **=**.

Worked Example

13. Using the calculator, find the probability of no more than 5 successes from 10 trials, with a probability of 0.2 of success in each trial.

    Enter Menu 7, Binomial CD, then 2 for variable.

    Entering $x = 5$, $N = 10$ and $p = 0.2$ gives:
    $P = 0.9936...$
    $P = 0.994$ (3 s.f.)

14.5 Summary

If n is the number of trials of an experiment, p is the fixed probability of success for each trial and x is the number of successes, then:
$$P(X = x) = {}^nC_x \, p^x \, (1-p)^{n-x}$$

The terms in the binomial expansion:
$$(a+b)^n = \binom{n}{0}a^n + \binom{n}{1}a^{n-1}b + \binom{n}{2}a^{n-2}b^2 + \cdots + \binom{n}{n}b^n$$

give the probabilities in the binomial distribution with n trials and probabilities of success and failure a and b, respectively. In general, the term $\binom{n}{x}a^{n-x}b^x$ is the probability of x successes.

To find $P(X \leq x)$, the Binomial Cumulative Distribution Function table given in the formula booklet can be used.

Some calculators have functions allowing calculation of $P(X = x)$ and $P(X \leq x)$.

Answers

Exercise 1A

1.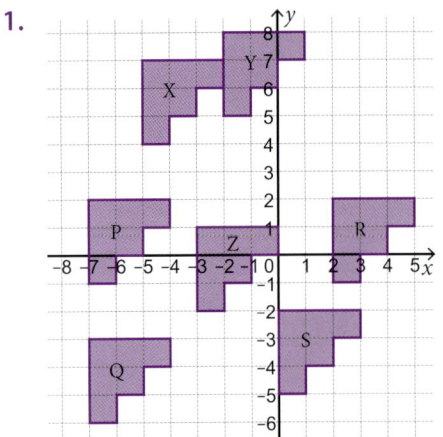

2. (a) Hypotenuse = 5.8;
   Angle = 31.0°
   (b) Hypotenuse = 5.8;
   Angle = 59.0°
   (c) Hypotenuse = 7.1;
   Angle = 45°
   (d) Hypotenuse = 3.6;
   Angle = 56.3°

Exercise 1B

1. (a) The ball is a particle. Gravity is a constant 9.8 m s^{-2}.
   (b) The pot is a particle (since air resistance is negligible). The table is rough (since there must be friction to slow the pot down).
2. (a) (i) It is valid to model the man as a particle (which has no size), but in a more sophisticated model, the man's size and shape could be considered. In this way the model could work out whether any part of his body touches the bar. Additionally, air resistance could then be considered. (ii) It is invalid to treat the pole as a rod, since the pole flexes. (iii) It is valid to model the acceleration due to gravity as a constant 9.8 m s^{-2}.
   (b) (i) It is valid to model the toboggan and two children as a single particle. The effects of air resistance can be ignored. (ii) Although the effects of friction will be small, it is important to include the effects of friction in the model. Without friction, the toboggan would keep moving indefinitely as it slides across the field. (iii) It is valid to model the acceleration due to gravity as a constant 9.8 m s^{-2}.
   (c) (i) It may not be valid to assume the astronaut's leg bones are rods, which are rigid, i.e. unbreakable. Since the model is studying the stresses on the bones, it may be better to treat them as non-rigid bodies. (ii) The second assumption is invalid. The acceleration due to gravity is lower than 9.8 m s^{-2} for objects further away from the surface of the earth. For an object in orbit, it may be valid to assume there is no gravity.

Exercise 1C

1. (a) 30 000 m (b) 26 kg
   (c) 13.9 m s^{-1} (d) 25 000 kg m^{-3}
   (e) 0.0075 m s^{-1} (f) 0.12 kg m^{-2}
   (g) 970 m s^{-1} (h) 0.046656 s or 0.0467 s (3 s.f.)

Exercise 1D

1. (a) Scalar (b) Scalar (c) Scalar
   (d) Scalar (e) Vector (f) Scalar
   (g) Vector (h) Vector
2. (a) 130 km h^{-1}
   (b) 7.28 m s^{-2} (3 s.f.)
   (c) 25 N
   (d) 40.4 km (3 s.f.)
3. (a) 12.8 m s^{-1} (b) 51.3°
4. (a) $(-4\mathbf{i} + \mathbf{j})$ km
   (b) 4.12 km (3 s.f.)
   (c) 15.7 km (3 s.f.)
5. $12\mathbf{i} + 5\mathbf{j}$

Exercise 2A

1. 30 km h^{-1}
2. (a) $d = \dfrac{C}{\pi}$ (b) $a = \dfrac{2A}{h} - b$
   (c) $b = \pm\sqrt{a^2 - c^2}$
3. (a) −2 (b) 180 (c) ±4.2

Exercise 2B

1. (a) $13\mathbf{i} + 20\mathbf{j}$ m (b) $6\mathbf{i} + 57\mathbf{j}$ m
   (c) $\begin{pmatrix}-1.5\\1\end{pmatrix}$ m s^{-1} (d) 2.5 s
   (e) 5 m s^{-1} (f) ⅓ s
2. 5 m s^{-1}
3. 3 s
4. (a) 7.5 m s^{-1}
   (b) $(4 + 6t)\mathbf{i} + (20 - 4.5t)\mathbf{j}$
   (c) 3 seconds (d) 2 m s^{-1}
5. (a) $(28800 - t)\mathbf{i} + (15000 + 45t)\mathbf{j}$ km (b) 2:00 am
   (c) 340 km (nearest km)
6. (a) $18jt$ km (b) $12\mathbf{i}t$ km
   (c) $(-12\mathbf{i} + 18\mathbf{j})t$ km
   (d) 21.6 km (e) 146°
7. (a) F: $5t\mathbf{i} + (490 + 5t)\mathbf{j}$ m;
   S: $(630 - 4t)\mathbf{i} + 12t\mathbf{j}$ m
   (b) $\mathbf{i}$-components:
   $5t = (630 - 4t) \Rightarrow t = 70$
   $\mathbf{j}$-components:
   $(490 + 5t) = 12t \Rightarrow t = 70$
   Since the $\mathbf{i}$ components and $\mathbf{j}$ components are equal at the same time, the ferry and ship collide at this time.
   (c) $350\mathbf{i} + 840\mathbf{j}$ m
8. (a) A: $(2 + 3t)\mathbf{i} + (5 - 2t)\mathbf{j}$ km
   B: $(6 - 2t)\mathbf{i} + (-3 + 6t)\mathbf{j}$ km
   (b) Position vector of B relative to A is $((6 - 2t)\mathbf{i} + (-3 + 6t)\mathbf{j})$
   $- ((2 + 3t)\mathbf{i} + (5 - 2t)\mathbf{j})$
   $= (4 - 5t)\mathbf{i} + (-8 + 8t)\mathbf{j}$ km
   (c) $\mathbf{i}$-components:
   $(2 + 3t) = (6 - 2t)$
   $\Rightarrow 5t = 4 \Rightarrow t = 0.8$
   $\mathbf{j}$-components:
   $(5 - 2t) = (-3 + 6t)$
   $\Rightarrow 8t = 8 \Rightarrow t = 1$
   The times at which the $\mathbf{i}$-components are equal and $\mathbf{j}$-components are equal are different. Therefore, the ships do not collide.
   (d) 10 km

Exercise 2C

1. (a) 7 m s^{-1} (b) 17 m
2. (a) 3.2 m s^{-2} (b) 131.95 m

ANSWERS – EXERCISE 3C

3. (a) 24 m s⁻¹ (b) 12 m s⁻²
4. (a) (i) 3 m s⁻² (ii) 885.5 m
   (iii) 514.5 m (b) 5.18 m s⁻²
5. (a) 0.4 m s⁻² (b) 600 m
   (c) 0.1875 m s⁻² (d) 105 seconds
6. (a) 10 m (b) Neither. Object A reaches point Y in 2 seconds. Object B also reaches point Y in 2 seconds.

Exercise 2D

1. (a) 2.0 m (2 s.f.)
   (b) 14 m s⁻¹ (2 s.f.)
2. (a) 22.5 m (b) 21.6 m s⁻¹ (c) 4.34 s
3. (a) 68 m (b) 41 m s⁻¹
4. (a) 27.7 m s⁻¹ (b) 5.90 s
5. (a) 78 m (b) 39 m s⁻¹
   (c) The twig is treated as a particle / air resistance is ignored.
6. (a) 9.8 m s⁻¹ (b) 19.6 m
7. (a) 7.3 m (b) 1.2 s (c) 10 m (2 s.f.)
   (d) 2.3 s
8. (b) 180 m (2 s.f.) (c) 9 s
9. (a) 44 m s⁻¹ (b) 4.3 s (c) The stone is modelled as a particle / air resistance is ignored.
10. (a) 5.3 m (b) 1.9 s (c) 1.3 m
11. (a) 1.6 s (b) 14 m s⁻¹ (c) 2.5 m

Exercise 2E

1. (a) 8 s (b) 160 m
2. (a) 7.7 m s⁻¹ (b) 1.5 s
3. (a) 17.7 s (b) 33.4 m
4. 5.95 m
5. (a) 1.3 s (b) 4.6 m
6. (a) 4 km (b) Speed of motorbike: 28.3 m s⁻¹. Speed of car: 14.1 m s⁻¹. The motorbike is breaking the speed limit.
7. (b) $\frac{110}{9}$ m
8. ⅓ m
9. (b) $\frac{H(2u^2 - gH)}{2u^2}$
10. T = 3.71 s; x = 33.4 m

Exercise 2F

1. (a) (7**i** + 7**j**) m s⁻¹ (b) 158 m
2. (−8.5**i** + 2.75**j**) m s⁻¹
3. (a) (8**i** + 2**j**) m s⁻¹
   (b) (32**i** + 8**j**) m
4. (a) (1 + t)**i** + (−1 + 2t)**j** m s⁻¹

(b) t = 3 (c) 15.3 m s⁻¹
5. (a) $\binom{13}{9}$ m (b) $\binom{5}{5}$ m (c) A's speed is $\sqrt{5}$ m s⁻¹; B's speed is $\sqrt{17}$ m s⁻¹.
6. (b) (7 + 0.2t)**i** + (5 + 0.15t)**j** m
   (c) (0.1 + 0.02t)**i** + (−0.3 − 0.02t)**j** m s⁻¹
   (e) 11**i** + 8**j** m (f) From part (b), Caitlyn's position vector is: (7 + 0.2t)**i** + (5 + 0.15t)**j**. When t = 20, this is: (7 + 0.2 × 20)**i** + (5 + 0.15 × 20)**j** = 11**i** + 8**j** m
   Therefore, the two boats are in the same position at the same time, so collide at this point.

Exercise 3A

1. (a) 20 m s⁻¹ (b) 24 km
   (c) 4 m s⁻² (d) 19.9 m s⁻¹ (3 s.f.)

Exercise 3B

1. (a)

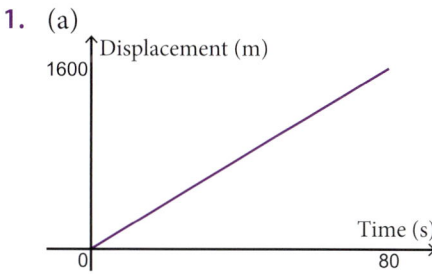

   (b) 20 m s⁻¹
2. (a) Between 10 and 11 a.m.
   (b) 10 km (c) He had returned home. (d) He remained at the same distance from home for half an hour, for example he may have been taking half an hour at one delivery point, or taking a break.
3.

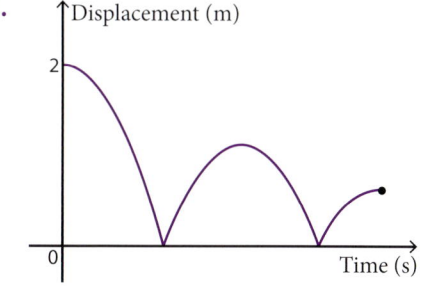

4. (a) 2 m s⁻¹ (b) −60 m
   (c) 1.64 m s⁻¹ (3 s.f.)
5. (a) uT metres
   (b) 2T seconds

(c)

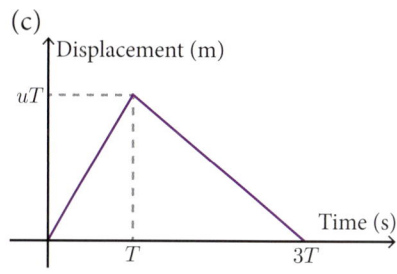

6. (a) 6 seconds; 1.2 metres.
   (b)

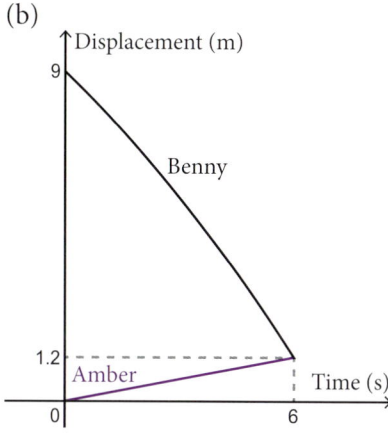

7. (a) $\frac{400}{u}$ seconds (b) 400 metres
   (c)

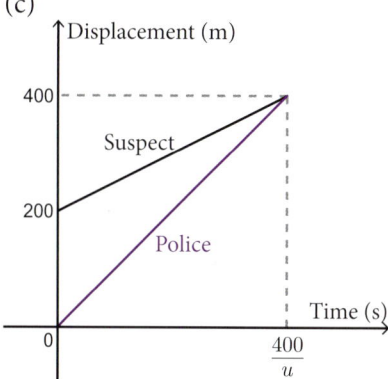

8. (a) $T_1 = 12$, $T_2 = 42$, $T_3 = 78$
   (b) 30 m

Exercise 3C

1. (a)

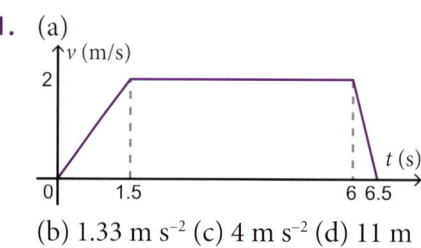

   (b) 1.33 m s⁻² (c) 4 m s⁻² (d) 11 m

ANSWERS – EXERCISE 3C

2. (a) 2.4 m s^{-2}
   (b)
   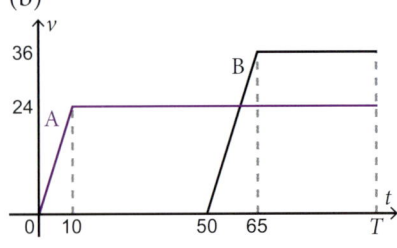
   (c) 162.5 s
3. (a) 30 m s^{-1} (b) 36 m s^{-1}
   (c)
   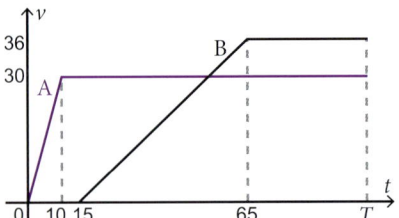
   (d) 215 seconds (3 minutes, 35 seconds) (e) 6300 m (6.3 km)
4. (a) 0.25 m s^{-2} (b) 15 s
   (c) T seconds is the time at which Caitlyn catches up with Alfie.
   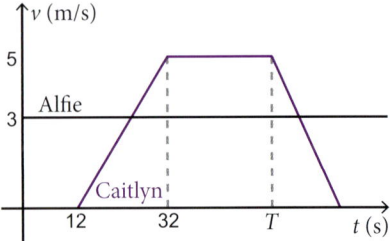
   (d) 55 s (e) 165 m (f) 12 s
   (g) Alfie is 7.5 m ahead
5. (a)
   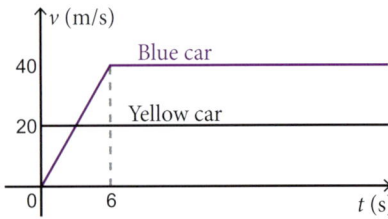
   (b) The yellow car takes 1500 seconds to reach Town C. The blue car takes 1253 seconds to reach Town C. The blue car arrives first.
6. (a) 5 m s^{-1} (b) 36 m s^{-1} (c) 65 s
   (d) 1 m s^{-2} (e) 2.4 m s^{-2}
   (f) Particle B comes to rest after 80 seconds. Particle A continues to move with a constant velocity of 20 m s^{-1}.

   (g) During the first 80 s: particle B travels 2610 m; particle A travels 1487.5 m. Particle B travels further.
   (h) After 136 seconds.
7. (a)
   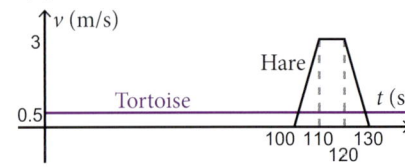
   (b) The tortoise wins the race
   (c) 10 seconds
8. (a) 24 m s^{-1} (b) 81 m
   (c) 8 seconds
9. (a)
   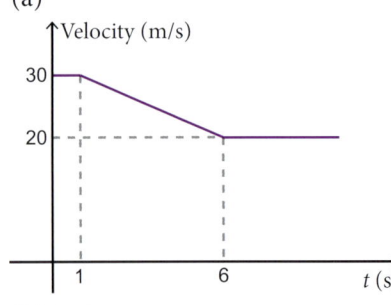
   (b) 155 m
10. (a) 10 m s^{-1} (b) 25 s
    (c)
    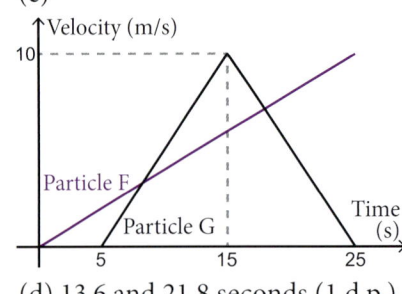
    (d) 13.6 and 21.8 seconds (1 d.p.)
11. (a)
    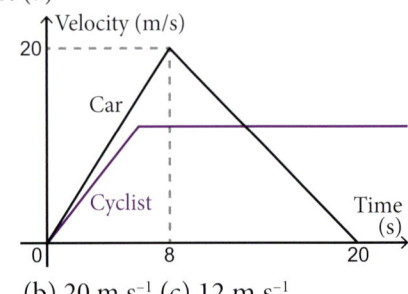
    (b) 20 m s^{-1} (c) 12 m s^{-1}
    (d) 5 s (e) 19.1 s

Exercise 3D

1. (a) The object's displacement from a reference point is zero. Usually this means the object has returned to its starting position.

   (b) The object's velocity is zero, i.e. the object is stationary.
   (c) A graph that lies above the time axis indicates that displacement is positive, i.e. the object is one side of its starting point, e.g. to the right. A graph that lies below the time axis indicates that displacement is negative. The object has moved to the other side of its starting point, e.g to the left.
   (d) The object's velocity has become negative, i.e. the object is moving in the opposite direction from those sections where the graph lies above the time axis.
   (e) The object's velocity is zero, i.e. the object is stationary.
   (f) The object is travelling at a constant velocity. It is still moving.
2. (a) The cat jumps when $t = 0$ seconds. Its height above ground increases for about 0.7 seconds, until it reaches its maximum height, around 1.1 metres above the ground. It then falls slightly, until it lands on something which is 1 metre high at time $t = 1$ s.
   (b)
   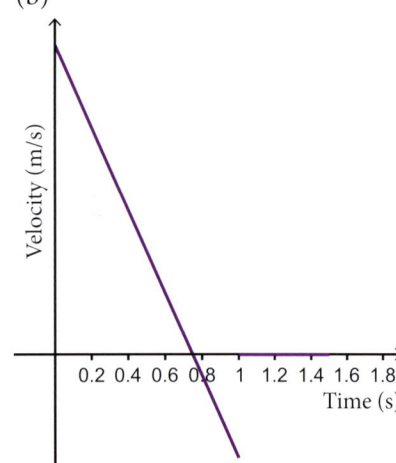

Exercise 4A

1. (a) $(7\mathbf{i} + \mathbf{j})$ (b) $\begin{pmatrix} 5 \\ -9 \end{pmatrix}$
2. (a) 5.83 cm (b) 31.0°
3. (a) The value of g is roughly 9.8 m s^{-2}.
   (b) All objects, when dropped near to the Earth's surface in a

vacuum, accelerate towards the Earth with the same acceleration g, or roughly 9.8 m s^{-2}. So, the apple and the feather accelerate downwards at the same rate and reach the ground at the same time. The acceleration does not depend on an object's mass.
(c) If the two objects were both dropped in air, the feather would experience more air resistance and it would fall more slowly.

Exercise 4B

1. (a) Forward thrust, lift, weight, air resistance (b) Forward thrust, lift (c) Tractive force (driving force), resistance, weight, normal reaction (d) Tension in the rope, friction, weight, normal reaction (e) Weight, normal reaction, friction (small) (f) Weight, air resistance

Exercise 4C

1. (a) $R_H = 3.42$ N; $R_V = 9.40$ N
 (b) $R_H = 31.7$ N; $R_V = 27.6$ N
 (c) $R_H = 6.00$ N; $R_V = 3.61$ N
 (d) $R_H = 12.2$ N; $R_V = 4.45$ N
 (e) $R_H = 23.0$ N; $R_V = 9.77$ N
 (f) $R_H = 17.0$ N; $R_V = 10.6$ N
2. (a) (i)

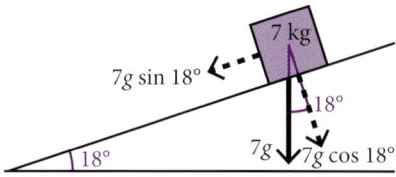

(ii) Perpendicular component: $7g \cos 18° = 65.2$ N. Parallel component: $7g \sin 18° = 21.2$ N
(b) (i)

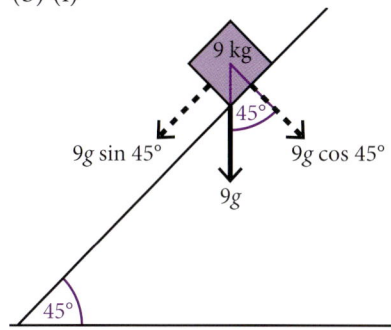

(ii) Perpendicular component: $9g \cos 45° = 62.4$ N. Parallel component: $9g \sin 45° = 62.4$ N
(c) (i)

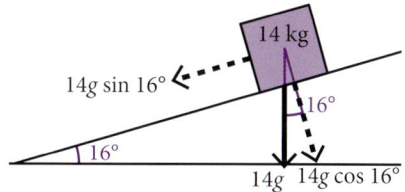

(ii) Perpendicular component: $14g \cos 16° = 132$ N. Parallel component: $14g \sin 16° = 37.8$ N
(d) (i)

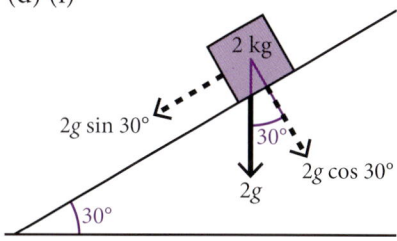

(ii) Perpendicular component: $2g \cos 30° = 17.0$ N. Parallel component: $2g \sin 30° = 9.8$ N
(e) (i)

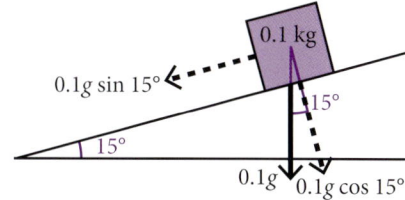

(ii) Perpendicular component: $0.1g \cos 15° = 0.947$ N.
Parallel component: $0.1g \sin 15° = 0.254$ N
3. 26.7 N

Exercise 4D

1. 5 N to the right
2. 3.61 N acting at 56.3° below the negative horizontal
3. 15.4 N acting at 0.765° above the positive horizontal
4. No. The resultant force in the direction of the canal is 26 548.6 N.
5. 81.9 N acting at 47.8° above the positive horizontal
6. (a)

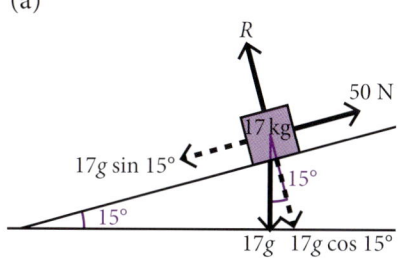

(b) Magnitude 6.88 N; direction parallel to and up the plane
7. (a) 20.3 N (b) 34.8 N

Exercise 4E

1. (a)

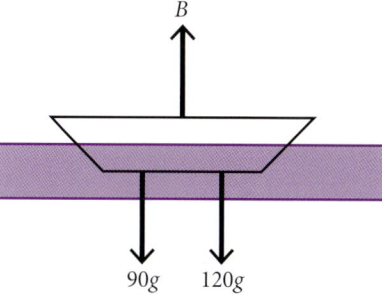

(b) The buoyancy force $B = 210g = 2060$ N (3 s.f.)
2. (a) $F = -2\mathbf{i} - 3\mathbf{j}$ N
 (b) $F = -2\mathbf{i} + 4\mathbf{j}$ N
 (c) $F = -3\mathbf{i} + 15\mathbf{j}$ N
3. 7.41 N, direction 25.2° above horizontal
4. (a)

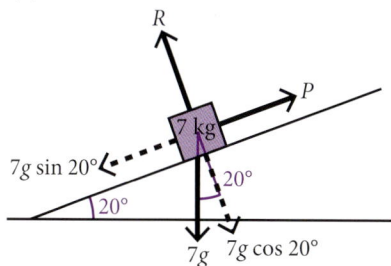

(b) 64.5 N (c) 23.5 N
5. 9.30 N at an angle of 17.2° below the horizontal
6. $J = 9.03$ N; $G = 16$ N
7. $S = 12.4$ N; $T = 9.72$ N

Exercise 5A

1. (a) 24 m s^{-1} (b) 9.6 s (c) 115 m (3 s.f.)
2. (a) 25 N, 53.1° below positive horizontal (b) 16.8 N, 72.6° below negative horizontal
3. $(-11.2\mathbf{i} - 0.670\mathbf{j})$ N

ANSWERS – EXERCISE 5B

Exercise 5B

1. The rocket starts at rest. It remains in this state until an external force causes its velocity to change. The external force is provided by combustion of the rocket's fuel.

2. (a) The upwards force (lift) is equal to the downwards force (the kite's weight). There is no wind, so the kite is not moving in the horizontal. (b) A gust of wind exerts a force upon the kite. By Newton's first law, the kite's velocity changes because of the additional force. It no longer remains in equilibrium.

3. (a) The weight force and air resistance. (b) This increases the air resistance. (c) The acceleration will be lower because of the air resistance. An object falling through air reaches a **terminal velocity**. It then stays at this velocity because the air resistance force becomes equal to the weight force. (d) The parachute increases the air resistance further. The terminal velocity is lower when the parachute is open (roughly 4 m s⁻¹ for a parachute with a diameter of 12 metres).

4. (a) 25 N (b) There is now a resultant force acting on the particle to the right. By Newton's first law, the particle cannot continue to move at a constant velocity. It accelerates to the right.

5. $\mathbf{P} = \begin{pmatrix} -4 \\ 2 \end{pmatrix}$ N

6. (a) $-4600\mathbf{i} + 3000\mathbf{j}$ N
   (b) No, the plane does not maintain the same constant velocity. There is now a resultant force acting on the plane. It will begin to accelerate in the direction of the resultant force.

7. (a)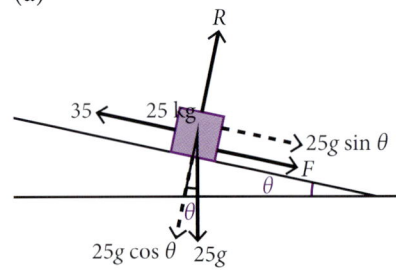

   (b) 7.9° (c) The component of the weight force acting down the slope remains unchanged at 27 N. The only force acting up the slope is the 15 N tension in the rope. Therefore, there is now a resultant force and the box will accelerate down the slope.

8. (a)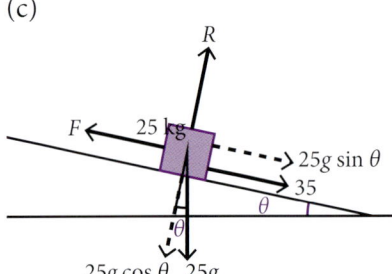

   (b) 22.8 N
   (c)

   47.2 N
   (d) The lawnmower is modelled as a particle.

Exercise 5C

1. (a) 10 N (b) 2.5 m s⁻² (c) 38.4 N (d) 22.5 N
2. (a) 90 000 N (b) 20 m
3. 1.5 m s⁻² to the right
4. (a) 20.8 m s⁻² (b) Use of fuel will decrease the overall mass and increase the acceleration. (c) The mass of the driver and any resistive forces have been ignored.
5. (a) 4.9 m s⁻² (b) 4.43 m s⁻¹

6. (a)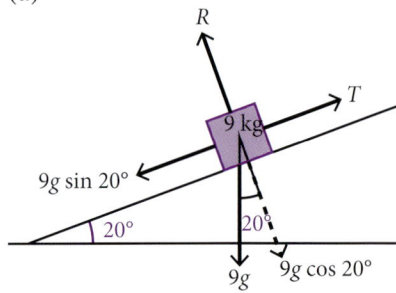

   (b) 1.09 m s⁻² (c) (i) 0.9 N up the plane (ii) 0.9 N down the plane
7. (a) 0.0503 N at an angle 63.6° below the positive horizontal
   (b) 0.503 m s⁻² (c) 22.6 cm

8. (a)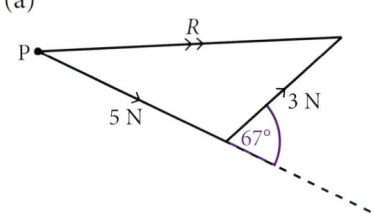

   (b) 6.76 N (c) 1.35 m s⁻²
   (d) 2.70 m

9. Venus

10.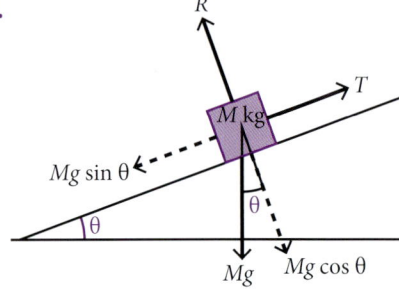

    Situation 1. The acceleration is a and the tension is T. Then:
    $F = ma$ parallel to the slope.
    $\Rightarrow T - Mg \sin\theta = Ma$
    $\Rightarrow T = Mg \sin\theta + Ma$
    Situation 2. The acceleration increases by k to $a + k$. Let the tension be T_2. Then:
    $T_2 - Mg \sin\theta = M(a + k)$
    $\Rightarrow T_2 = Mg \sin\theta + Ma + Mk$
    $T_2 = T + Mk$
    i.e. the tension increases by Mk newtons.

Exercise 5D

1. (a) 5 m s⁻² (b) 36.9°
2. (a) $(-0.6\mathbf{i} - 0.8\mathbf{j})$ m s⁻²
   (b) 53.1° (1 d.p.) (c) 2.1° (1 d.p.)

3. (a) $\begin{pmatrix} -2 \\ 2 \end{pmatrix}$ N (b) $\begin{pmatrix} 0.5 \\ -0.5 \end{pmatrix}$ m s^{-2}
4. (a) 5 (b) $(16\mathbf{i} + 12\mathbf{j})$ N
   (c) 10 m s^{-1}
5. (a) $(4\mathbf{i} + 47\mathbf{j})$ N
   (b) $(-6\mathbf{i} - 42\mathbf{j})$ N
6. (a) $(18\mathbf{i} - 11.5\mathbf{j})$ N
   (c) $(48\mathbf{i} - 21\mathbf{j})$ m s^{-1}
7. (a) $(-3\mathbf{i} + \mathbf{j})$ m s^{-1}
   (b) $(-3\mathbf{i} + 14\mathbf{j})$ m
   (c) $\left(-\dfrac{3\sqrt{10}}{10}\mathbf{i} + \dfrac{\sqrt{10}}{10}\mathbf{j}\right)$
8. Position vector of Q: $6\mathbf{i}$ m, due east of O.
9. (a) −13 (b) 22.4 m s^{-1}
10. (a) $(5\mathbf{i} + \mathbf{j})$ m s^{-2} (b) 25.5 m s^{-1}
    (c) 079°

Exercise 5E

1. (a) 90 m s^{-2} (b) 45 m (c) 90 m s^{-1}
   (d) 450 m (e) 19.5 s (3 s.f.)
2. (a) 5.25 m s^{-2}
   (b) Possible answers:
   - The model rocket accelerates for a shorter time (because it carries less fuel). Its acceleration must be higher for it to gain a significant height.
   - The mass of the space shuttle is far greater, leading to a lower acceleration.
   - The acceleration of the shuttle has to be lower because of the astronauts inside. Humans could not survive inside an object accelerating at 90 m s^{-2}.
3. (a) 5160 N (b) 8
   (c) Any two from:
   - The lift cable is light.
   - The lift cable is inextensible.
   - The lift and any passengers are particles.
4. 3910 N (3 s.f.)
5. (a) 44.4 m s^{-1} (b) 4.22 s
   (c) 6.47 m s^{-2} (d) 36.1 m s^{-1}
6. (a) 49 m s^{-1} (b) Air resistance ignored and/or object modelled as a particle. (c) 0.096 m (9.6 cm)

Exercise 5F

1. (a) This is false. The two forces are equal and opposite. The man can survive this force, but the cockroach cannot.
   (b) This is true. The two forces are equal in size.
   (c) This is false. The reaction force from the ground is equal to my weight force whether I stand on one leg or two legs.
   (d) This is true. This is how a rocket gains its thrust.
   (e) This is true. This is how a helicopter gains its lift.
2. 45 N
3. (a) 200 N (b) 200 N
4. (a) 800 N (b) 800 N
   (c) 250 N (d) 250 N
5. (a) By Newton's third law, these forces are equal in magnitude.
   (b) Newton's second law is $F = ma$, which can be rearranged to give $a = \dfrac{F}{m}$. The forces are equal, but the comet will experience a greater acceleration, since its mass is smaller.
6. (a) One foot pushes backward against the blocks. According to Newton's third law, the blocks give the sprinter an equal and opposite force forward.
   (b) 7 m s^{-2} (c) 0.14 m
7. She could choose a heavy tool from her toolbelt and push it as hard as possible away from the module. According to Newton's third law, the tool would exert an equal and opposite force upon her body, sending her back towards the module. The force provided would not be very large, but it may be enough to get her moving in the right direction at a low speed. Then, according to Newton's first law, she would continue moving at this speed until she reached the module, since no external forces would be acting on her body.

Exercise 6A

1. 100 m
2. $R_H = 19.0$ N; $R_V = 28.2$ N
3. $F = 13.5$ N; $G = 6.95$ N
4. 4 kg

Exercise 6B

1. (a) Sean is wrong. The object experiences a friction force, which slows it down. If there really were no forces acting the object, it would continue at the same speed until it reaches the end of the track. (The track could then be described as smooth.)
   (b) Amy is wrong. Once the rocket has escaped from the Earth's gravitational pull, on its journey to the moon the capsule experiences no friction forces and no other external forces. To continue moving at a constant speed, the rocket engines are not needed. An object will continue to move at a constant speed unless it is acted upon by some force.
2. (a) 0.2 m s^{-2} (b) The puck travels 22.5 m, so it does not reach the other end of the rink.
3. 0.91 s (2 s.f.)
4. (a)

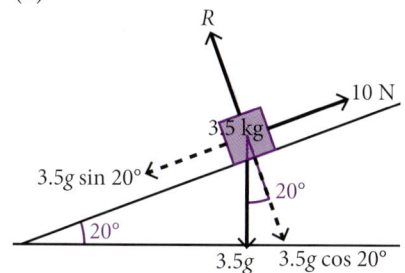

   (b) 0.49 m s^{-2} (2 s.f.)
   (c) 6.1 s (2 s.f.)
5. (a)

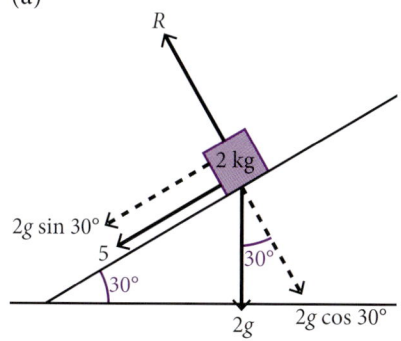

   (b) Using $F = ma$:
   $-2g \sin 30 - 5 = 2a$

ANSWERS – EXERCISE 6B

$a = -\dfrac{g+5}{2}$

Then using $v^2 = u^2 + 2as$ with

$s = ?;\ u = 4;\ v = 0;\ a = -\left(\dfrac{g+5}{2}\right)$

$0 = 16 - 2\left(\dfrac{g+5}{2}\right)s$

$s = \dfrac{16}{g+5}$

(c)

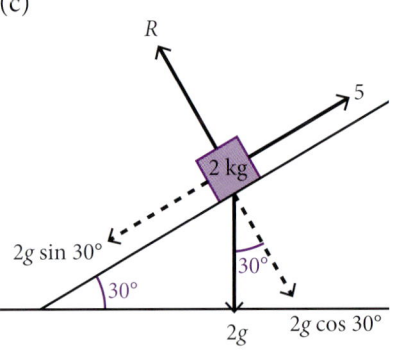

(d) $t = \dfrac{8}{\sqrt{g^2 - 25}}$

Exercise 6C

1. *Suggested* answers: (a) An ice hockey puck sliding across ice: 0.05 (b) A tomato in a Teflon frying pan: 0.1 (c) A book sliding across a polished wooden table: 0.3 (d) A chair being dragged across a bedroom carpet: 0.6 (e) Bare feet on a tiled bathroom floor: 0.7 (f) Feet wearing socks on a tiled bathroom floor: 0.4
2. 0.340
3. Matchbox yes; Car no; Skier yes; Brick no
4. (a)

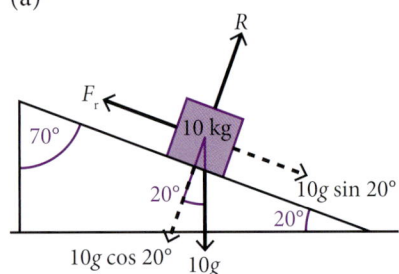

(b) 1.05 m s⁻²
(c) Final speed 2.51 m s⁻¹, greater than the maximum safe speed.

5. (a)

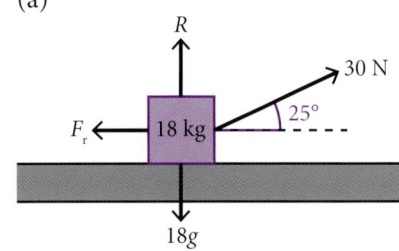

(b) 0.150 (c) The package is modelled as a particle; the rope is modelled as being light and inextensible. (d) Increasing the angle would increase the vertical component of the tension. Since this component and the reaction force add up to the package's weight, the reaction force becomes smaller. Since $F_r = \mu R$, the friction force also becomes smaller.

6. (a)

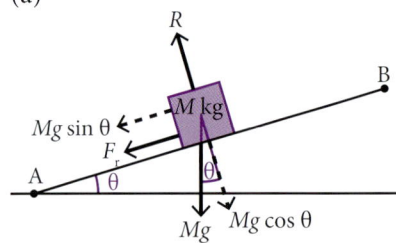

(b) $U = \left(\dfrac{24\mu + 7}{25}\right)gT$ (c) The box is being modelled as a particle.

7. 15 kg (2 s.f.)
8. (a)

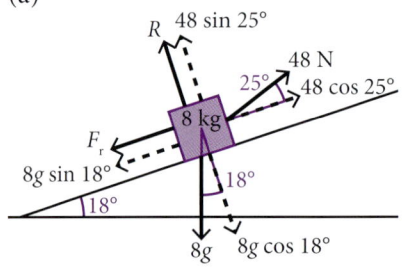

(b) 0.306 m s⁻² (c) 0.612 m

9. (a)

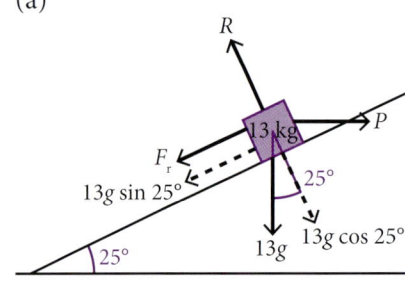

(b) 116.6 N or 120 N (2 s.f.)

10. (a) $F_r = \mu R = 0.45 \times 1000 \times 9.8 = 4410$

$F = ma \Rightarrow a = -\dfrac{4410}{1000} = -4.41$

$v^2 = u^2 + 2as \Rightarrow$

$s = \dfrac{(v^2 - u^2)}{2a} = \dfrac{(0 - 900)}{2 \times -4.41}$

= 102 m. The car stops in a distance of 102 m, so does not collide with the stationary car.
(b) 131 m

Exercise 6D

1. (a) 0.4 N (b) (i) 0.6 N (ii) 0.3 (c) (i) 0.6 N (ii) 0.1 m s⁻²
2. 21.2 N
3. (b) 498 N (c) 0.1 m s⁻¹ (d) 0.55 m
4. (a)

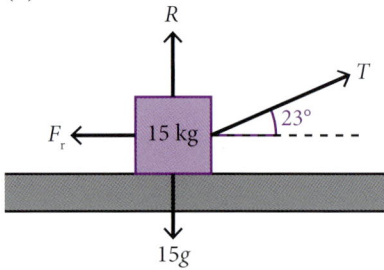

(b) 0.0776 (c) 35.7 N

5. (a)

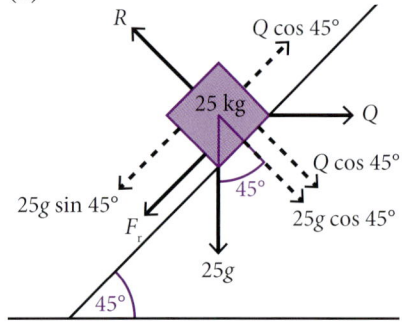

(b) Parallel to plane:
$Q \cos 45 = 25g \sin 45 + F_r$
Perpendicular to plane:
$R = 25g \cos 45 + Q \cos 45$
Eliminating $Q \cos 45$:
$R - 25g \cos 45 = 25g \sin 45 + F_r$
Using $F_r = \mu R$:
$R - \mu R = 25g \sin 45 + 25g \cos 45$
$\Rightarrow R(1 - \mu) = 25g\sqrt{2}$
$\Rightarrow R = \dfrac{25g\sqrt{2}}{1 - \mu}$

(c) $25g\left(\dfrac{1+\mu}{1-\mu}\right)$

ANSWERS – EXERCISE 7D

6. (a)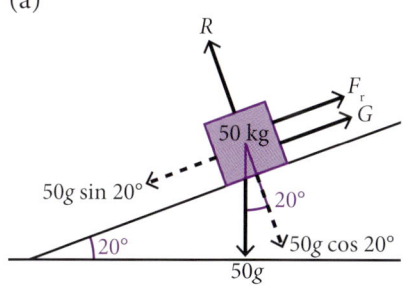
   (b) 98.5 N (3 s.f.)
   (c)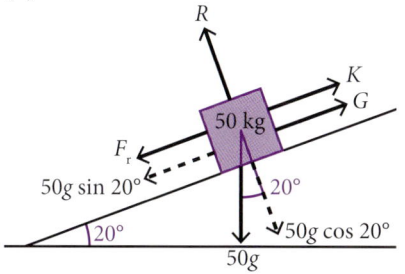
   138 N
   (d) Any two of: the rope is light; the rope is inextensible; the crate is a particle.
7. (b) $\tan^{-1}\left(\dfrac{17}{20}\right)$

Exercise 7A

1. (a) (i) No external forces are acting. According to Newton's first law, a body experiencing no forces, or no resultant force, will continue to move at a constant velocity. (ii) The engines will burn some fuel, causing the spacecraft to experience a resultant force. Its velocity will change because of this force. (It will probably decelerate as it approaches its destination.)
   (b) 281 kN (c) 240 N
2. (a) 0.98 m (b) 2.4 seconds

Exercise 7B

1. (a)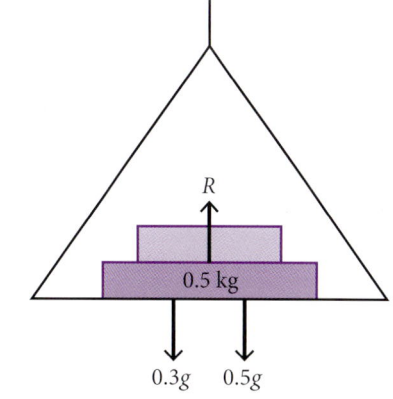
   (b) $0.3g$ or 2.94 N
   (c) $0.8g$ or 7.84 N
   (d) $0.8g$ or 7.84 N
2. Tension in top string 3N; tension in lower string 1N
3. (a) 29.4 N (b) 68.6 N

Exercise 7C

1. (a) 580 N (2 s.f.) (b) 700 N (2 s.f.)
2. 740 N; 690 N; 620 N (2 s.f.)
3. (a)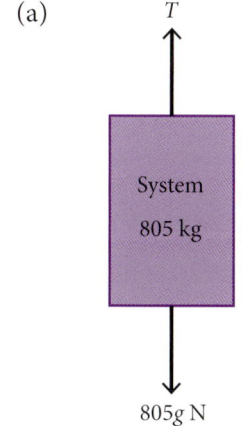
   (b) 8140 N (3 s.f.)
   (c)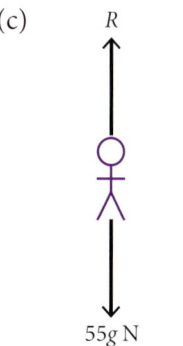
   (d) 526 N (3 s.f.)

4. (a)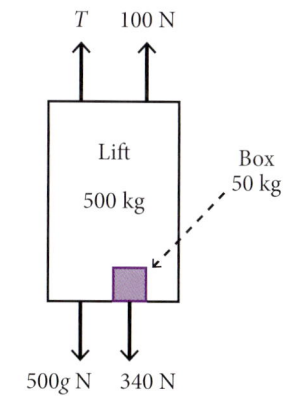
   (b)

   340 N ↑

   ↓ 50g N

   (c) 3 m s^{-2} (d) 3640 N
5. (a) 8.2 N (1 d.p.) (b) 3.1 N (c) 8.2 N
6. (a) 6000 N (b) 7 (c) The lift cable is light; the lift cable is inextensible; the lift is treated as a particle.

Exercise 7D

1. (a)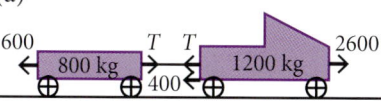
   (b) $F = ma$ for whole system
   $\Rightarrow 1000 - 100 = 1000a$
   $\Rightarrow a = 0.9$ m s^{-2} (c) 280 N
2. (a)

   600 ← 800 kg — T T — 1200 kg → 2600
                        400

   (b) 0.8 m s^{-2} (c) 360 m
   (d) 0.75 m s^{-2} (e) 384 m (f) 62 s
3. (a) Tow truck 3600 kg; car 1800 kg (b) 17 000 N (c) The mass of the towbar is neglected since it is light. The towbar does not stretch, since it is inextensible, meaning that the acceleration of both vehicles is the same.
4. (a) Car: 54 N; trailer 12 N

ANSWERS – EXERCISE 7D

(b)
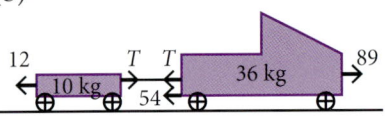

(c) 0.5 m s⁻² (d) 17 N (e) 3 s
(f) 66 N (g) 12 N

5. (a) 0.25 m s⁻²
(b)
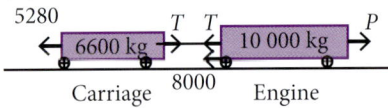

(c) 6930 N (d) 17 400 N (3 s.f.)

6. (a)
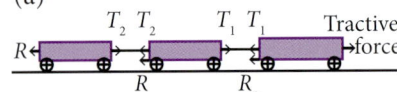

(b) Let m be the mass of a carriage, a be the acceleration and R be the resistance force for each carriage. (Also marked on the diagram is R_e, which is the resistance force to the motion of the engine, but this is not used in the working.)
$F = ma$ for carriage 1:
$T_1 - T_2 - R = ma$ (1)
$F = ma$ for carriage 2:
$T_2 - R = ma \Rightarrow R = T_2 - ma$ (2)
Sub (2) into (1):
$T_1 - T_2 - (T_2 - ma) = ma$
$T_1 - T_2 - T_2 + ma = ma$
$T_1 - 2T_2 = 0$
$T_1 = 2T_2$

Exercise 7E

1. (a)
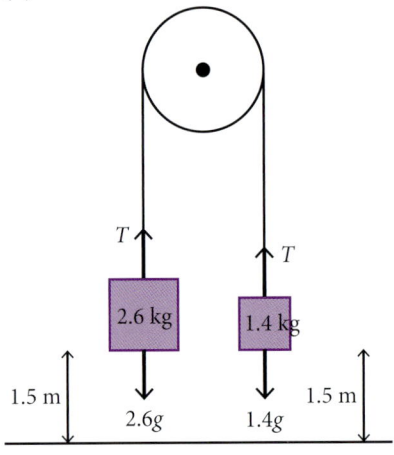

(b) $a = 2.9$ m s⁻²; $T = 18$ N (both to 2 s.f.)

(c) 1.0 s (2 s.f.) (d) 3.5 m (2 s.f.)
2. (a) 37.5 N (b) 3 kg
3. (a) ²⁵⁄₁₆ m s⁻²
(b)
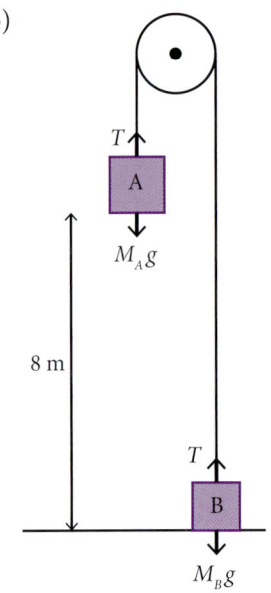

(c) 4 kg (d) 2.9 kg (e) The objects are treated as particles.
4. (a)
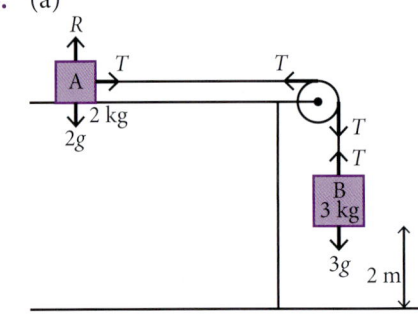

(b) $a = 5.89$ m s⁻²; $T = 11.8$ N
(c) 16.6 N (d) 0.825 s (e) The two blocks are particles.
5. (a)
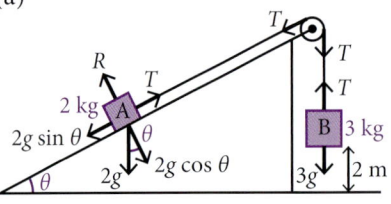

(b) $a = 4.1$ m s⁻²; $T = 17.1$ N (both to 1 d.p.)
(c) 4.0 m s⁻¹ (1 d.p.)
(d) 29.2 N in the direction 58.5° below the horizontal (both to 1 d.p.)

6. (a)
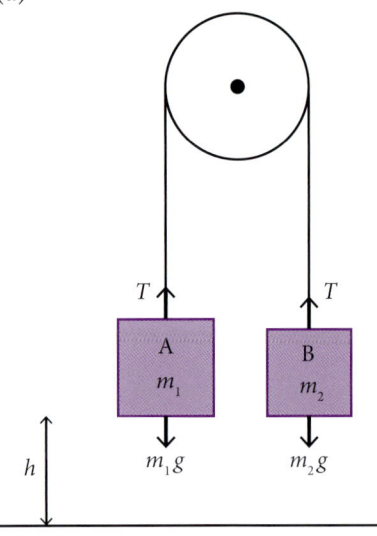

$F = ma$ for block A:
$m_1 g - T = m_1 a$ (1)
$F = ma$ for block B:
$T - m_2 g = m_2 a$ (2)
(1) + (2)
$\Rightarrow m_1 g - m_2 g = m_1 a + m_2 a$
$\Rightarrow (m_1 - m_2)g = (m_1 + m_2)a$
$a = \left(\dfrac{m_1 - m_2}{m_1 + m_2}\right)g$

(b) From (2):
$T = m_2 a + m_2 g$
$= m_2 \left(\dfrac{m_1 - m_2}{m_1 + m_2}\right)g + m_2 g$
$= m_2 \left(\dfrac{m_1 - m_2}{m_1 + m_2}\right)g + m_2 \left(\dfrac{m_1 + m_2}{m_1 + m_2}\right)g$
$= m_2 g \left(\dfrac{m_1 - m_2}{m_1 + m_2} + \dfrac{m_1 + m_2}{m_1 + m_2}\right)$
$= m_2 g \left(\dfrac{m_1 - m_2 + m_1 + m_2}{m_1 + m_2}\right)$
$= m_2 g \left(\dfrac{2m_1}{m_1 + m_2}\right)$
$= \dfrac{2m_1 m_2 g}{m_1 + m_2}$

(c) $v^2 = u^2 + 2as$
$v^2 = 0 + 2\left(\dfrac{m_1 - m_2}{m_1 + m_2}\right)gh$
$v = \sqrt{2\left(\dfrac{m_1 - m_2}{m_1 + m_2}\right)gh}$

(d) $v = u + at \Rightarrow t = \dfrac{v - u}{a} = \dfrac{v}{a}$
(since $u = 0$)

$t = \sqrt{2\left(\dfrac{m_1 - m_2}{m_1 + m_2}\right)gh} \div \left(\dfrac{m_1 - m_2}{m_1 + m_2}\right)g$

ANSWERS – EXERCISE 7F

$$= \sqrt{\frac{2(m_1 - m_2)gh}{(m_1 + m_2)} \times \frac{(m_1 + m_2)}{(m_1 - m_2)g}}$$

$$= \sqrt{\frac{2(m_1 - m_2)gh}{(m_1 + m_2)} \times \sqrt{\frac{(m_1 + m_2)^2}{(m_1 - m_2)^2 g^2}}}$$

$$= \sqrt{\frac{2(m_1 - m_2)gh}{(m_1 + m_2)} \times \frac{(m_1 + m_2)^2}{(m_1 - m_2)^2 g^2}}$$

$$t = \sqrt{\frac{2(m_1 + m_2)h}{(m_1 - m_2)g}}$$

Exercise 7F

1. (a)

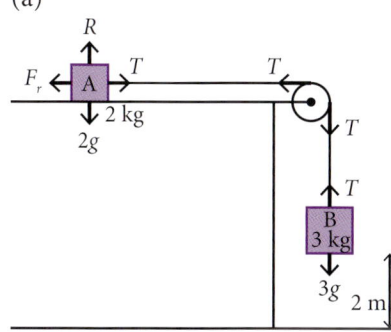

(b) 0.89 s (c) (i) 7.2 m (ii) 3.2 s

2. (a)

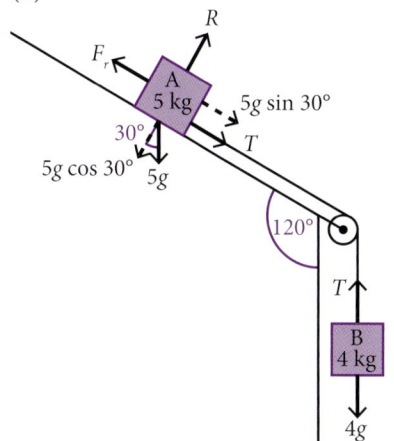

(c) $\frac{1}{18}(13 - 3\sqrt{3})g$ m s^{-2}

(d) Modelling the objects as particles means they have no size. Therefore, they are not affected by an air resistance force.

3. (a) 2.2 m s^{-2} (2 s.f.)
(b) 4.2 m s^{-1} (2 s.f.) (c) 7.0 m s^{-2}
(d) While the string is taut distance travelled is 4 m. While the string is slack distance travelled would be 1.25 m. Total distance travelled is 5.25 m,

but this is not possible, because the pulley is only 5 m from P's starting position. Therefore, P does collide with the pulley.

4. (a)

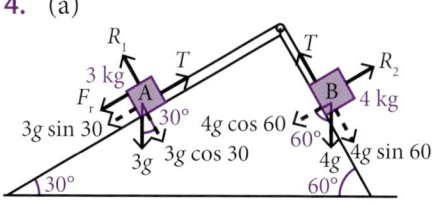

(b) Perpendicular to plane for Box A: $R_1 = 3g \cos 30 = 15\sqrt{3}$
$F_r = \mu R_1 = 0.6 \times 15\sqrt{3} = 9\sqrt{3}$
$F = ma$ parallel to the plane for Box A: $T - F_r - 3g \sin 30 = 3a$
$F = ma$ parallel to the plane for Box B: $4g \sin 60 - T = 4a$
Add, and substitute for F_r:
$4g \sin 60 - 9\sqrt{3} - 3g \sin 30 = 7a$
$20\sqrt{3} - 9\sqrt{3} - 15 = 7a$
$a = \frac{1}{7}(11\sqrt{3} - 15)$

5. (a)

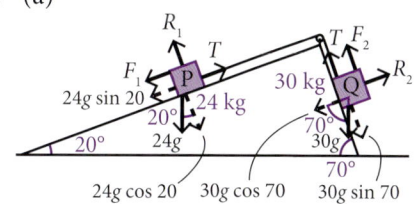

(b) $a = 2.1$ m s^{-2}; $T = 200$ N (both to 2 s.f.) (c) 280 N (2 s.f.)

6. (a)

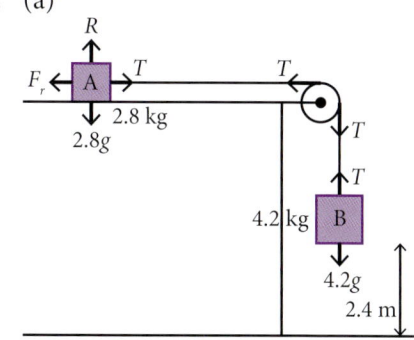

(b) 0.45 (c) In total Box A travels 4.5 metres towards the pulley, so it does not reach the pulley.
(d) The rope is light / The rope is inextensible.

7. (a)

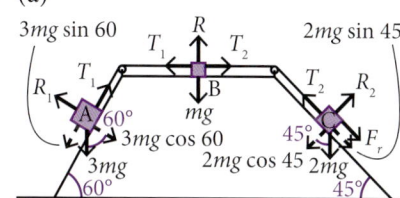

(b) (i) $a = \dfrac{2P - 5mg}{6m}$

(ii) $T = \dfrac{2P + mg}{3}$

(iii) $\sqrt{2}\left(\dfrac{2P + mg}{3}\right)$

8. (a)

(b) $F = ma$ for A:
$3mg \sin 60 - T_1 = 3ma$ (1)
$F = ma$ for B:
$T_1 - T_2 = ma$ (2)
$F = ma$ for C:
$T_2 - F_r - 2mg \sin 45 = 2ma$ (3)
Add:
$3mg \sin 60 - F_r - 2mg \sin 45 = 6ma$ (4)
$F_r = \mu R$ for C:
$F_r = \dfrac{1}{\sqrt{2}} 2mg \cos 45 = mg$

Substitute in (4):
$\dfrac{3mg\sqrt{3}}{2} - mg - \sqrt{2}mg = 6ma$

$6a = g\left(\dfrac{3}{2}\sqrt{3} - 1 - \sqrt{2}\right)$

$a = \dfrac{g}{12}(3\sqrt{3} - 2\sqrt{2} - 2)$ m s^{-2}

(c) $T_1 = \dfrac{mg}{4}(3\sqrt{3} + 2\sqrt{2} + 2)$ and

$T_2 = \dfrac{mg}{6}(3\sqrt{3} + 4\sqrt{2} + 4)$

(d) 1.7 m s^{-1} (2 s.f.)

9. (c) $\sqrt{\dfrac{2h(m_1 + m_2)}{g(m_2 - m_1\mu)}}$ (d) μg

ANSWERS – EXERCISE 8A

Exercise 8A
1. Mean 35.75; Median 36; Mode 39; Range 18
2. No box for zero people; No box for >4 people; First two categories overlap. Improved question:

   How many people have mobile phones in your family?
   - ☐ 0 people
   - ☐ 1–2 people
   - ☐ 3–4 people
   - ☐ >4 people

Exercise 8B
1. (a) In a census every unit of the population is observed. (b) An example: census of citizens of a country. In the UK this is carried out once every 10 years.
2. Sampling involves observing or measuring a certain number of the units in the population, not the entire population.
3. This is a census, since every member of the population is included in the survey.
4. (a) A sampling frame is a list of all units within the population.
   (b) An increase in a population causes the sampling frame to be a longer list of units.
   (c) An increase in the variability within a population would have no effect on the sampling frame. It remains a list of all the units within the population.
5. (a) Quantitative and discrete
   (b) Quantitative and discrete
   (c) Qualitative
   (d) Quantitative and continuous
   (e) Qualitative
   (f) Quantitative and discrete

Exercise 8C
1. (a) Advantages of a sample:
   1. A sample requires fewer resources (time, money, etc).
   2. Results are obtained more quickly since fewer units of the population have to be observed (e.g. fewer people have to respond, etc).
   3. Overall the data processing is easier because there is less data to deal with.
   (b) A census should be taken when complete accuracy is required about a population, or when the population is small enough that the census does not become too time consuming or costly.
2. A census would not be appropriate when testing the pallets because testing a pallet would cause its destruction. As small a sample as possible should be used.
3. (a) Census. Information is required on every citizen.
   (b) A relatively small sample should be sufficient, for example a sample of size 50, since the bee population should be relatively uniform.
   (c) A large sample should be taken. A census would be too time-consuming and costly, but the population of Belfast is fairly diverse, so a small sample may not be sufficient. In addition, the team has enough resources (10 people) to conduct a large survey.
   (d) Census. It would not be too costly to measure the lengths of the necks of all the giraffes in both zoos.
   (e) Census. It should be possible to record the temperature every day during the month of July. The data collection can be done automatically, so no manpower is required.
   (f) A relatively small sample. It would be unrealistic to take a census (measuring or asking every man in Banbridge) since this would be too costly and time-consuming. Only one student is working on the study, so a small sample is probably required.
   (g) Census. In this case accurate figures are required. In addition, the data should be readily available, so the census does not become too costly or time-consuming.
   (h) Small sample. To test a bulb, the time must be measured until it burns out. This test is a destructive test – the bulb cannot be used or sold afterwards. It would be inappropriate to use a census, since this would involve destroying every light bulb. A sample should be used instead, and it would be wise to use as small a sample as possible to save costs.
4. (a) A census would not be appropriate when testing the carrier bags because testing a bag causes its destruction.
   (b) (i) The claim does not seem to have enough evidence to support it. In the test only 3 out of 5 (60%) of the bags were strong enough to hold 20 kg. (ii) The test could be improved by increasing the sample size.

Exercise 8D
1. (a) 30 000 (b) Random number sampling may be possible here if each household is listed in a spreadsheet with a unique numerical ID.
2. The simplest approach is to stir the pot and, without looking, choose 30 marbles. This is lottery sampling. (The less simple way, in this case, would be to give each marble a number and select 30 random numbers on the calculator.)

Exercise 8E
1. The sampling interval is $\frac{200}{25} = 8$. A random number m is chosen between 1 and 8. This is the first clock-in number chosen. Then every 8th number after that is chosen. For example, if $m = 4$

the manager would choose clock-in numbers 4, 12, 20, 28, etc.
2. (a) 3 (b) The number of sites required for the sample is 20% of 15, which is 3.
The sampling interval $k = \dfrac{15}{3} = 5$
A random number is chosen between 1 and 5, e.g. 4. The 15 roadworks sites should be given a unique ID from 1 to 15. Then sites 4, 9 and 14 are selected as the sample.
(c) Simple random sampling may be simpler than systematic sampling in this case, as the population size and sample size are both small.

Exercise 8F

1. (a)

P1	P2	P3	P4	P5	P6	P7
3	3	2	3	3	4	3

(b) Each year group is fairly represented in the sample.
2. (a) Stainberry's 6; Priceland 4; Superstuff 3; Scrounders 2
(b) It would not be possible to use a sample stratified by supermarket as the original population of shoppers can no longer be divided into non-overlapping groups. It may be possible to stratify by gender, age, or some other variable.
3. (a) 1 soprano, 3 alto, 2 tenor, 1 baritone (b) Each voice type is fairly represented in the Christmas choir, so the overall sound of the choir should remain the same as the original larger group.

Exercise 8G

1. Possible reasons:
• Only men are questioned – will miss all female staff.
• Only Wednesdays – may miss some staff who don't work on Wednesdays.
• Only after 9 am – may miss some staff who work night shifts

• Many of the people questioned may not be members of the police force, e.g. visitors, delivery workers, etc.
2. (a) Assuming the membership numbers are from 1 to 50, random number sampling could be used, selecting 10 numbers from 1 to 50. The disadvantage of this method is that each of the activities may be over- or under-represented in the sample.
(b) A more suitable method would be stratified sampling. In the sample there would be 4 pool users, 3 gym users and 3 pilates class members.
3. (a)

	M	W	T
Poor	10	3	13
Satisfactory	3	3	6
Good	1	4	5
Very Good	1	5	6
Total	15	15	30

(b) Annie concludes that 13 out of the 30 in the sample rated the service as poor, so the retraining is not necessary.
(c) Random sampling can lead to bias. In this case, women are over-represented in the sample (a half of the sample, but only one third of the population).
(d) Brad concludes that 15 out of 30 customers, exactly half, rated service as poor. Since the manager had specified **more than half**, Brad concludes that retraining is not necessary.
(e) Reason 1: With 12 women in the sample of 30, women are slightly over-represented. Reason 2: Brad works through the alphabetical list, choosing every 5th customer until he has 30 names. The 30th customer chosen is the 150th customer in the list. Therefore, customers in the second half of the list cannot be chosen. In a fair sample, all customers have an equal chance of being included in the sample.
(f) The number of women chosen for the sample is 10, number of men chosen for the sample is 20.
(g)

	M	W	T
Poor	14	2	16
Satisfactory	4	2	6
Good	2	3	5
Very Good	0	3	3
Total	20	10	30

(h) Cormac's conclusion is that more than half of customers rate service as poor (16 out of 30) and the manager should pay for the retraining.
(i) The best sampling methodology in this case is Cormac's stratified sampling. This is because men and women are represented fairly in the sample. A better approach still would be to stratify the customer base by age groups, as well as by gender, and ensure that each age group is represented fairly within the sample.
4. (a) Fionnuala should calculate the sampling interval $k = \dfrac{600}{20} = 30$. Then she should choose a random number between 1 and 30. If this is 14, for example, she would then choose the photos with filenames 14.JPG, 44.JPG, 74.JPG, etc.
(b) No type of random sample produces the best results here. Instead, Joy should probably choose what she thinks are the best 20 photos of the 600.
5. The sampling interval $k = \dfrac{250}{10} = 25$
Taking every 25th working day, gives a dataset containing 10 of the same day of the week, e.g. 10 Mondays or 10 Fridays. There would be a bias in the sample, since, for example, there may be more people taking their annual leave on Fridays and fewer on

ANSWERS – EXERCISE 8G

Tuesdays. Ideally, the sample should include a mixture of different days of the week. Alternative sampling strategies would be:
- Simple random sampling. Using this approach, 10 days would be picked at random from the 250 days in the dataset.
- Stratified sampling. Using this approach, the population of 250 days would be split into 5 strata – one for each day of the working week. Then 2 days could be selected at random from each of the 5 strata to obtain the 10 days required for the sample.

Exercise 9A
1. (a) 16 (b) 30 (c) 0.933 mm (3 s.f.)

Exercise 9B
1. (a)

Mass rounded to nearest kg	Mid-point	Class limits	Class boundaries
35 – 37	36	35, 37	34.5, 37.5
38 – 40	39	38, 40	37.5, 40.5
41 – 43	42	41, 43	40.5, 43.5
44 – 46	45	44, 46	43.5, 46.5

(b) 3 kg (c) 43.5 kg is rounded to 44 kg and appears in the 44 – 46 kg group.

Exercise 9C
1. (a) 24 (b) 15 (c) 28 (d) ⅓
2. (a) **Method 1**: Consider the mean and standard deviation for each location. The means are: Belfast £1.42; Omagh £1.48 (both to the nearest penny). Compare the standard deviation of the prices in each location. From the calculator, for Belfast $\sigma = 9.69$; for Omagh $\sigma = 9.66$ The standard deviations are very similar.
Method 2: The medians are: Belfast £1.43; Omagh £1.47. Compare the interquartile range of the prices in each location.

Belfast: 18p; Omagh 18p. The interquartile ranges are the same.
(b) Petrol is more expensive in Omagh. It is often more expensive in locations where the fuel must be transported a greater distance from its place of origin, because of the transportation costs.

Exercise 9D
1. (a) 21 minutes (b) 23 minutes
(c) • The Douglas Dynamos have a lower median time (23 mins for DD; 25 mins for RR).
• The Douglas Dynamos have a lower interquartile range (5 mins for DD; 6 mins for RR).
• The Douglas Dynamos have slightly faster runners overall. The Ramsey Racers have a slightly wider spread of abilities.

2. (a)

Distance travelled d (miles)	Freq.	Cumulative frequency
$350 \leq d < 400$	2	2
$400 \leq d < 450$	5	7
$450 \leq d < 500$	9	16
$500 \leq d < 550$	20	36
$550 \leq d < 600$	3	39
$600 \leq d < 650$	1	40

(b) (i) Roughly 510 miles
(ii) Roughly 475 miles and 530 miles
(c)

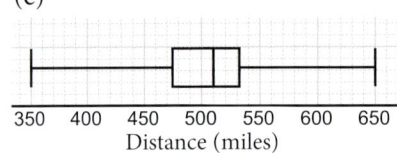

3. (a) Grey (b) The first box plot represents the red squirrels, the second the greys.
(c)

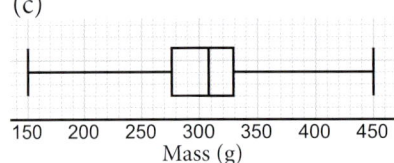

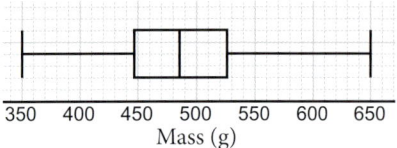

(d) The grey squirrels. Although the range is roughly equal for each group, the interquartile range is greater for the grey squirrels.

Exercise 9E
1. (a) $a = 1, b = 2, c = 3, d = 4$
(b)

Weekly income (£)	Frequency
150 –	20
170 –	120
210 –	160
250 –	60

(c) 360 (d) 50 (e) It assumes that exactly a quarter of the 120 households in the £170 – £210 class are between £170 and £180.

2. (a) 120 (b) 7/60 (c) 37/120 (d) (i) 7/30 (ii) ⅙ (e) 17/60 (f) The calculation in (d) assumes the pupils in the 140 – 150 cm group are uniformly distributed, with half of them above 145 cm, half of them below.

3. (a)

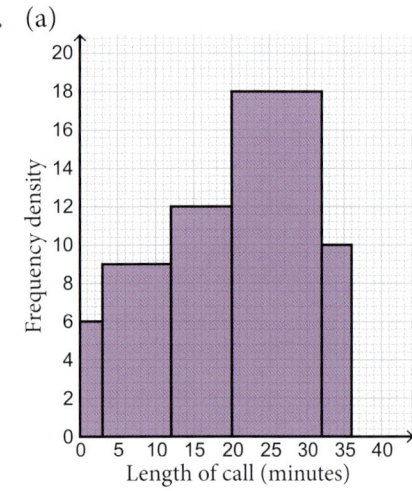

(b) A sample in which each group is represented proportionately, according to its size. (c) 3
4. Width: 2 cm, height: 1.1 cm
5. (a) 1.5 cm (b) 6 cm

ANSWERS – EXERCISE 10B

6. (a), (b) and (d)

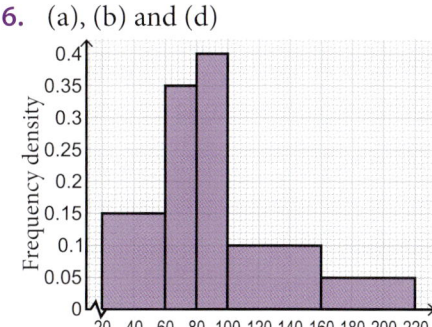

(c) Converting the information in the histogram to a frequency table gives:

Time (mins)	Midpoint x	f	fx
20 – 60	40	6	240
60 – 80	70	7	490
80 – 100	90	8	720
100 – 160	130	p	$130p$
160 – 220	190	3	570

$\dfrac{\Sigma fx}{\Sigma f} = \text{mean}$

$\dfrac{240 + 490 + 720 + 130p + 570}{6 + 7 + 8 + p + 3} = 93\dfrac{1}{3}$

$\dfrac{2020 + 130p}{24 + p} = \dfrac{280}{3}$

$6060 + 390p = 6720 + 280p$

$110p = 660$

$p = 6$

(d) See histogram above

Exercise 9F

1. (a)

Number of people in car	Frequency
1	50
2	25
3	17
4	6
5	2

(b)

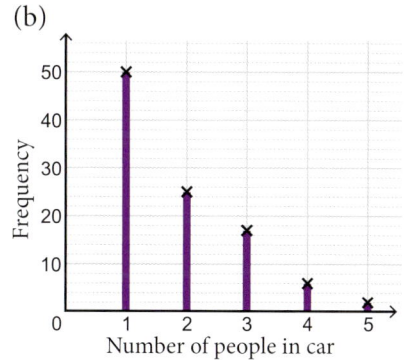

2. (a)

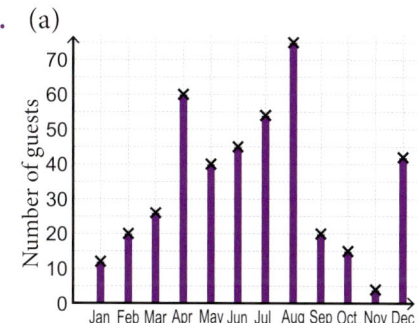

(b) The graph shows an increase in bookings during the summer months: June, July and August. There is also a peak in April (during the Easter holidays) and during the Christmas holidays in December.
(c) A vertical line chart is appropriate because it is not possible to interpolate between any pair of data points given.

Exercise 9G

1. Box plot
2. Cumulative frequency diagram
3. • Overlapping categories, e.g. which group does a day with 2 people go in? • Frequencies should add up to 31.
4. • There should not be an axis break on the y-axis. The y-axis should always start at zero on a histogram so that the full area of each bar is visible. • The numbering should be given on the y-axis. • There should not be gaps between the bars of a histogram. • The x-axis should be labelled and units given.
5. • There is no key. • There are only 9 numbers on the "Home" side. • The smaller numbers should be adjacent to the stem on the "Home" side.
6. (a) Box plots show the range and interquartile range clearly. (b) The stem and leaf diagram includes all the raw data. In a box plot the data are summarised.
7. (a) The first histogram shows data with a larger mean (the bars are further towards the right) and slightly larger spread (roughly 60 cm compared with 50 cm).
(b) The first histogram may represent older children.

Exercise 10A

1. (a) 7.85 cm (b) 7.65 cm
2. (a) 14 (b) 14 (c) 14

Exercise 10B

1. (a)

Distance cycled, d (km)	Freq-uency, f	Mid-point, x km	fx
$0 \leq d < 10$	5	5	25
$10 \leq d < 20$	14	15	210
$20 \leq d < 30$	24	25	600
$30 \leq d < 40$	15	35	525
$40 \leq d < 50$	3	45	135

(b) $\Sigma f = 61; \Sigma fx = 1495$
Estimate of mean
$= \dfrac{\Sigma fx}{\Sigma f} = \dfrac{1495}{61} = 24.5$ (3 s.f.)

(c) $\dfrac{Q_2 - 20}{30 - 20} = \dfrac{30.5 - 19}{43 - 19}$
$\Rightarrow Q_2 = 24.8$ (3 s.f.)

(d) The data are given in a grouped frequency table; therefore, the raw data are not known.

2. Mean for Sunday to Friday is 61 560 ÷ 6 = 10 260. Saturday's figure of 12 000 is higher than 10 260, so the additional value increases the mean.

3. 13 minutes 45 seconds

4. (a) The mean exam score gives a good indication how Jane is doing overall.
(b) Mode may be a good choice. It is always one of the data values

ANSWERS – EXERCISE 10B

and gives the shop an indication of which size to keep most of in stock.
(c) For salaries, the median is often used. The mean is distorted by a small number of very large salaries.
5. 50
6. 8
7. (a) 9 (b) 8.5 (c) 8.16 (3 s.f.)
(d) The mode is an actual data value and gives the manufacturer information on the most common size worn or purchased.
(e) Yes. There appears to be a peak at size 6, which probably relates to the modal size for women and another peak at size 9, probably the mode for men.

Exercise 10C

1. Mean £320; Standard deviation £53.23
2. (a) Midpoints: 10, 20, 30, 40, 50
   $\Sigma f = 54$; $\Sigma fx = 1620$;
   $\Sigma fx^2 = 55200$;
   Estimate of mean = 30
   Estimate of standard deviation = 11.1 (3 s.f.)
   (b) Since this is a random sample, it is likely the mean and standard deviation will be different for a different random sample the following week.
3. Midpoints: 24.5, 74.5, 124.5, 174.5, 250
   $\Sigma f = 156$; $\Sigma fx = 19391$;
   $\Sigma fx^2 = 3\,503\,829.5$
   Estimate of mean = 124 (3 s.f.)
   Estimate of standard deviation = 83.7 (3 s.f.)
4. (a) 4 (b) 17.9 minutes (c) 16.7 minutes (d) 6.45 minutes (e) The data are presented in a grouped frequency table; therefore, we do not know the raw data. We only know that 2 people took between 2 and 8 minutes, etc.
5. Estimate of mean = 17.7 cm (3 s.f.).
   Estimate of sample standard deviation = 8.93 cm (3 s.f.)
6. Mean = 17; standard deviation = 5.31 (3 s.f.)
7. Midpoints: 17.7, 17.9, 18.25, 18.75, 20.1, 20.3, 20.7; $\Sigma f = 50$;
   $\Sigma fx = 968.9$; $\Sigma fx^2 = 18812.175$
   Mean = 19.6 (3 s.f.); standard deviation = 0.761 (3 s.f.).
8. Mean: 48.6 minutes; standard deviation 5.96 minutes (both to 3 s.f.)
9. (a) $\mu = 51$; $\sigma = 3.77$ (3 s.f.)
   (b) The mean is unaffected since one value is 14 above the mean and one value 14 below the mean.
10. Johnny's mean time is 25.1 mins; Deborah's mean is 27 mins. Johnny's standard deviation is 0.917 mins; Deborah's is 0.742 mins. Johnny is faster, on average, but Deborah's lower standard deviation tells us she is more consistent in her times.

Exercise 11A

1. The older a car, the lower its value.
2. (a) Graph 3 – a positive correlation (b) Graph 1 – a negative correlation
   (c) Graph 2 – no correlation.
3. Gradient –2.2; y-intercept 1.56

Exercise 11B

1. Strong positive correlation; when it is sunny in Alfreton, it is usually sunny in Belper; when it's cloudy in Alfreton it's usually cloudy in Belper.
2. (a) No correlation (b) A larger number of flights for Máire is not necessarily associated with a larger number for John. (c) Probably not. If they worked together, they would probably largely attend the same meetings. There would be a positive correlation between the number of meetings they attended.
3. (a) No correlation (b) Negative correlation (c) For phones less than 18 months old, the age does not affect the battery life. For phones over 18 months old, the older the phone the shorter is its battery life.
4. (a) Positive correlation (b) For cars over 20 years old, an older car has a higher value.

Exercise 11C

1. (a) 0.850 (3 s.f.) (b) There is a strong positive correlation between the unemployment rate and the number of house repossessions. As unemployment increases, house repossessions rise too.
2. (a) 0.561 (b) There is a positive correlation between the two judges' scores. If Judge P gives a high score, Judge Q is likely to give a high score as well.
3. (a) –0.545 (b) There is a negative correlation. If a pupil scores well in chemistry, they are likely to score poorly in geography and vice versa.
4. (a) 0.0869 (b) There is almost no correlation between shoe size and collar size. If a customer has large shoes, it makes it no more or less likely that they will have a large collar size. (This result is slightly unexpected – perhaps a larger sample is required.)
5. (a) 0.789 (b) There is a fairly strong positive correlation between the number of hours of practice and the exam score. The more you practice, the better you get!
6. (a) $r = 0.586$ (b) There is a positive correlation between p and w. This means that if a student does well in programming, they are more likely to do well in web design as well.
7. (a) 0.07 (b) The value of r indicates almost no correlation between the level of UK interest rates and the value of the pound. It is true that higher interest rates make the pound a more attractive

investment, but there are many other factors that affect the value of the pound, for example, it fell sharply when the UK voted to leave the European Union. Interest rates were cut to record low levels during the coronavirus pandemic, but this did not affect the pound, since many other countries had to take similar measures.

Exercise 11D

1. (a) 3.5 (b) 4
2. (a) Yes. The points follow a straight line fairly closely and there appears to be a positive correlation. (b) Yes. The points follow a straight line fairly closely and there appears to be a negative correlation. (c) No. There may be a negative correlation, but the points do not closely follow a straight line. (d) No. There is not enough data.
3. (a) 60.0 cm (3 s.f.)
   (b) 26.5 g (3 s.f.)
   (c) 0.2305 is the gradient of the straight line. For every extra gram of mass added, the string extends by 0.2305 cm. 43.89 is the y-intercept of the straight line. This is the length of the spring when a mass of 0 g is attached, i.e. the spring's natural length.
   (d) No. This would be extrapolation, which can be dangerous. The straight line relationship holds for masses up to 100 g in weight, but it may not hold outside this weight range.
4. (a) 140 (b) 45 (c) Probably not. The number of guests is also limited by the number of rooms available, so extrapolation is unsafe. (d) The number 6 is the gradient of the regression line and it represents how many extra guests the hotel can accommodate when the number of staff increases by 1.

Exercise 11E

1. (a) The size of a house is the independent variable and its value is the dependent variable.
   (b) The length of Mr Walker's daily walk is the independent variable and the time he takes for it is the dependent variable.
   (c) The age of a child is the independent variable and the child's height is the dependent variable.
2. The principal **may** be correct, but she must be careful. The causation could be the other way around: perhaps a lack of success at school for some pupils makes them more likely to watch TV.
3. There is no direct causation between CO_2 and obesity levels. Instead there is a third factor causing both the increase in CO_2 and higher obesity levels: an increase in wealth. Richer populations tend to eat more food, leading to higher levels of obesity. Richer populations also generate more CO_2 emissions through increased use of vehicles and other activities.
4. A higher number of colds and flu causes patients to take more cold and flu medication. It would be wrong to conclude that the causation is working the other way around (taking the medication is causing the illnesses).
5. This is an area of ongoing research, but it seems likely that the causation works both ways: sleeping less than 7 hours each night can cause ill health, but in some cases ill health can also cause a lack of sleep, e.g. for patients suffering chronic pain.
6. The causation works in the opposite direction: a larger number of sick people in the city calls for a larger hospital with more staff.

Exercise 12A

1. Q_1 = £10250; Q_2 = £20250; Q_3 = £27000; IQR = £16750
2. Mean: 1.69; Sample standard deviation: 0.950 (3 s.f.)
3. 6.11 m

Exercise 12B

1. (a) The median may stay the same, or it could be reduced.
   (b) The standard deviation will be reduced.
2. (a) Mean: £313.36 (nearest penny); standard deviation £54.92 (nearest penny)
   (b) Ordered list of values is:
   189, 247, 266, 299, 324, 334, 344, 347, 363, 363, 371. $n = 11$
   Position of lower quartile is:
   $\frac{n+1}{4} = 3 \Rightarrow Q_1 = 266$
   Position of upper quartile is:
   $3\left(\frac{n+1}{4}\right) = 9 \Rightarrow Q_3 = 363$
   IQR = 363 − 266 = 97
   1.5 × IQR = 1.5 × 97 = 145.5
   Critical values are:
   266 − 145.5 = 120.5
   and: 363 + 145.5 = 508.5
   There are no salary figures lying outside the range 120.5 − 508.5. Therefore there are no outliers in this dataset.
3. (a) (78, 14)
   (b) If the league position is p, then: $\sum p = 68, n = 12$,
   Mean $= \frac{68}{12} = 5.666\ldots$
   $\sum p^2 = 528, \sigma = 3.448\ldots$
   The mean + 2 standard deviations
   $= 5.666 + 2 \times 3.448 = 12.562$.
   14 > 12.562, therefore a league position of 14 is an outlier.
   (c) The equation gives a league position of 4 (rounded up from 3.95).
   (d) There is a lot of variability in the data; the points do not follow the regression line closely. Therefore, we should not have

ANSWERS – EXERCISE 12B

too much confidence in the prediction.
(e) Weak negative correlation. There is clearly a pattern, but the points do not follow a straight line closely.

4. (a) The age 110 is erroneous
(b) The erroneous value of 110 could be replaced with an average of the other 10 values; or the erroneous value could be removed from the dataset before the mean is calculated. Either way, the mean age for the cleaned data is 10.3 years.

5. (a) $Q_1 = 13$; $Q_3 = 38 \Rightarrow$ IQR = 25
$Q_3 + 1.5$ IQR $= 38 + 1.5 \times 25 = 75.5$
$165 > 75.5 \therefore 165$ is an outlier
(b) The outlier could be removed from the dataset. Alternatively, it could be replaced with an average of the other 10 values.
(c) 22.9 years (d) Including 65 increases the mean.
(e) (i) Including 65 increases the standard deviation. (ii) The interquartile range also increases, since Q_3 increases from 37 to 38.

6. 104%. Yes, the teacher was justified in removing the fifth score. It must have been an error, since it was above 100%

7. (a) The age of 9 is incorrect.
(b) If the mean of the 19 ages is 30, the total is 570. Using the uncorrected data, the total is 560. Therefore, the person whose age is recorded as 9 should be 19.
Median: 28; IQR: 18

8. (a)

Destination	Approx. distance (miles)	Frequency
Dunmurry	4.7	13
Lisburn	8.5	24
Hillsborough	11.9	11
Dromore	16.4	8
Banbridge	23.5	17
Newry	36.8	15
Dublin	99.7	11
Vladivostok	7460	1

(b) Vladivostok should be removed from the dataset. This was either an error or somebody playing a joke.
(c) Roughly 26.0 miles

9. (a) 82 (b) Mean = 47.2, Standard deviation = 19.7

10. From both diagrams it can be seen that the highest value in the dataset is 20 kg. The second box plot treats this value as an outlier; the first does not. To determine whether this is an outlier, calculate the upper critical value: From the first plot, using all 23 values in the dataset, $Q_3 = 11$ and $Q_1 = 6$.
$Q_3 + 1.5$ IQR $= 11 + 1.5(11 - 6)$ $= 18.5$. The highest value of 20 is greater than this critical value; therefore, this value is an outlier. The second box plot includes this value as an outlier, so this is the correct plot.
Note: the quartiles are re-calculated from the remaining 22 data values, after the outlier is removed. The lower quartile and median are unchanged, but the upper quartile is different.

Exercise 12C

1. (a) 23 mm (b) (i) The median would either stay the same (at 33 mm) or increase. (ii) The standard deviation would increase.

2. (a) 78 and 36 (b) $\sum x = 1173$; $n = 30$; $\mu = 39.1$

3. (a) For the purposes of plotting a graph, it is acceptable to remove the missing record from the dataset. There are enough data points to plot a graph, which should show a clear trend.
(b) 132 cm (3 s.f.)

4. (a) The second strategy would result in a lower standard deviation. Assuming one data point to be the mean reduces the average distance from the mean.
(b) 35.1 (3 s.f.)

5. (a) £218 000 (b) £225 000
(c) Assuming the missing price was £100 000, the mean becomes £206 200. The error in the mean would be 225 000 − 206 200 = £18 800. Assuming the missing price was £450 000, the mean becomes £241 200. The error in the mean would be 241 200 − 225 000 = £16 200. The largest possible error in the mean is £18 800.
(d) Assuming the missing price was £100 000, the mean becomes £224 000. The error in the mean would be 225 000 − 224 000 = £1000. Assuming the missing price was £450 000, the mean becomes £227 000. The error in the mean would be 227 000 − 225 000 = £2000. The largest possible error in the mean is £2000.
(e) The estate agent concludes that house prices may have risen slightly in the area, but there is a lot of uncertainty in the April 2020 figures. For that month, the mean is very dependent on the one missing value because the dataset is small. She can be a lot more confident about the mean for the April 2021 figures.

6. (a) Air temperatures between 95 and 100°C are not possible.
(b) The buoy appears to have broken down between 1600 and 1700. From 1700 until 2300 it reports 99.99°C. This is probably an error code.
(c) The readings from 1700 to 2300 should be treated as missing data and removed from the dataset. If a daily mean is to be calculated, the error codes could be replaced with average (mean) figures for the relevant times of day. The daily mean should not be calculated using only the 17 good readings from that day, since this

would be biased towards daytime temperatures.
7. (a) Mean 15.5; variance 5.85
   (b) Mean 16.2; variance 3.96
8. (a) £1 250 000.
   (b) Removing the data from the dataset would create a large error in the total wage bill.
   (c) Possible alternative strategies: 1. Taking mean figure for all employees. 2. Taking mean of upper and lower figure for IT department. 3. Taking lower figure for IT department.
   Strategy 1. £25 000. 2. £28 500. 3. £22 000.
   (d) If s is Faye's true salary, then:
   $1250000 + (s + 2545)$
   $= 1.002(1250000 + s)$
   $1252545 + s = 1252500 + 1.002s$
   $0.002s = 45$, giving $s = £22\,500$
   (e) Lower. The error of £2545 would be smaller in relation to the overall wage bill.

Exercise 13A
1. (a) $\frac{2}{9}$ (b) $\frac{4}{9}$ (c) $\frac{2}{3}$ (d) 0
2. GGG, GGB, GBG, GBB, BGG, BGB, BBG, BBB
3. (a) $\frac{25}{216}$ (b) $\frac{11}{36}$ (c) $\frac{125}{216}$

Exercise 13B
1. (a)

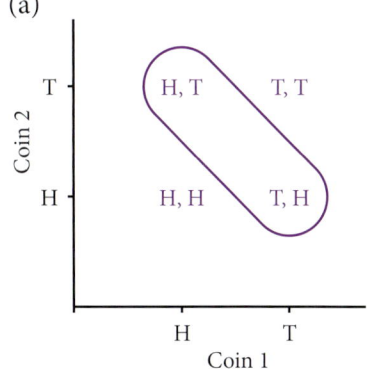

   (b) $P(\text{one head and one tail}) = \frac{2}{4} = \frac{1}{2}$

2. (a)

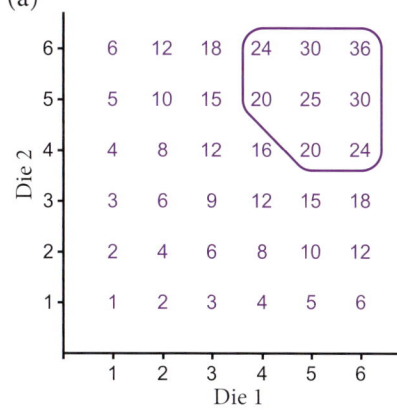

   Line shown for part (b)(ii)

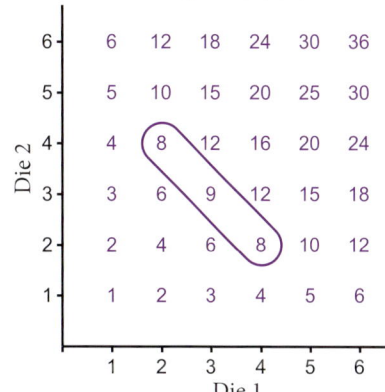

   Line shown for part (b)(iii)
   (b) (i) $P(X = 18) = \frac{2}{36} = \frac{1}{18}$
   (ii) $P(X > 18) = \frac{8}{36} = \frac{2}{9}$
   (iii) $P(8 \leq X \leq 9) = \frac{3}{36} = \frac{1}{12}$
3. (a) 0.27 (b) 0.84 (c) 0.53
4. $x = 5$
5. (a) $\frac{1}{5}$ (b) $\frac{13}{60}$

Exercise 13C
1. (a)

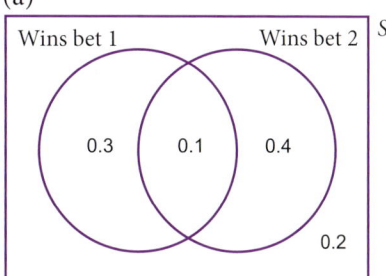

   (b) 0.2 (c) 0.7

2. (a)

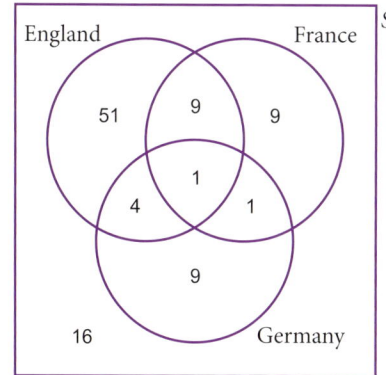

   (b) $\frac{3}{20}$ (c) $\frac{4}{25}$
3. (a) 0.15 (b) 0.2 (c) 14 (d) 0
4. (a) $x = 22$; $y = 16$ (b) 0
5. (a) 0.2 (b) 0.5 (c) 0.7 (d) 0.1
6. (a) $a = 0.04$; $b = 0.24$; $c = 0.1$
   (b) 5
   (c) No. There is no intersection between the red and green circles.

Exercise 13D
1. (a)

	Meal	No meal	Total
Car passengers	76	38	114
Foot passengers	30	56	86
Total	106	94	200

   (b) 0.53

2. (a)

	Rocky planets	Gas giants	Total
Has at least one moon	2	4	6
Has no moons	2	0	2
Total	4	4	8

   (b) (i) $\frac{1}{4}$ (ii) 0 (c) $\frac{2}{3}$

3. (a)

	Girls	Boys	Total
School lunch	24	26	50
Packed lunch	36	33	69
Home for lunch	0	1	1
Total	60	60	120

ANSWERS – EXERCISE 13D

(b) $^{12}/_{25}$ (c) $^{11}/_{20}$ (d) $^{3}/_{10}$

Exercise 13E

1. 0.48
2. 0.4
3. (a) 0.0338 (b) 0.3562
4. (a)

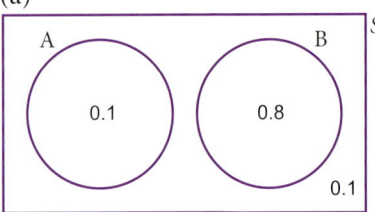

(b) 0.9 (c) 0.1

5. (a)

	UV protection	No UV protection	Total
Polarisation	16	14	30
No polarisation	42	28	70
Total	58	42	100

(b) (i) Not mutually exclusive, since $P(P \cap U) = \frac{16}{100} = \frac{4}{25} \neq 0$

(c) $P(P) = \frac{30}{100} = \frac{3}{10}$;

$P(U) = \frac{58}{100} = \frac{29}{50}$

$P(P) \times P(U) = \frac{3}{10} \times \frac{29}{50}$

$= \frac{87}{500} \neq P(P \cap U)$

Therefore, events P and U are not independent.

6. Let C be the event 'Wins at coconut shy' and W be the event 'Wins at whack-a-mole'. From the diagram:
$P(C) = 0.15 + 0.1 = 0.25$;
$P(W) = 0.1 + 0.3 = 0.4$
$\therefore P(C) \times P(W) = 0.25 \times 0.4 = 0.1$
From the diagram:
$P(C \cap W) = 0.1$
$P(C) \times P(W) = P(C \cap W)$
Therefore independent events.

7. (a)

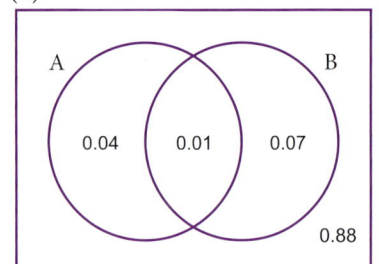

$P(A \cup B) = 0.04 + 0.01 + 0.07$
$= 0.12$
(b) (i) A and B are not mutually exclusive. Mutually exclusive events cannot both happen. On a Venn diagram there is no intersection between the two circles.
(ii) $P(A) \times P(B) = 0.05 \times 0.08$
$= 0.004; P(A \cap B) = 0.01$
$P(A) \times P(B) \neq P(A \cap B) \therefore A$ and B are not independent.

8. (a) No. For two events to be mutually exclusive there is no intersection. Independent events always have an intersection.
(b) Yes. The Venn diagram shows two such events.

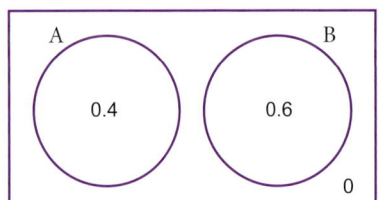

9. There is an outcome that is common to both events, namely rolling two threes. *Alternative answer:* On a Venn diagram Event A's circle intersects Event B's circle. In the intersection is the outcome two threes.

10. (a) Yes, events F and H are mutually exclusive. There is no intersection between the F and H circles, so these two events never happened at the same time (in the sample space of 100 nights).

(b) $P(F) = \frac{10 + 15}{100} = \frac{1}{4}$;

$P(G) = \frac{15 + 21 + 24}{100} = \frac{60}{100} = \frac{3}{5}$

$\Rightarrow P(F) \times P(G) = \frac{1}{4} \times \frac{3}{5} = \frac{3}{20}$

$P(F \cap G) = \frac{15}{100} = \frac{3}{20}$

$P(F) \times P(G) = P(F \cap G)$
Therefore, events F and G are independent.

(c) $P(G) = \frac{3}{5}$;

$P(H) = \frac{24 + 13}{100} = \frac{37}{100}$

$\Rightarrow P(G) \times P(H) = \frac{111}{500}$

$P(G \cap H) = \frac{24}{100} = \frac{6}{25}$

$P(G) \times P(H) \neq P(G \cap H)$
Therefore events G and H are not independent.

(d) Freddie probably has his own room. Events G and H are not independent, meaning that one occurring affects the probability of the other occurring. This probably happens because Grace and Harriet disturb each other when they waken.

11. $p = 0.4; q = 0.1$
or $p = 0.1; q = 0.4$

12. (a) 0.1 (b) (i) $P(A \cap B) \neq 0 \therefore A$ and B are not mutually exclusive.
(ii) $P(A) \times P(B) = 0.2$
and $P(A \cap B) = 0.1$
$\therefore P(A) \times P(B) \neq P(A \cap B) \therefore A$ and B are not independent.
(c) $P(A') = 0.6$
(d) $P(A \cup B') = 0.6$
(e) $P(A' \cup B) = 0.7$

Exercise 13F

1. (a) $^{7}/_{15}$ (b) $^{1}/_{18}$ (c) $^{7}/_{15}$
2. $^{1}/_{40}$

ANSWERS – EXERCISE 14C

3. (a)

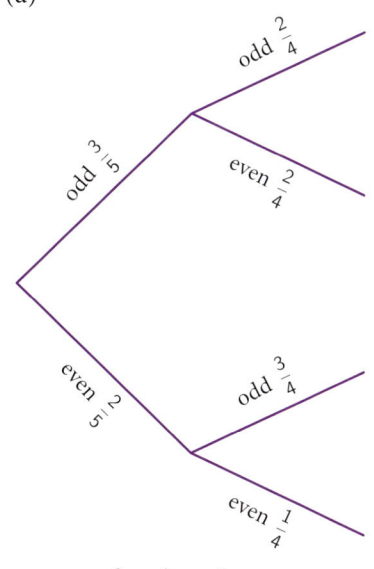

$P(EE) = \dfrac{2}{5} \times \dfrac{1}{4} = \dfrac{1}{10}$

(b) $P(EO \text{ or } OE) =$

$\left(\dfrac{3}{5} \times \dfrac{2}{4}\right) + \left(\dfrac{2}{5} \times \dfrac{3}{4}\right) = \dfrac{6}{20} + \dfrac{6}{20} = \dfrac{3}{5}$

4. (a) 0.144 (b) 0.416
5. (a) ⅚ (b) ²⁵⁄₂₁₆
6. (a)

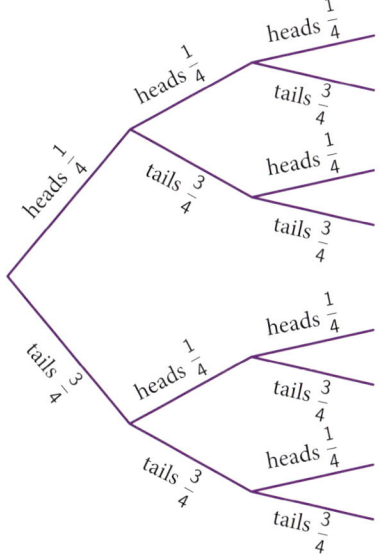

(b) $P(HHH) = \dfrac{1}{4} \times \dfrac{1}{4} \times \dfrac{1}{4} = \dfrac{1}{64}$

(c) $P(\text{one head}) =$
$3 \times \dfrac{1}{4} \times \dfrac{3}{4} \times \dfrac{3}{4} = \dfrac{27}{64}$

(d) $P(TTT) = \dfrac{3}{4} \times \dfrac{3}{4} \times \dfrac{3}{4} = \dfrac{27}{64}$

$P(\text{Either 3 heads or 3 tails in one experiment}) = \dfrac{1}{64} + \dfrac{27}{64} = \dfrac{28}{64}$

$P(\text{Either 3 heads or 3 tails in both experiments}) = \dfrac{28}{64} \times \dfrac{28}{64} = \dfrac{49}{256}$

7. 91/216 (3 s.f.)

8.

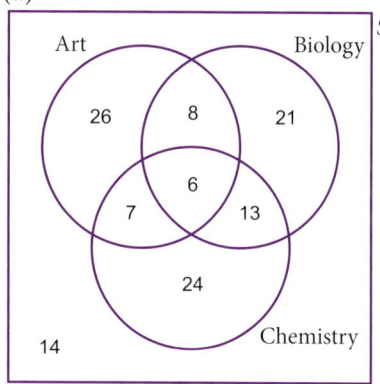

$P(\text{Late}) = (0.4 \times 0.6) + (0.6 \times 0.4)$
$= 0.48$. The supervisor is wrong. The bus driver is late 48% of the time.

Exercise 13G

1. ¹⁄₁₂₈
2. 6
3. ¹⁶⁄₃₇
4. ³⁄₂₅
5. (a) 0.05 (b) Not independent, since $0.7 \times 0.2 = 0.14 \neq 0.05$
6. (a)

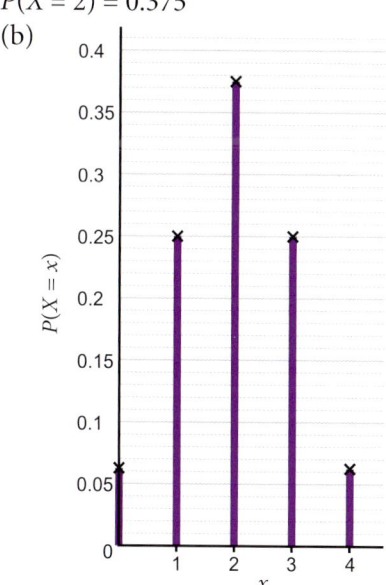

(b) ²⁴⁄₁₁₉ (c) ⁷⁹⁄₁₁₉

(d) $P(B) = \dfrac{48}{119}$; $P(C) = \dfrac{50}{119}$;

$P(B) \times P(C) = \dfrac{2400}{14161}$

$= 0.169$ (3 s.f.)

$P(B \cap C) = \dfrac{19}{119} = 0.160$ (3 s.f.)

Therefore, studying Biology and studying Chemistry are not independent events.

7. (a) 0.2 (b) 0.5 (c) 0.2 (d) 0.5
8. ¹⁄₁₂

Exercise 14A

1. (a) 24 (b) 1680
2. (a) 35 (b) 20
3. $8x^3 + 36x^2 + 54x + 27$
4. ⁴⁄₂₅

Exercise 14B

1. Any two of:
- Exactly 2 possible outcomes for each trial.
- Fixed number of trials.
- Probability of success the same for each trial.
- Trials are independent of each other.

2. (a) No, he is not correct. $P(X = 2) = 0.375$
(b)

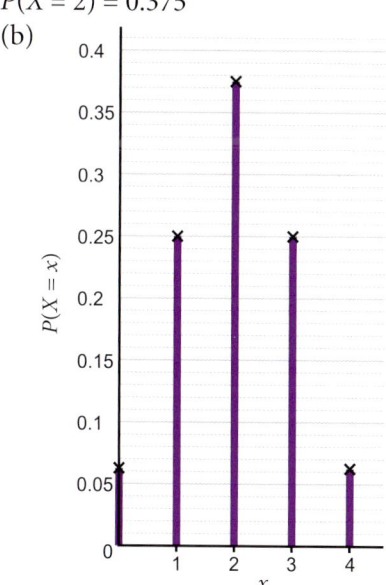

3. 0.000475
4. (a) 0.0625 (b) 0.938 (3 s.f.)
5. 0.0036
6. (a) 0.125 (b) 0.343
7. (a) ¹⁸⁄₃₇ (b) 0.385 (c) 0.865
8. (a) 0.410 (3 s.f.) (b) 0.154 (3 s.f.) (c) 0.998 (3 s.f.)
9. 0.0956 (3 s.f.)
10. (a) $P(X = 1) = 9p(1 - p)^8$
(b) $P(X = 2) = 36p^2(1 - p)^7$
(c) 0.2 (d) 0.176 (3 s.f)
11. (a) ¹⁄₁₆ (b) ¹⁄₁₆ (c) 0.311 (3 s.f.)

Exercise 14C

1. $(p + q)^3 = p^3 + 3p^2q + 3pq^2 + q^3$
2. (a) $(a + b)^6 = a^6 + 6a^5b +$

$15a^4b^2 + 20a^3b^3 + 15a^2b^4$
$+ 6ab^5 + b^6$
(b) (i) 0.262 (3 s.f.)
(ii) 0.345 (3 s.f.)
3. (a) $(a + b)^5 = a^5 + 5a^4b + 10a^3b^2 + 10a^2b^3 + 5ab^4 + b^5$
(b) ⅕ (c) (i) 0.328 (3 s.f.)
(ii) 0.00032 (iii) 0.672 (3 s.f.)
4. (a) $(a + b)^{10} = a^{10} + 10a^9b + 45a^8b^2 + 120a^7b^3 + 210a^6b^4 + 252a^5b^5 + 210a^4b^6 + 120a^3b^7 + 45a^2b^8 + 10ab^9 + b^{10}$ (b) $\dfrac{1}{1024}$

Exercise 14D

1. (a) $P(X \leq 2) = 0.679$ (3 s.f.)
   (b) $P(X > 3) = 1 - P(X \leq 3)$
   $= 1 - 0.8862 = 0.114$ (3 s.f.)
2. (a) 0.253 (3 s.f.) (b) 0.276 (3 s.f.)
   (c) 6
3. 0.0781
4. (a) 0.147 (3 s.f.) (b) 8
5. (a) $P(X = 2) = 190p^2(1 - p)^{18}$
   (b) $P(X = 3) = {}^{20}C_3 p^3(1 - p)^{20-3}$
   $= 1140p^3(1 - p)^{17}$
   Then: $P(X = 3) = 4P(X = 2)$
   $\therefore 1140p^3(1 - p)^{17}$
   $\qquad = 4 \times 190p^2(1 - p)^{18}$
   $1140p^3(1 - p)^{17}$
   $\qquad = 760p^2(1 - p)^{18}$
   $3p^3(1 - p)^{17} = 2p^2(1 - p)^{18}$
   $3p = 2(1 - p)$
   $3p = 2 - 2p$
   $5p = 2$
   $p = 0.4$
   (c) 0.996 (3 s.f.)
6. (a) 0.372 (3 s.f.) (b) 0.969 (3 s.f.)
   (c) 0.0307 (3 s.f.)
7. (a) 0.3 (b) The probability it rains on a Monday is a fixed, constant value. The probability of it raining one Monday is independent of it raining on all the other Mondays.
8. 6
9. (a) 0.4 (b) 3
10. (a) False (b) True (c) False
    (d) False